# VLSI PHYSICAL DESIGN AUTOMATION
## Theory and Practice

# VLSI PHYSICAL DESIGN AUTOMATION
## Theory and Practice

## SADIQ M. SAIT

*Department of Computer Engineering*
*King Fahd University of Petroleum and Minerals*
*Saudi Arabia*

## HABIB YOUSSEF

*Department of Computer Engineering*
*King Fahd University of Petroleum and Minerals*
*Saudi Arabia*

## McGRAW-HILL BOOK COMPANY

**London** · New York · St Louis · San Francisco · Auckland · Bogotá · Caracas
Lisbon · Madrid · Mexico · Milan · Montreal · New Delhi · Panama · Paris
San Juan · São Paulo · Singapore · Sydney · Tokyo · Toronto

Published by
McGRAW-HILL Book Company Europe
Shoppenhangers Road, Maidenhead, Berkshire, SL6 2QL, England
Telephone 01628 23432
Facsimile 01628 770224

---

**British Library Cataloguing in Publication Data**

Sait, Sadiq M.
  VLSI Physical Design Automation : Theory
  and Practice
  I. Title II. Youssef, Habib
  621.395

ISBN 0-07-707742-3

1234 BL 9765

Typeset by MFK Typesetting Ltd., Hitchin, Herts.
Printed and bound in Great Britain by Biddles Ltd., Guildford, Surrey.

# DEDICATION

To the memory of my mother, Amina Omer Sait and to my father, Muhammad Iyoob Sait

Sadiq M. Sait

To the memory of my father, Muhammad Youssef and to my mother, Fatima Braham

Habib Youssef

# CONTENTS

# FOREWORD

Teachers of advanced undergraduate and graduate courses in VLSI design have for several years lamented the non-existence of a book like this one. There exist books on VLSI algorithms for the student whose primary interest is algorithms and not VLSI. These books reduce the scope of problems below what must be implemented in CAD tools while, at the same time, pushing mathematics prerequisites beyond those usually possessed by students in Computer Engineering. Those books which have treated the issues of placement and routing at the engineering level are now out of date and with passing years have come to appear superficial. This book treats current issues at the appropriate depth.

Sadiq M. Sait and Habib Youssef have limited the scope of this volume to chip layout. Treatment begins at a point after synthesis and netlist manipulation are complete. Simulation can be accomplished before and after, but is not one of the issues addressed herein. Thus, the authors are able to provide a comprehensive analysis of every classic chip layout level topic from partitioning and floorplanning to routing and compaction. In every chapter there is space for treatment of all competing approaches.

This book will be essential reading for the next generation of CAD tool developers and will strengthen the hand of the more casual student who must choose intelligently among available vendor supplied tools. The book contains more than enough material for a one semester graduate course. It will serve as the primary text for a course limited to layout. It should become a resource for any course on VLSI design. As soon as the book becomes available, my colleague and I will integrate it into our Computer Engineering course on CAD algorithms at the University of Arizona.

I have known Sadiq M. Sait since the time of his PhD dissertation. I endorsed at its inception the project that led to this volume and predicted solid results. My initial confidence has been more than justified.

Dr Frederick J. Hill, Professor
University of Arizona
Tucson 85721 AZ
USA

# PREFACE

## MOTIVATION

This book is intended as a text for senior undergraduates and first-year graduate students in Computer Engineering, Computer Science and Electrical Engineering. It is also a good reference book for CAD practitioners.

VLSI design is now recognized as an important area of Computer/ Electrical Engineering and Computer Science. VLSI design is a very complex process. The design of a VLSI system is unthinkable without the use of numerous computer aided design (CAD) tools, which automate most of the design tasks. Except for the initial specification of the system, every other step of the design process has been either automated or made easy through user-friendly CAD tools/programs. The first paper on CAD appeared in 1955. CAD started being recognized as indispensable as early as 1960. Now for more than a decade, CAD of digital systems has become a mature area. Almost every engineering school offers at least one graduate-level course in VLSI design and VLSI design automation. Computer Science departments offer courses in VLSI computation and design automation.

There are three general aspects of CAD: (1) synthesis, (2) verification, and (3) design management. This book deals with the first aspect, that is, the synthesis aspect. Synthesis is known as the problem of obtaining a lower level representation of a digital system from a higher level representation of the same system. When the higher level representation is a behavioural description and the lower level is a structural description, the synthesis process from the higher level to the lower level is known as high-level synthesis. When both the higher and lower levels are structural representations, the process is called physical synthesis. In this book, we are mostly concerned with physical synthesis, i.e., partitioning, floorplanning, placement, and routing. However, high-level synthesis aspects are also addressed in the context of silicon compilation.

To our knowledge, this is the first textbook suitable for both first year graduate and senior undergraduate education in VLSI Physical Design Automation. In fact, very few *textbooks* on VLSI/CAD exist. Moreover, most are intended for PhD students, researchers, and engineers from the CAD industry. Others are outdated and/or mix physical design and verification (simulation) problems. All existing books are mostly edited and assume a significant amount of theoretical background on the part of the reader. For instance, the reader is expected to know elements of graph theory, algorithm design, and analysis of algorithms. This assumption is far from true in our context.

This book fills the void for senior undergraduate students and first year graduate students who may not have been exposed to many of the mathematical ideas from graph theory and algorithms. The book makes up for this lack of exposure through the use of a large number of illustrative examples and solved exercises. The only prerequisite assumed is the knowledge of basic logic design and computer architecture.

## Organization of the book

The book is organized into nine chapters. The chapters are organized in a sequence similar to the physical design process itself.

Chapter 1, the introductory chapter, motivates the student towards a study of physical design automation of integrated circuits. Layout is examined in the backdrop of the entire design automation process. Basic terminology is introduced, and the important subproblems of layout are identified. The fact that many layout subproblems are 'hard' is brought out with illustrative examples.

Chapters 2 to 7 examine the different subproblems in IC layout. Chapter 2 covers the problem of circuit partitioning. Chapter 3 discusses floorplanning. In Chapter 4 we present the problem of module placement. Chapters 5 to 7 are on wiring. We discuss grid/maze routing, global routing, and channel routing in Chapters 5, 6 and 7 respectively.

In Chapter 8, we consider the problem of silicon compilation and automatic generation of cells. Three different cell styles are examined, namely, Standard-cell, Gate Matrix and PLA. Algorithms for automatic generation of layout in the above cell styles are covered.

In Chapter 9, layout editors are considered and techniques for hand-drawing of layouts are examined. The importance of *compaction* in hand layouts is explained, and the two main compaction approaches together with their related algorithms are described.

In order to achieve uniformity in treatment, each of these topics is examined in the following light. *Problem Definition* introduces the reader to the essence of the layout subproblem, its graph theoretic formulation, and the notation associated with the treatment of the problem. Under *Cost Functions and Constraints*, we examine the problem as a constrained optimization problem and explain the possible cost functions which must be minimized subject to practical constraints. Next, we examine the various *approaches* that researchers and engineers in CAD have adopted to solve the concerned optimization problem. Such optimization algorithms include graph theoretic, numerical, and stochastic techniques. The algorithms are illustrated with examples and exercises. Finally, a section on latest developments in the problem area guides the reader/student to open research problems. *Exercises* are provided at the end of each chapter, followed by *References*. The exercises are classified into *routine exercises* which typically include applying an algorithm to a problem instance. *Challenging exercises* include programming projects and research-oriented problems (indicated by (*)).

The book contains an appendix which covers an overview of combinatorial optimization, and some basic definitions in algorithms and graph theory, as they apply to the book.

A Solutions Manual is available and can be obtained from the publisher or authors.

## How to use this book

The book can serve as a text for a one semester first year graduate course on CAD of digital systems. It will be difficult to cover all material in detail in the 15 weeks of the semester. Our experience using an early manuscript of the text was that the first seven chapters can be well covered. One week is required to introduce the topic of CAD in general and physical design in particular (Chapter 1). Approximately two weeks per chapter are spent on Chapters 2 to 7. In each chapter, several heuristics are described. Instructors may choose to cover all topics of the book by discussing only selected sections from each chapter.

As an undergraduate text, the depth and pace of coverage will be different. Only portions from each chapter could be covered. The instructors may decide as to which portions to cover. Material covered may preferably be from early portions of each chapter.

## Acknowledgements

The book took about two years to write. Many people supported the book. We thank all those who provided comments, corrections, and reviews for the book. We wish to thank Dr Muhammad S. T. Benten, Dean, College of Computer Sciences and Engineering, King Fahd University of Petroleum and Minerals (KFUPM), for his tremendous support. We apologize for the many times we had to call him late at night to fix system problems. He was a wonderful consultant on matters related to UNIX and LATEX. We also thank Dr Samir H. Abdul-Jauwad, Chairman, Computer Engineering Department, Dr Ala H. Al-Rabeh, Dean, College of Graduate Studies, and Dr M. Y. Osman, all at KFUPM, for their encouragement and support.

We are especially indebted to our teachers, Dr Eugene Shragowitz (Department of Computer Science, University of Minnesota), Dr Manzer Masud (formerly with Department of Computer Engineering, KFUPM, and now Engineering Vice-President of Attest Software Incorporated, Santa Clara, California), who attracted us to the general area of VLSI DA and shaped/influenced many of our views.

We also gratefully acknowledge Dr Gerhard Beckhoff (Department of Electrical Engineering, KFUPM), Dr Farouk Kamoun, and Dr Muhammad Ben-Ahmed (formerly Professors with Department of Informatics, Faculty of Science, Tunis), and Dr Valdis Berzins (Professor, Naval Post Graduate School, Monterey, CA), who played a key role during our formative years.

Thanks are due to Asjad M. T. Khan and Dr C. P. Ravikumar for providing material and exercises for parts of the book. Asjad spent an extraordinary amount of time and energy on this project and was a consultant on such matters as computer systems, text editing, typesetting etc. Dr Ravikumar provided the Appendix. He also helped in the writing of Chapter 1 and edited early drafts of some chapters (Partitioning, Placement and Grid Routing).

We used early drafts of this book to teach a graduate course in VLSI Physical Design Automation. Our thanks and appreciation to all those students who contributed with corrections to the text, solutions to the problems, etc., namely, Talal Al-Kharroubi, Talhah Al-Jarad, Maher Al-Sharif, Munir Zahra, Usamah Abed, Muhammad Al-Akroush, Emran Ba-Abbad, Shahid Khan, Shah Edreess, and Khaled Nassar. Special thanks to: (1) Talal Al-Kharroubi for his meticulous reading of the entire manuscript and for detecting many technical errors and subtle typos, and (2) Shahid

Khan, Khalid A. A., and Masud-ul-Hasan, for their patience in making and correcting all illustrations.

Any book writing requires a sizeable amount of reference material. Mr Shoaib A. Quraishi of KFUPM library was helpful in getting all requested material in record time.

The editors and staff of McGraw-Hill Book Company Europe were extremely helpful in providing timely feedback on reviewers' comments. We particularly thank Ms Camilla Myers, Engineering Editor, Ms Fiona Sperry, and Ms Rosalind Hann, Production Editor, for their help and cooperation during the writing of this book.

The design on the cover page is the layout of the Static Control Superscalar Processor (StaCS), courtesy of MASI laboratory, Paris, France. The StaCS processor has 875,000 transistors and was designed and implemented using the Alliance CAD system. We thank Professor Alain Greiner of MASI laboratory for providing the photograph of the layout and granting permission to reproduce it.

Lastly, we are very grateful to our wonderful wives (Sumaiya Sadiq and Leila Youssef) and children (Ifrah and Aakif Sadiq, Elias and Osama Youssef), without their patience and sacrifices such a major undertaking and its successful completion would not have been possible.

The authors also acknowledge the support provided by King Fahd University of Petroleum and Minerals under Project Code COE/AUTOMATION/163.

The manuscript was produced in LATEX at King Fahd University of Petroleum and Minerals using the computing facilities of the College of Computer Sciences and Engineering.

e-mail: facyøø9@saupmøø.bitnet          Sadiq M. Sait

e-mail: facyøø16@saupmøø.bitnet         Habib Youssef

# INTRODUCTION

## 1.1 VLSI DESIGN

VLSI, or Very Large Scale Integration refers to a technology through which it is possible to implement large circuits in silicon—circuits with up to a million transistors. The VLSI technology has been successfully used to build microprocessors, signal processors, systolic arrays, large capacity memories, memory controllers, I/O controllers, and interconnection networks.

Present-day VLSI technology permits us to build entire systems with hundreds of thousands of transistors on a single chip. For example, the Intel 80286 microprocessor has over 100,000 transistors, the 80386 has 275,000 transistors, the 80486 has approximately 1,000,000 transistors. The RISC processor from National Semiconductor NS32SF641 has over a million transistors. The Pentium processor of Intel has over three million transistors (Alpert and Avnon, 1993).

Integrated circuits of the above complexity would not have been possible without the assistance of computer programs during all phases of the design process. These computer programs automate most of the design tasks. Designing a VLSI chip with the help of computer programs is known as CAD, or Computer Aided Design. Design Automation (DA), on the other hand, refers to entirely computerized design process with no or very little human intervention. CAD and DA research has a long history of over three decades. Some of the earliest CAD software dealt with placement of logic modules on printed circuit boards (PCBs) and finding short electrical paths to wire the interconnections. Logic minimization was also an important facet of electronic design, since eliminating even a handful of logic gates resulted in significant cost savings. As technology has changed from Small Scale Integration (SSI) to Very Large Scale Integration (VLSI), the demand for design automation has escalated; the *types* of design automation tools have also multiplied due to changing needs. For example, in the LSI and VLSI domains, it is important to simulate the behaviour of a circuit before the circuit has been manufactured; this is because it is impractical to breadboard a circuit of LSI complexity in order to verify its behaviour. The rapidly changing technology has also radically transformed design issues. For instance, in the LSI/VLSI technologies, it is not very important to save on transistors; the cost reduction through logic minimization is unlikely to be significant when the total number of transistors is in the order of a million. On the other hand, it is important to save on interconnection costs, since wires are far more expensive in VLSI than transistors.

As a result of sustained research and development efforts by a number of groups all over the world for over three decades, a number of sophisticated design tools are available today for designing integrated circuits, and we are briskly moving towards complete design automation. In this book, we are concerned with algorithms for VLSI design automation, with an emphasis on physical design automation. Physical design of an integrated circuit refers to the process of generating the final layout for the circuit. Needless to say, physical design is of vital importance, and a lion's share of design automation research has gone into developing efficient algorithms for automating the layout process.

## 1.2 THE VLSI DESIGN PROCESS

Since the complexity of VLSI circuits is in the order of millions of transistors, designing a VLSI circuit is understandably a complex task. In order to reduce the complexity of design process, several intermediate levels of abstractions are introduced. More and more details about the new design are introduced as the design progresses from highest to lowest levels of abstractions. Typical levels of abstractions together with their corresponding design steps are illustrated in Figure 1.1. As indicated in Figure 1.1 the design is taken from specification to fabrication step by step with the help of various CAD tools. Clearly it is not possible to sit down with paper and pencil to design a million-transistor circuit (or chip). A human engineer can reason about a handful of objects at best. It is easy for a human engineer to think in terms of larger *circuit modules* such as arithmetic units, memory units, interconnection networks, and controllers.

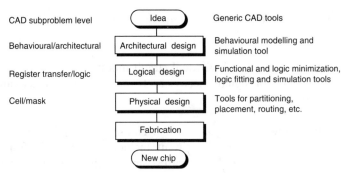

CAD subproblem level

Behavioural/architectural

Register transfer/logic

Cell/mask

Generic CAD tools

Behavioural modelling and simulation tool

Functional and logic minimization, logic fitting and simulation tools

Tools for partitioning, placement, routing, etc.

Idea

Architectural design

Logical design

Physical design

Fabrication

New chip

**Figure 1.1** Levels of abstraction and corresponding design step.

Designing a circuit at this level of abstraction is known as *architectural design*.

### 1.2.1 Architectural design

Architectural design of a chip is carried out by expert human engineers. Decisions made at this stage affect the cost and performance of the design significantly. Several examples of decisions made during the architectural design of a microprocessor are given below.

1. What should be the instruction set of the processor? What memory addressing modes should be supported? Should the instruction set be compatible with that of another microprocessor available in the market?
2. Should instruction pipelining be employed? If so, what should be the depth of the pipeline?
3. Should the processor be provided with an on-chip cache? How big should the cache memory be? What should be the organization of the cache? Should instruction cache be separated from data cache?
4. Should the arithmetic unit be designed as a bit-serial unit or as a bit-parallel unit? If bit-serial arithmetic is used, one saves on hardware cost but loses on performance.
5. How will the processor interface to the external world? Are there any international standards to be met?

Architectural design cannot be done entirely by a computer program. However, computer programs can aid the system architect in making important decisions. For instance, an architect can tune a parameter (such as the size of the cache) through simulation. Simulators and performance prediction tools are very useful to a computer architect who is experimenting with an innovative idea.

Once the system architecture is defined, it is necessary to carry out two things:

(a) Detailed logic design of individual circuit modules.
(b) Derive the control signals necessary to activate and deactivate the circuit modules.

The first step is known as *data path design*. The second step is called *control path design*. The *data path* of a circuit includes the various functional blocks, storage elements, and hardware components to allow transfer of data among functional blocks and storage elements. Examples of functional blocks are adders, multipliers, and other arithmetic/logic units. Examples of storage elements are shift registers, random access memories, buffers, stacks, and queues. Data transfer is achieved using tristate busses or a combination of multiplexers and demultiplexers.

The *control path* of a circuit generates the various control signals necessary to operate the circuit. Control signals are necessary to initialize the storage elements in the circuit, to initiate data transfers among functional blocks and storage elements, and so on. The control path may be implemented using hardwired control (random logic) or through micro-programmed control.

**Example 1.1**  It is required to design an 8-bit adder. The two operands are stored in two 8-bit shift registers $A$ and $B$. At the end of the addition operation, the sum must be stored in the $A$ register. The contents of the $B$ register must not be destroyed. The design must be as economical as possible in terms of hardware.

SOLUTION  There are numerous ways to design the above circuit, some of which are listed below.

1. Use an 8-bit carry look-ahead adder.
2. Use an 8-bit ripple-carry adder.
3. Use two 4-bit carry look-ahead adders and ripple the carry between stages.
4. Use a 1-bit adder and perform the addition serially in 8 clock cycles.

Since it is specified that the hardware cost must be minimum, it is perhaps best to select the last option, namely the serial adder. The organization of such an adder is shown in Figure 1.2. Let $A_k$ and $B_k$ indicate the $k$th significant bit of register $A$, and $B$ respectively, $k = 0, 1, \cdots, 7$. The basic idea in the serial adder is to use a full-adder to add $A_k$ and $B_k$ and the carry $C_{k-1}$ during the $k$th clock cycle. The

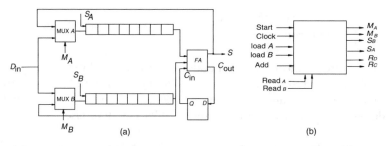

**Figure 1.2** Organization of a serial adder. FA is a full-adder. (a) Data path. (b) Control path block diagram.

carry generated during the $k$th clock cycle is saved in a D flip-flop for use in the next iteration. The output of the D flip-flop is initially set to 0 with a RESET operation, hence $C_{-1} = 0$.

Each shift register has a serial-input pin, and a serial-output pin. The bits $A_k$ and $B_k$ are available on the serial-output pins of the registers $A$ and $B$ at the beginning of the $k$th clock cycle. The serial-input pin of $A$ is fed from the output of a multiplexer which selects either the sum bit or the input signal.

The data path of the serial adder consists of two 8-bit shift registers, a full-adder, a D flip-flop, and two multiplexers. In addition, a 3-bit counter is required to count the number of times bit-wise addition is being performed. The relevant control signals are tabulated below.

| | |
|---|---|
| $S_A$ | Shift the register $A$ right by one bit |
| $S_B$ | Shift the register $B$ right by one bit |
| $M_A$ | Control multiplexer $A$ |
| $M_B$ | Control multiplexer $B$ |
| $R_D$ | Reset the D flip-flop |
| $R_C$ | Reset the counter |
| $START$ | A control input, which commences the addition |

The control algorithm for adding $A$ and $B$ is given below.

**forever do**
        **while** (START = 0) **skip**;
        Reset the D flip-flop and the counter;
        Set $M_A$ and $M_B$ to 0;
        **repeat**
                Shift registers A and B right by one;

counter = counter + 1;
**until** counter = 8;

The control path of the serial adder consists of hardware required to implement the above control algorithm, i.e., to generate the control signals tabulated above. We leave it as an exercise to the reader to design the control path for the serial adder (see Exercise 1.1).

Several observations can be made by studying the example of the serial-adder. First, note that designing a circuit involves a tradeoff between cost, performance, and testability. The serial adder is cheap in terms of hardware, but slow in performance. It is also more difficult to test the serial adder, since it is a sequential circuit. The parallel 8-bit carry look-ahead adder is likely to be fastest in terms of performance, but also the costliest in hardware.

All the different ways that we can think of to build an 8-bit adder constitute what is known as the *design space* (at that particular level of abstraction). Each method of implementation is called a point in the design space. There are advantages and disadvantages associated with each design point. When we try optimizing the hardware cost, we usually lose out on performance, and vice versa. We have mentioned hardware cost, performance, and testability as three important design aspects; there are many more, such as power dissipation, fault tolerance, ease of design, and ease of making changes to the design. A circuit specification may pose *constraints* on one or more aspects of the final design. For example, when the specification says that the circuit must be capable of operating at a minimum of 15 MHz, we have a constraint on the timing performance of the circuit.

Given a specification, the objective is to arrive at a design which meets all the constraints posed by the specification, and optimizes on one or more of the design aspects. This problem is also known as *hardware synthesis*. Computer programs have been developed for data path synthesis as well as control path synthesis. The automatic generation of data path and control path is known as *high-level synthesis* (Camposano et al., 1990).

## 1.2.2 Logic design

The data path and control path (derived automatically or manually) will have components such as arithmetic/logic units, shift registers, multiplexers, buffers, and so on. Further design steps depend on the following factors.

1. How is the circuit to be implemented, on a PCB or as a VLSI chip?
2. Are all the components readily available as off-the-shelf integrated circuits or as predesigned modules?

If the circuit must be implemented on a printed-circuit board using off-the-shelf components, then the next stage in design is to select the components so as to minimize the total cost and at the same time maximize the performance. Following the selection procedure, the IC chips are placed on one or more circuit boards and the necessary interconnections are established using one or more layers of metal deposits. A similar procedure may be used in case the circuit must be implemented on a VLSI chip using predesigned circuit components from a *module library*. The predesigned modules are also known as *macro-cells*. The cells must be placed on the layout surface and wired together using metal and polysilicon (poly) interconnections.

## 1.2.3 Physical design

Physical design of a circuit is the phase that precedes the fabrication of a circuit. In most general terms, physical design refers to all synthesis steps succeeding logic design and preceding fabrication. These include all or some of the following steps: logic partitioning, floorplanning, placement, and routing. The performance of the circuit, its area, its yield, and its reliability depend critically on the way the circuit is physically laid out. To begin with, consider the effect of layout (placement and routing) on the timing performance of a circuit. In an integrated circuit layout, metal and polysilicon are used to connect two points that are electrically equivalent. Both metal and poly lines introduce wiring impedances. Thus a wire can impede a signal from travelling at a fast speed. The longer the wire, the larger the wiring impedance, and the longer the delays introduced by the wiring impedance. When more than one metal layer is used for layout, there is another source of impedance. If a connection is implemented partly in metal layer 1 and partly in metal layer 2, a *via* is used at the point of layer change. Similarly, if a connection is implemented partly in poly and partly in metal, a *contact* becomes necessary to perform the layer change. Contacts and vias introduce a significant amount of impedance, once again contributing to the slowing down of signals.

Layout affects critically the area of a circuit. There are two components to the area of an integrated circuit—the functional area, and the wiring area. The area taken up by the active elements in the circuit is known as the functional area. For instance, in the example of the serial adder of the previous section, the functional modules are the full-adder, the registers, the multiplexers, the D flip-flop, the counter, and the logic circuits necessary to implement the control path. The area occupied by these modules constitutes the functional area of the circuit. The wires used to interconnect these functional modules contribute to the wiring area.[†] Just as they affect the performance of the circuit, long wires and *vias* also affect the area of the circuit. A good layout should have strongly connected modules placed close together, so that long wires are avoided as much as possible. Similarly, a good layout will have as few *vias* as possible.

The area of a circuit has a direct influence on the *yield* of the manufacturing process. We define *yield* to be the number of chips that are defect-free in a batch of manufactured chips. The larger the chip area, the lower the yield. A low yield would mean a high production cost, which in turn would increase the selling cost of the chip.

The reliability of the chip is also influenced by the layout. For instance, *vias* are sources of unreliability, and a layout which has a large number of *vias* is more likely to have defects. Further, the width of a metal wire must be chosen appropriately by the layout program to avoid *metal migration*. If a thin metal wire carries a large current, the excessive current density may cause wearing away of metal, tapering the wire slowly, resulting in an open circuit.

## 1.3 LAYOUT STYLES

In this section, we describe the layout approaches used to generate physical representations of circuits. These approaches differ mainly in the structural constraints they impose on the layout elements as well as the layout surface. The various layout approaches belong to two general classes described below.

1. The full-custom layout approach where layout elements are handcrafted and can be placed anywhere on the layout surface (no constraints imposed).
2. The semi-custom approaches which impose some structure on the layout elements and surface in order to reduce the complexity of the layout tasks.

---

[†]Some authors treat data routing hardware such as multiplexers, demultiplexers, and busses as *interconnects* and add their areas to the wiring area as well. Some authors treat transistors alone as functional elements in the circuit and consider only the contribution of transistors towards functional area.

Current layout styles are:

1. Full-custom;
2. Gate-array design style;
3. Standard-cell design style;
4. Macro-cell (building block layout);
5. PLA (programmable logic array); and
6. FPGA (field programmable gate-array) layout.

Next, we discuss each of these layout styles in detail.

## 1.3.1 Full-custom layout

*Full-custom layout* refers to manual layout design, where an expert artwork designer uses a layout editor to generate a physical layout description of the circuit. Layout editors are discussed in Chapter 9 of this book. Full-custom design is a time-consuming and difficult task. However, it gives full control to the artwork designer in placing the circuit blocks and interconnecting them. As a result, an expert can often achieve a high degree of optimization in both the area and performance of the circuit. For instance, the layout designer can control the width-to-length ratio of individual transistors to tune the performance of the circuit. Similarly, the shape of the layout can be controlled more easily in a full-custom approach. Using this approach, it may take several man-months to lay out a VLSI chip manually. As a result, the full-custom approach is used only for circuits that are to be mass produced, e.g., a microprocessor. For a circuit which will be reproduced in millions, it is important to optimize on the area as well as performance. Full-custom design pays off in this situation, since the high design cost is amortized over the large number of copies manufactured.

For the full-custom layout style the designer productivity can be greatly improved with the availability of a good layout editor. A layout editor is more than a drawing tool. It can aid the artwork designer in several ways. For example, a layout editor can perform on-line design rule check (DRC). Design rules are a set of precautions that must be taken while drawing a layout. There are two types of design rules, width rules and spacing rules. A *width rule* specifies the minimum width of a feature. For example, in a particular CMOS technology, it may be required that a metal wire be at least 4 μm thick. If a metal line is thinner than 4 μm, the metal line may not be continuous when the chip is actually manufactured, due to the tolerances in mask alignment and other manufacturing processes. A *spacing rule* specifies the minimum distance that must be allowed between two features. Again, in a particular CMOS technology, it may be specified that the minimum spacing between two metal wires is 6 $\mu$m. If two metal wires are indeed placed closer than recommended, it is possible that they are short-circuited due to tolerances of manufacturing processes.

The width rules and spacing rules differ for different manufacturing technologies. Mead and Conway simplified these design rules and created what are popularly known as $\lambda$-based design rules (Mead and Conway, 1980). The key idea is to characterize a manufacturing technology with a single parameter $\lambda$. All the width and spacing rules are then specified in terms of the parameter $\lambda$. Thus, consider four design rules which call for a minimum spacing of 4, 5, 6.5, and 8 $\mu$m. If we select $\lambda = 2$ $\mu$m, these spacings can be specified as $2\lambda$, $2.5\lambda$, $3.25\lambda$, and $4\lambda$ respectively. Mead and Conway recommended the use of integral multiples of $\lambda$ for spacings; if we permit real coefficients for $\lambda$, it beats our purpose of achieving simplicity in stating the design rules. Thus, it is meaningless to state that the spacing between metal lines should be $1.0125\lambda$; we might as well state the actual spacing. Keeping this in mind, we can round off the coefficients of $\lambda$ in the above example of four design rules and restate the spacings as $2\lambda$, $3\lambda$, $4\lambda$, and $4\lambda$. The price to be paid for achieving this simplicity is, of course, that the design rules are somewhat 'overcautious'—a minimum spacing of 8 $\mu$m is specified even though 6.5 $\mu$m is adequate. Thus, a layout which follows Mead–Conway design rules can be expected to be somewhat larger than necessary.

Layout editors such as Magic (see Chapter 9) employ Mead–Conway design rules to perform DRC on-line (Ousterhout et al., 1984). Mead–Conway rules simplify the process of design rule verification. The layout editor can treat the layout surface as a grid, where the separation between two vertical (or horizontal) lines is $\lambda$. Further, the design rules (stated in terms of $\lambda$) can be read from a 'technology file'; this allows the same editor to be used for creating layouts for different manufacturing technologies such as 2 $\mu$m CMOS, 1.5 $\mu$m CMOS, 3 $\mu$m $n$MOS, and so on.

Skilled full-custom designers are not likely to use layout editors based on Mead–Conway rules, since optimizing the chip area is of utmost importance in a full-custom design. A specialized layout editor which uses the distance rules specific to technologies will be more suitable for full-custom design.

Full-custom design is prohibitively expensive for circuits which are unlikely to be manufactured in large numbers. A class of circuits, known as Application-Specific Integrated Circuits (ASICs), falls into this category. An ASIC is a circuit which performs a specific function in a particular application. Unlike a microprocessor, an ASIC is not programmable for different applications. As a result, an ASIC has a limited market. A digital filter is an example of an ASIC. Similarly, a chip which performs a specific image processing function is an example of an ASIC. Optimizing the area and performance of the chip are not the important issues in designing an ASIC. It is more important to reduce the *time to market*, i.e., the sum total of design time, manufacturing time, and testing time. Since manufacturing time is out of the purview of a designer, it is only possible for the designer to reduce the other two, namely, the time to design and test the chip.

A high degree of automation is required in order to reduce design and test times. In turn, automation can only be achieved by standardizing the design and testing process. Standard test techniques such as *scan based design* are commonly employed in ASICs; such techniques reduce the time for test generation. Standard layout architectures are used to reduce design time. Gate-arrays, sea-of-gates, standard-cells, and programmable logic arrays are examples of standard layout architectures.

## 1.3.2 Gate-array layout

A gate-array† consists of a large number of transistors which have been prefabricated on a wafer in the form of a regular two-dimensional array. A single wafer can consist of many arrays. Initially the transistors in an array are not connected to one another. In order to realize a circuit on a gate-array, metal connections must be placed using the usual process of masking. This process of adding metal wires to a gate array is called *personalizing* the array. After personalization, the wafer can be diced and individual gate-arrays can be separated, packaged, and tested.

Since all the processing other than personalization is identical to all gate-arrays, irrespective of the circuit to be implemented, a foundry can stock up a large number of wafers which have been prefabricated up to metalization. Therefore, it takes a very short time to get a gate-array chip fabricated. Gate-arrays are also called Mask Programmable Gate-Arrays (MPGAs). The cost of producing a gate-array chip is low due to the high yield. This is because there are only a small number of processing steps involved in a personalization—only four masking steps are necessary, one each for the two metal layers and two contact layers.

Personalization involves two types of interconnections—intra-cell wiring and inter-cell wiring. A cell is a small circuit module, such as a two-input NAND gate, which can be implemented by connecting a group of transistors in a local neighbourhood of the gate-array. Implementing a cell on a gate-array is straightforward. A cell library can be maintained, in which an interconnection pattern is stored for each cell. Thus intra-cell wiring is also independent of the circuit being implemented on the gate-

---

† Sometimes gate-arrays are also referred to as *master slice* circuits.

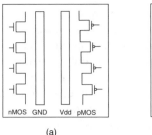 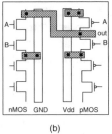

(a)                                  (b)

**Figure 1.3** (a) Example of a basic cell in a gate-array. (b) Cell personalized as a two-input NAND gate.

array. Inter-cell wiring, however, is specific to the circuit and is handled by the layout software. A typical gate-array cell personalized as a two-input NAND gate is shown in Figure 1.3.

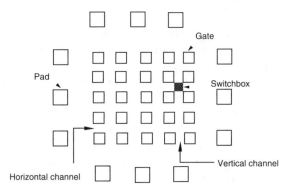

**Figure 1.4** A gate-array floor plan.

In order to carry out inter-cell wiring in a systematic fashion, the gate-array is structured as a regular array of basic cells (see Figure 1.4).

Each square in the figure represents a cell and encloses a group of transistors in the gate-array. The inter-cell wiring is carried out in the regions called *channels* shown in Figure 1.4. The floorplan of the gate-array chip can be likened to that of a township which has a number of buildings (cells) and streets (channels) to carry traffic (wires) from one building to another. A cross-street, the area where a horizontal channel intersects a vertical channel, is called a *switchbox*. A fixed number of horizontally (vertically) running wires can be placed in any horizontal (vertical) channel. This number is called the horizontal (vertical) track density of the channel.

Due to the limited amount of wiring space, gate-arrays present difficulties to an automatic layout generator. If the layout program attempts to avoid long inter-cell wires by placing strongly connected cells close together in the array, the local congestion may make the layout unrouteable. In other words, the number of wires to be routed across a channel exceeds its capacity. The floorplan of a gate-array chip also presents limitations. Since the cells are of the same size (width and height) the cells in the library must be specially designed to meet this requirement.

### Sea of gates (channel-less gate-arrays)

A special case of the gate-array architecture is when routing channels are very narrow, or virtually absent. Thus, the chip consists of a closely packed array of transistors. Since the channels are too narrow, they cannot be used to route wires. Wires must therefore be routed *over the transistors*. This is sometimes called *over the cell* routing. The architecture is called *sea of gates* to suggest the absence of routing streets.

### 1.3.3 Standard-cell layout

A standard-cell, known also as a polycell, is a logic block that performs a standard function. Examples of standard-cells are two-input NAND gate, two-input XOR gate, D flip-flop, two-input multiplexer, and so on. A cell library is a collection of information pertaining to standard-cells. The relevant information about a cell consists of the name of the cell, its functionality, its pin structure, and a layout for the cell in a particular technology such as 2 μm CMOS. Cells in the same library have standardized layouts, that is, all cells are constrained to have the same height.

Example 1.2 Consider the description of a cell named i1s shown in Figure 1.5. The description has been taken from a standard-cell library associated with the OASIS design automation system (OASIS, 1992). The cell description is that of an inverter whose input is a and output is q. We are only interested in the profile, termlist, and siglist statements; the other details of the circuit, most of which are self-explanatory, are unimportant here. The cell is rectangular in shape (all standard-cells in OASIS are) and has dimensions 16 μm ×58 μm in 2.0 μm CMOS technology. The area of the cell is 928.0 μm$^2$. The unplaced cell has its lower left corner at $(-1, -1)$ and the top right corner at $(15, 57)$. The input a is available both on top and bottom; the pin

```
cell begin i1s generic=i1 primitive=INV area=928.0 transistors=2
        function="q = INV((a))"
        logfunction=invert
 profile top (-1,57) (15,57);
 profile bot (-1,-1) (15,-1);
 termlist
  a (1-4,-1) (1-4,57)
  pintype=input
  rise_delay=0.35 rise_fan=5.18
  fall_delay=0.28 fall_fan=3.85
  loads=0.051 unateness=INV;
  q (9-12,-1) (9-12,57)
  pintype=output;
 siglist
 GND Vdd a q
 ;
 translist
 m0 a GND q length=2000 width=7000 type=n
 m1 a Vdd q length=2000 width=13000 type=p
 ;
 caplist
 c0 Vdd GND 2.000f
 c1 q GND 5.000f
 c2 Vdd a 2.000f
 c3 a GND 11.000f
 ;
 cell end i1s
```

**Figure 1.5** Description of an inverter logic module named i1s. Focus only on the profile, termlist, and siglist statements.

position is from 1–4 in both the top and bottom. Similarly, the signal q is available both on top and bottom at coordinates 9–12. The siglist statement lists the signals associated with the cell; these are the power signal Vdd, the ground signal GND, the input signal a, and the output signal q. The CMOS layout of this cell is given in Figure 1.6.

## Cell-based design

The reader is likely to be familiar with the process of designing a circuit using SSI and MSI level components. Cell-based design is identical to this process, except for the implementation details; instead of off-the-shelf SSI or MSI components, we have to select components from a cell library. And, instead of placing the components on a PCB, we place the cells in silicon. The advantage of designing with a cell library is, of course, that designs can be completed rapidly. Since cell layouts are readily available, a layout program will only be concerned with:

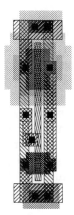

**Figure 1.6** Layout of standard-cell i1s.

1. the location of each cell; and
2. interconnection of the cells.

Placement and routing is again simplified using a standard floorplan (see Figure 1.7). The layout is divided into several rows. A row consists of cells

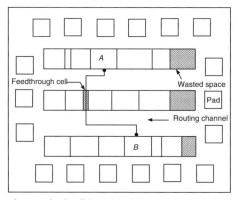

**Figure 1.7** Floorplan of a standard-cell layout.

placed next to each other. Note that the height of a row is the same as the height of any cell in the row, since all the cells are predesigned to have the same height. Rows are separated by horizontal routing channels. Cells within the same row, or cells from two facing rows can be interconnected by running wires through the adjacent channel. If two cells in non-adjacent rows have to be connected, a more elaborate technique is called for. Figure

1.7 illustrates the point. Here, cell $A$ in row 1 has a connection to cell $B$ in row 3. A special type of cell, called a *feedthrough cell* is placed in row 2. A feedthrough cell has nothing else but one or more vertical wires running straight. The cell $A$ is first connected to one of the wires in the feedthrough from the top side; the bottom connection of the same wire is then connected to cell $B$.

Compared to gate-array layout, standard-cell layout offers more flexibility. In a standard-cell chip, the wiring space is not fixed beforehand. Moreover, the cells can have varying widths. The disadvantage of standard-cells in relation to gate-arrays is, of course, that all the fabrication steps are necessary to manufacture the chip.

### 1.3.4 Macro-cell layout

Both gate-array design and standard-cell design impose restrictions on the cells that are used to design the circuit. For example, the cells in a standard-cell layout must have the same height. If this restriction is removed, the cells can no longer be placed in a row-based floorplan as explained in Section 1.3.3. Even if we force a row-based floorplan, it would be very

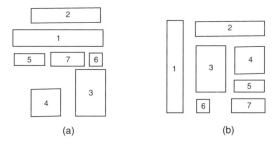

(a)                    (b)

**Figure 1.8** (a) Cells of varying heights and widths placed in a row-based floorplan. (b) A more compact floorplan for the same circuit.

inefficient in terms of layout area. Figure 1.8(a) illustrates the point. In Figure 1.8(b), the same set of cells are arranged in a better more compact floorplan. The design style which permits cells to vary in both dimensions is called *macro-cell* design style, or *building-block* design style. The main advantage of macro-cell design is that cells of significant complexity are permitted in the library. Thus, registers, register files, arithmetic-logic units, memories, and other architectural building blocks can be accommodated in the library.

There is an advantage in storing blocks such as arithmetic-logic units in a cell library. Such blocks can often be designed to have efficient layout characteristics. Consider, for example, an 8-bit array multiplier. Due to its regular structure, it permits an efficient layout. If the same multiplier has to be designed using simple cells such as logic gates, there is no guarantee of how the cells will be finally arranged by the layout program. If we wish to maintain the array topology of the multiplier, the only way is to store its layout as a building-block.

Building-block layout (BBL) comes closest to full-custom layout. Like standard-cell layout style, all the processing steps are required to manufacture a BBL chip. As you can guess, it is much more difficult to design an automatic layout program for the BBL design style. This is because there is no standard floorplan to adhere to. As a result, the routing channels are not predefined either. Floorplanning and channel definition are additional steps required in a BBL layout system.

## Module generation

The concept of storing cell layouts in a library is an attractive one, since it can save plenty of design effort. The library-based approach is applicable to gate-array, standard-cell, and building-block design styles. There is, however, a disadvantage with this approach. A cell library is strongly dependent on a manufacturing technology. Thus, separate libraries are required for 1.5 μm CMOS and 2.0 μm CMOS technologies. If a site is using 2.0 μm CMOS libraries today, and wants to upgrade to 1.5 μm technology, a considerable amount of effort is needed to redesign the cells for the new technology. Libraries can also be bought from design houses, but tend to be very expensive.

An alternative approach is to use a module generator that can compile the layout of a cell starting from a specification of the cell. This specification may be a *functional description*, such as truth table, or a *structural description*, such as a netlist. The required characteristics of the layout, such as the height of the cell, can be specified to the module generator. Cell compilation is gaining rapid acceptance in the industry. Chapter 8 considers layout generation in more detail.

## 1.3.5 Programmable logic arrays

Recall that any combinational logic function $Z$ can be written in the sum-of-products (SOP) form. For instance, suppose $Z$ is defined as a function of three inputs $A_0, A_1, A_2$, such that $Z$ is true if and only if two or more inputs are true. The reader may verify that

$$Z = A_0 \cdot A_1 + A_0 \cdot A_2 + A_1 \cdot A_2 \tag{1.1}$$

is a minimal SOP representation for $Z$. A SOP expression can be realized using a two-level logic. The products (AND terms) are formed in the first level and their sum (the OR term) is computed in the second level. It is assumed that the inputs are available in both normal and inverted forms.

A programble logic array (PLA) is a convenient way to implement two-level sum-of-products expressions. A PLA consists of an AND plane to implement the product terms and an OR plane to implement the sums. For instance, consider the function $Z$ above and the function $Y$ below.

$$Y = A_0 \cdot A_1 + \overline{A_0} \cdot \overline{A_2} \tag{1.2}$$

These two functions can be implemented using the PLA shown in Figure 1.9. The AND plane of the PLA has two vertical lines corresponding to each input—one connected directly to the input, and the other to the inverted form of the input. There are as many horizontal lines in the PLA

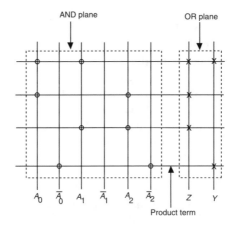

**Figure 1.9** A PLA to implement functions $Y$ and $Z$. There are four rows corresponding to four products terms. There are six columns in the AND plane, two for each of the inputs. There are two columns in the OR plane, one for each output.

as there are product terms. Looking at the equations for $Y$ and $Z$, there are four unique product terms. A circle is placed in the AND plane at the intersection of a row $i$ and a column $j$ if the product term $i$ contains the input $j$. For instance, in the row $A_0 \cdot A_1$, there are two circles where the row intersects the columns $A_0$ and $A_1$. The OR plane contains as many vertical lines as there are outputs (two in this case). A cross is placed in the OR plane at the intersection of row $i$ and column $j$ if the output corresponding to column $j$ contains the product $i$. For example, a cross is placed where the

$Y$ column intersects the row $A_0 \cdot A_1$. A circle in row $i$ and column $j$ of the AND plane represents a switch which will be turned ON if the input $j$ is true. A product line $i$ is ON if all the switches in row $i$ of the AND plane are turned ON. Similarly, a cross in row $i$ and column $j$ of the OR plane represents a switch that will be turned ON if the product term $i$ is ON. An output line $j$ is turned ON if any one of the switches on line $j$ is ON.

## Merits and limitations

Due to its fixed architecture, it is easy to automate the generation of PLA layouts. Mead and Conway (1980) popularized the use of PLAs in VLSI layouts. The use of PLAs is not restricted to combinational circuits alone. Recall that a finite state machine (FSM) has a combinational part and memory elements; some of the outputs of the combinational circuit are fed back as inputs to the combinational circuit through memory elements. If memory elements (such as D flip-flops) are separately available, PLAs can be used to implement the combinational part of an FSM.

PLAs are commonly used to implement the control path of a digital circuit, since control signals are often written as SOP expressions. For instance, suppose that the $LOAD$ control of a register $R0$ must be ON during the first clock cycle if the signal $NEG$ is true, during the seventh clock cycle if the signal $NEG$ is false, and unconditionally during the eighth clock cycle. Then

$$LOAD = NEG \cdot \phi_1 + \overline{NEG} \cdot \phi_7 + \phi_8 \qquad (1.3)$$

where $\phi_i$ signal indicates that the $i$th clock cycle is in progress.

PLAs are not well suited for implementing the data path of a circuit. Since PLAs can only implement two-level logic, it may become necessary to write data path expressions in an (unnatural) SOP format. (Try to implement a 4-bit adder as a PLA to appreciate the point.) Such expressions tend to contain:
1. product terms with a large number of inputs, and
2. sum terms with a large number of products.
The rise and fall delays of the output lines are severely affected by either of the above conditions.

## 1.3.6 FPGA layout

Similar to an MPGA, an FPGA (field programmable gate-array) also consists of a two-dimensional array of logic blocks. Each logic block can be programmed to implement any logic function of its inputs. Thus they are usually referred to with the name configurable logic blocks (CLBs). In addition to this, as shown in Figure 1.10, the channels or switchboxes between logic blocks contain interconnection resources. The interconnection resources (or simply *interconnect*) consist of wire segments of various lengths. These interconnects contain programmable switches that serve to connect the logic blocks to the wire segments, or one wire segment to another. Furthermore, I/O pads are confined to the array periphery and are programmable to be either input or output pads. The main design steps when using FPGAs to implement digital circuits are: (1) mapping of the initial logic description of the circuit into a netlist of CLBs (technology mapping), (2) assigning to each CLB in the netlist, a corresponding CLB in the array (placement), (3) interconnecting the CLBs of the array (routing), and finally, (4) generating the bit patterns to ensure that the CLBs perform the assigned function and are interconnected as decided by the routing step. In MPGAs the interconnection is done at the foundry by customizing the metallization mask to a specified digital system implementation. However in FPGAs, both the logic blocks, and the interconnects are *field programmable*.

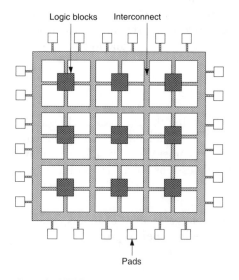

**Figure 1.10** Diagram of a typical FPGA.

FPGAs were first introduced in 1985 by Xilinx Company. Since then, many different FPGAs have been developed by a number of other companies such as Actel, Altera, etc. (Brown *et al.*, 1992). The judicious design of logic blocks coupled with that of the interconnection resources, facilitates the implementation of a large number of digital logic circuits. There are many ways to design FPGAs. Design issues include tradeoffs in the granularity and flexibility of both logic blocks and the interconnection resources.

Logic blocks can be fine grain modules such as two-input NAND gates or coarse grain modules consisting of complex structures such as multiplexers, look-up tables, PAL (programmable array logic), etc. Most FPGA logic blocks contain one or two flip-flops to aid the implementation of sequential circuits.

The structure and content of the interconnect in an FPGA is called the *routing architecture*. The routing architecture consists of wire segments and programmable switches. The programmable switches are constructed using pass-transistors (controlled by static RAM cells), anti-fuses, or EPROM/ EEPROM transistors. Similar to the logic block, the complexity of routing architecture can vary from simple connections between blocks to more complex interconnection structures.

The advantages of FPGAs over MPGAs are lower prototyping cost and shorter production times. The main disadvantages are their lower speed of operations and lower gate density. The programmable switches and associated programming circuitry require a large amount of chip area compared to the metal connections in gate-arrays. These programmable switches also have significant resistance and capacitance which account for the low speed of operation.

FPGAs are most ideally suited for prototyping applications, and implementation of random logic using PALs. They have also been successfully used in the implementation of ASICs. Reported examples of ASICs include controllers for FIFO, printers, graphics engine, network transmitter/receiver, etc. (Brown *et al.*, 1992).

## 1.4 DIFFICULTIES IN PHYSICAL DESIGN

From the preceding section, it must be clear that physical design is a complex optimization problem, involving several objective functions. A good layout is one which occupies minimum area, uses short wires for interconnection, and uses as few *vias* as possible. As importantly, a layout must meet all the *constraints* posed by the specification. For instance, if the target technology is a gate-array, then there is a constraint on the amount of wiring space available. The number of routing layers available is another constraint. Similarly, there may be constraints on the routing model, e.g., only horizontal and vertical wiring is permitted, layer changes are permitted only between two adjacent layers, power and ground wires must be in metal, power and ground wires must be wide enough to permit maximum current density, and so on. In addition to all this, we would like to be able to generate a correct and good layout as quickly as possible.

There are practical difficulties in trying to meet all the requirements stated above. First of all, it is difficult to model the physical design problem if we place so many constraints and wish to optimize so many objective functions. What makes the problem harder is the fact that some of these objective functions conflict with one another. To illustrate this point, consider the gate-array layout problem. If we attempt to minimize the total length of interconnection wiring by trying to place strongly connected components close together, we are likely to increase wiring congestion in some regions of the layout. It may then not be possible to route the circuit at all, because there is only a fixed number of tracks available in each wiring channel of the gate-array.

It is clear that we cannot write a single computer program to deal with the physical design problem. The fact that there are several layout styles makes the problem more difficult. Thus, different approaches are required for macro-cell placement, standard-cell placement, and gate-array placement. In a gate-array placement, there are constraints on wiring area, the number of channels, and the number of tracks per channel. Therefore, routability is of major concern in gate-array layout. In standard-cell layout, there is more flexibility in terms of wiring area, and hence the stress is on optimizing the wiring area. In addition, one attempts to minimize feedthrough cells so as to reduce the total area of the chip. Macro-cell placement must deal with cells of different sizes and shapes. Therefore, it is first necessary to design a floorplan for the chip and define the channels. The order in which these channels must be routed is also an important consideration.

### 1.4.1 Problem subdivision

Even if we restrict ourselves to a single layout style, physical design still remains a complex task. It is therefore customary to adopt a stepwise approach and subdivide the problem into more manageable subproblems. A possible subdivision is as follows.

1. Circuit partitioning.
2. Floorplanning and channel definition.
3. Circuit placement.
4. Global routing.
5. Channel ordering.
6. Detailed routing of power and ground nets.
7. Channel and switchbox routing.

Partitioning a circuit is necessary if it is too large to be accommodated on a single chip. Floorplanning and channel definition are required for a macrocell layout. Floorplanning includes finding the alignment and relative orientation of modules so that the total area of the chip is minimized. After a floorplan has been found, the routing region must be divided into channels and switchboxes. During placement, the exact positions of circuit components are determined, so as to reduce the estimated wiring area. Routing follows the placement phase, and is carried out in two stages— global routing and detailed routing. Global routing decides, for each net, a rough routing plan in terms of the *channels* through which the net will be routed. It is then necessary to select an order for routing the channels and switchboxes. This is because the description of one channel may depend on the routing of other channels. By a channel description we mean the exact ordering of pins on the top and bottom sides of the channel. When the complete description of a channel is available, the *detailed* routing of the nets within the channel can begin. Detailed routing involves actual assignment of wires to tracks. Power and ground nets are generally handled separately, due to the special constraints on their routing style, as explained earlier. It is a common practice to perform power and ground routing first and then deal with the remaining signal nets.

## 1.4.2 Computational complexity of layout subproblems

All of the aforementioned subproblems are constrained optimization problems. A constrained optimization problem consists of finding a feasible solution which satisfies a specified set of design constraints and optimizes a stated objective function. Examples of design constraints for the layout problems are: restricted number of wiring layers, cell sizes and/or shapes, available routing resources, geometric constraints, etc. Examples of objective functions could be: overall wiring length, wiring channel densities, circuit performance (wiring delays), or a combination of these.

Unfortunately these layout subproblems are NP-Hard.[†] Even simplified versions of these problems remain NP-Hard. Therefore, there are no known efficient algorithms that will find optimum solutions for these problems and it is very unlikely that such efficient algorithms will be found. For instance, a subproblem arising during the layout of a circuit with $n$ cells is the arrangement of these cells into a linear sequence so as to minimize the overall connection length. The search space contains $n!$ possible

[†]Appendix A gives a brief summary of NP-hardness and NP-completeness.

arrangements. A brute force approach is necessary to examine all $n!$arrangements in order to select the one with minimum connection length. However, this is an impractical solution approach. For example, if 1 $\mu$s is required per solution, then for $n = 20$, the brute force approach will identify the optimum in about 80,000 years! Of course, the existence of a large solution space does not imply that all of it *should* be searched in order to find the best solution. There might be, one could argue, a clever way to eliminate searching through large portions of the search space and curtail the actual search to $f(n)$ arrangements, where $f(n)$ is a polynomial of $n$. A polynomial function of $n$, such as $n^2 + 2n$, does not increase too fast for large $n$. Unfortunately, no such clever search technique is known thus far for this problem.

How do we solve layout problems then? There is little doubt that VLSI layout involves large problem sizes. There are hundreds to thousands of logic blocks to be placed, and hundreds to thousands of nets to be routed. Therefore, instead of optimal enumerative techniques, we must resort to *heuristic techniques*. An heuristic is a *clever* algorithm which will only search inside a subspace of the total search space for a 'good' (rather than the best) solution which satisfies all design constraints. Therefore, the time requirement of an heuristic algorithm is small. A number of heuristic algorithms have been developed over the past three decades for various layout problems. Many of these algorithms will be discussed in later chapters of this book.

It is natural to ask how good the solution generated by an heuristic really is. Assume that an heuristic algorithm $A$ has been developed for a minimization problem. If $S_A$ is the solution generated by the heuristic, and $S^*$ is the optimum solution, a measure of the error ($\epsilon$) made by the heuristic is the relative deviation of the heuristic solution from the optimal solution, that is,

$$\epsilon = \frac{S_A - S^*}{S^*} \tag{1.4}$$

Unfortunately, it is not easy to measure the error, since $S^*$ is not known. Therefore, we have to resort to other techniques for judging the quality of solutions generated by heuristic algorithms.

### 1.4.3 Solution quality

One method to tackle the above problem is to artificially generate test inputs for which the optimum solution is known a priori. For instance, in order to test an heuristic algorithm for floorplanning, we may generate the test input as follows. We start with a rectangle $R$ and cut it into smaller rectangles. If these smaller rectangles are given as input to the floorplanner, we already know the best solution—a floorplan which resembles the rectangle $R$. This method of testing, however, is not always feasible. It is difficult to generate such test inputs for channel routers, global routers, etc. Furthermore, an heuristic algorithm may perform well on artificial inputs, but poorly on real inputs and vice versa.

Test inputs comprising real circuits, called *benchmarks*, are used to compare the performance of heuristics. Generally, such benchmarks are universally recognized. Benchmarks are created by experts working in the field. For layout problems, there are two widely used sets of benchmarks: the Microelectronics Center of North Carolina (MCNC) benchmarks and the International Symposium on Circuits and Systems (ISCAS) benchmarks. Then alternative layout procedures are compared against the same benchmark tests (Brglez, 1993).

## 1.5 DEFINITIONS AND NOTATION

In this section, we shall introduce the terminology and notation used throughout the book.

A *cell* refers to a logic block which is useful in building larger circuits. Two- or three-input gates (AND, OR, NAND, NOR, and XOR), and flip-flops (D flip-flops, Set-Reset flip-flops, and JK flip-flops) are typical examples of a cell. The name *standard-cell* is used if the design of the cells has been standardized in some way. For instance, all the cells may have been specially laid out to have the same height and similar pin structure. Standard-cell layouts are usually stored in a database for ready access; such a database is known as the cell library. A *macro-cell*, or simply a *macro*, is a logic circuit composed of basic gates and possibly flip-flops. A macro is useful in building larger circuits. An example of a macro is a circuit which receives four inputs $A, B, C, D$ and computes $\overline{AB + CD}$. This particular macro is known as an AND-OR-INVERT *gate* for obvious reasons. Macro-cells are also known as *building blocks*, or simply, *blocks*. Macro-cell layouts can also be maintained in a library. But the layouts in a macro library may not adhere to any standard; their shapes and sizes may differ widely. Two particular shapes are popular in designing layouts for macro-cells—rectangular shapes and L-shapes. In this book, we will be predominantly concerned with rectangular shaped cells.

A rectangular cell is characterized by its height $h$ and width $w$. Alternatively, one may specify the coordinates of the lower left corner and the upper right corner. The *aspect ratio* of a rectangular cell is defined as the ratio $\frac{h}{w}$. For standard-cells, the aspect ratio is generally less than 1. No such restriction holds for macro-cells.

The word *logic module*, or simply *module*, is used when one wants to refer to either macro-cells or standard-cells. A module interfaces to other modules through *pins*. A pin is simply a wire (in either metal or polysilicon) to which another external wire can be connected. Pins may be provided on some or all sides of the module. In a standard-cell, pins are generally provided only on the top and bottom sides. In a macro-cell, pins are provided on all sides.

## 1.5.1 Nets and netlists

A *signal net*, or simply *net*, is a collection of pins which must be electrically connected. For example, in a sequential circuit, the *clock* pins of all the flip-flops must be connected. Suppose that there are three flip-flops in a circuit and their clock pins are named CK1, CK2, and CK3. The clock *net* consists of these three pins. A layout program must be told that the three pins must be connected; this is done by specifying the net CK = {CK1, CK2, CK3} to the program. The names of the pins are sometimes specified differently. In the above example, suppose that the flip-flop is called FF and the clock pin of the flip-flop is called CLK; the three *instances* of the flip-flop are termed FF[0], FF[1] and FF[2]. The pin names will then be FF[0].CLK, FF[1].CLK and FF[2].CLK. A list of all nets constitutes a netlist.

## 1.5.2 Connectivity information

The input to a layout program is a circuit description. Basically, one must specify the modules used in the circuit and how they are interconnected. The module information consists of the name of the module (such as FF[1] in the example above), its shape and size information, and the pin structure of the module. In particular layout styles, some of this information may be omitted. For example, in gate-array layout, all the modules are identical and there is no need to specify the size, shape and pin structure for each module separately.

The connectivity of modules may be described in more than one way. A *netlist* description is, as the name suggests, a list of all the nets in the circuit. For instance, consider how a netlist for the circuit of Figure 1.11 can be formed. This circuit generates the carry signal in a full-adder. There are three two-input AND gates in the circuit, which we shall name AND[1], AND[2], and AND[3] respectively. There is a three-input OR gate, which we shall denote as OR3[1]. The two-input AND gate has two input pins, IN1 and IN2, an output pin OUT, the power pin Vdd, and the ground pin GND. Similarly, the OR gate has three inputs IN1, IN2, IN3, the output

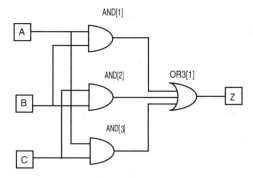

**Figure 1.11** Logic diagram of 'full-adder carry' circuit.

and the power and ground pins. Figure 1.12; shows an example of a complete netlist description for the carry circuit. Ground and power signals are omitted in Figure 1.11.

```
(A, AND[1].IN1)
(B, AND[1].IN2)
(C, AND[2].IN1)
(B, AND[2].IN2)
(A, AND[3].IN1)
(C, AND[3].IN2)
(AND[1].OUT, OR3[1].IN1)
(AND[2].OUT, OR3[1].IN2)
(AND[3].OUT, OR3[1].IN3)
(OR3[1].OUT, Z)
(AND[1].Vdd, AND[2].Vdd, AND[3].Vdd, OR3[1].Vdd)
(AND[1].GND, AND[2].GND, AND[3].GND, OR3[1].GND)
```

**Figure 1.12** Netlist for the 'full-adder carry' circuit.

For some layout problems, such as placement, it is customary to compose from the circuit netlist a connectivity graph model enclosing the information needed for the task. A graph is an abstract representation that is more convenient to work with than the original netlist. The connectivity graph of a circuit has one node corresponding to each module, input pad, and output pad. (For simplicity, we shall treat pads also as modules.) An edge is introduced between a node $i$ and a node $j$ if some pin of module $i$ is connected to a pin in module $j$. The connectivity graph for the previous example has eight nodes (three input pads, four gates, and one output pad). An edge is added from the node AND[1] to node OR3[1], to take care of the net (AND[1].OUT, OR3[1].IN1).

Thus the procedure to form a connectivity graph from the netlist description is straightforward; we look at each net of the form $(i, j)$ and draw an edge between the modules $m_i$ and $m_j$ to which the pins $i$ and $j$

belong. But how do we deal with multipin nets, i.e., nets with more than two pins? To illustrate, consider a three-pin net $(a, b, c)$. Let the pins $a, b, c$ belong to modules $m_a, m_b$, and $m_c$ respectively. We add three edges $(m_a, m_b)$, $(m_b, m_c)$, and $(m_c, m_a)$ to capture the information in the net $(a, b, c)$. In general, to handle a $k$-pin net, we add $\binom{k}{2}$ nets—a complete subgraph on the $k$ modules connected by that net.

The above procedure may give us multiple edges between two nodes $i$ and $j$ in the connectivity graph. We combine all these edges into a single edge which has an integral weight $c_{ij}$, where $c_{ij}$ is the number of multiple edges between nodes $i$ and $j$. We refer to $c_{ij}$ as the connectivity between modules $i$ and $j$. We also define the connectivity of a module $i$ as the sum of all terms of the form $c_{i*}$. If there are $n$ modules in the circuit, the connectivity $c_i$ of module $i$ is defined as follows,

$$c_i = \sum_{j=1}^{n} c_{ij} \qquad (1.5)$$

The connectivity information can be conveniently represented in the form of an $n \times n$ matrix $C$, where element $c_{ij}$ is the connectivity between modules $i$ and $j$. The matrix representation is especially convenient for computer programming. It may be worth noting that the connectivity matrix is symmetric, i.e.,

$$c_{ij} = c_{ji}, \quad i, j = 1, 2, \cdots, n \qquad (1.6)$$

Also, since it is never required to externally connect two pins of the same module,

$$c_{ii} = 0, \quad i = 1, 2, \cdots, n \qquad (1.7)$$

The connectivity graph and the connectivity matrix for the 'full-adder carry' circuit of Figure 1.11 is shown in Figures 1.13. and 1.14 respectively.

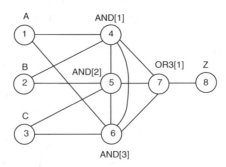

**Figure 1.13** The connectivity graph for the 'full-adder carry' circuit.

|   | 1 | 2 | 3 | 4 | 5 | 6 | 7 | 8 |
|---|---|---|---|---|---|---|---|---|
| 1 | 0 | 0 | 0 | 1 | 0 | 1 | 0 | 0 |
| 2 | 0 | 0 | 0 | 1 | 1 | 0 | 0 | 0 |
| 3 | 0 | 0 | 0 | 0 | 1 | 1 | 0 | 0 |
| 4 | 1 | 1 | 0 | 0 | 1 | 1 | 1 | 0 |
| 5 | 0 | 1 | 1 | 1 | 0 | 1 | 1 | 0 |
| 6 | 1 | 0 | 1 | 1 | 1 | 0 | 1 | 0 |
| 7 | 0 | 0 | 0 | 1 | 1 | 1 | 0 | 1 |
| 8 | 0 | 0 | 0 | 0 | 0 | 0 | 1 | 0 |

**Figure 1.14** The connectivity matrix for the 'full-adder carry' circuit.

Exercises 1.7–1.10 deal with computer programs to derive the connectivity matrix for a given circuit and verify a given connectivity matrix.

### 1.5.3 Weighted nets

Consider the netlist for the carry circuit, shown in Figure 1.12. As explained earlier, each net of the form $(i_1, i_2, \cdots, i_k)$ represents an electrical connection between the pins $i_1, i_2, \cdots, i_k$. Sometimes, it may be necessary to associate more information with a net. For example, it may be important to realize the connection specified by a certain net using as short wires as possible. In such a case, a positive *weight* is associated with the net to indicate how critical the net is.

The procedure to form the connectivity matrix from a weighted netlist is similar to what was discussed earlier. When forming the connectivity graph from the netlist, we draw an edge of weight $w_{ij}$ between modules $i$ and $j$ when we come across a net of weight $w_{ij}$ involving the two modules. Finally, to compute $c_{ij}$, we combine the weights on all the edges between nodes $i$ and $j$.

### 1.5.4 Grids, trees, and distances

For the purpose of routing the signal nets, it is convenient to superimpose an imaginary grid on the layout surface. Wires are routed either horizontally or vertically along the lines of the grid. This style of routing, where wires can only make 90° turns, is known as Manhattan routing. More recently, wires with 45° turns are also permitted.

The length of the wiring required to implement a two-pin net is measured by taking the *Manhattan distance* between the two pins. If the two pins are located at coordinates $(x_1, y_1)$ and $(x_2, y_2)$, the Manhattan distance between them is given by

$$d_{12} = |x_1 - x_2| + |y_1 - y_2| \qquad (1.8)$$

Note that $d_{12}$ is the shortest possible length of wire required to connect the two pins using the Manhattan style of routing. It may or may not be possible to route the net along the shortest path.

When a multipin net is under consideration, there are several ways to interconnect the pins. For instance, consider a three-pin net $(A, B, C)$. One possible realization is to split the three-pin net into two two-pin nets and realize them separately, e.g., $(A, B)$ and $(B, C)$. Alternatively, one can connect pins $A$ and $B$ and then connect $C$ to *any* point on the wire segment $AB$. The former way to realize the three-pin net is an example of a *rectilinear spanning tree*. The latter technique is an example of a *Steiner tree*.

Given a $k$-pin net, we can treat each pin as a node in a graph $G$. We then draw weighted edges from each node in the graph to every other node. The weight of the edge from node $i$ to node $j$ is the length of wire required to connect pin $i$ and pin $j$. A spanning tree of the graph $G$ is a set of $k - 1$ edges which form a tree; such a tree connects (or spans) all the $k$ nodes. The routing of a net can follow the construction of a spanning tree; we look at each edge $(i, j)$ in the spanning tree and connect the pins $i$ and $j$ using a Manhattan path. The resulting wiring pattern is called a *rectilinear spanning tree*.

The *cost* of a spanning tree is the sum of the weights on all the $k - 1$ edges of the tree. A spanning tree of minimum cost is known as the minimum spanning tree (MST). The rectilinear minimum-cost spanning tree, also abbreviated as RMST, is a good way of implementing a multipin net. However, a RMST does not give the shortest wirelength; a minimum-cost *Steiner tree* is required for that purpose.

To construct a *Steiner tree* which connects the $k$ pins of a net, we are permitted to add additional points called Steiner points. Suppose that we add $r$ Steiner points; a Steiner tree implementation of the $k$-pin net is a rectilinear spanning tree on the $k + r$ points. For example, consider a three-pin net $(A, B, C)$. Let the coordinates of the three pins be $(5, 2)$, $(5, 12)$, and $(10, 7)$. All spanning trees of this net have the same cost, namely, 20 units. Thus, one possible RMST results by joining $A$ to $B$ and $B$ to $C$ as shown in Figure 1.15(a). However, a Steiner tree can be constructed by joining $A$ and $B$ (10 units of wire) and then dropping a vertical wire segment from $C$ to the line $AB$. The point $(5, 7)$ is a Steiner point. The cost of the Steiner tree is 15 units (see Figure 1.15(b)). Constructing a minimum-cost Steiner tree on $k$ points is a hard problem. On the other hand, constructing a minimum spanning tree on $k$ points is solvable in polynomial-time. It has been shown that the length of a RMST is no more than 1.5 times the length of a minimum-cost Steiner tree (Du and Hwang, 1992); so it is reasonable to route a multipin net in the form of an RMST.

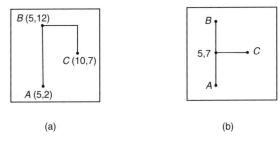

(a)                                   (b)

**Figure 1.15** (a) A minimum spanning tree. (b) Steiner tree.

## 1.6 SUMMARY

This chapter has introduced the reader to basic concepts of VLSI design and design automation. VLSI design is a complex process and is best dealt with in a hierarchical fashion; the use of automatic design tools is also crucial in rendering VLSI design feasible. In this book, we are concerned with physical design and the automation of physical design.

Physical design is the final step in the VLSI design process, and involves mainly the placement of circuit modules and routing of the nets. From the circuit layout, masks can be extracted for manufacturing the chip. VLSI layouts can be handcrafted with little CAD support, where an artwork designer uses a layout editor to draw the layout. This approach is tedious and error-prone. Automatic layout methodologies rely on the extensive use of computer programs during all design phases, from functional/logic/timing verification to artwork generation. In order to automate the layout procedure, it is common to impose restrictions on the layout architecture. Several layout architectures are popular. The gate-array design style uses a regular two-dimensional array architecture. A gate-array consists of a large number of uncommitted transistors that have been fabricated on a wafer. A cell, such as a logic gate, can be created by grouping the transistors in a local neighbourhood of the gate-array and connecting the transistors appropriately. The interconnection patterns for different cells are stored in a library. Inter-cell routing is specific for the circuit being designed. The standard-cell design style is also based on a cell library approach, but requires all the fabrication steps to manufacture the chip. However, the standard-cell architecture is different, consisting of rows of cells, all of which have the same height; the rows are separated by horizontal routing channels. Short vertical wires are permitted within a routing channel to connect cells in opposite rows. A net that connects two cells in non-consecutive rows must use feedthrough cells. A programmable logic array (PLA) is useful in implementing two-level sum-of-products expressions. A PLA has a regular architecture consisting of an AND plane

| Design style | Application | Design time | Fab effort | Cost | Performance |
|---|---|---|---|---|---|
| Full-custom | Chips for high-volume production, e.g., microprocessors | High(4) | High(2) | High(4) | High(4) |
| Gate-array | ASICs | Low(2) | Low(1) | Low(2) | Low(4) |
| Standard-cell | ASICs | Low(3) | High(2) | Low(3) | High(1) |
| Macro-cell | General | High(1) | High(2) | Low(3) | High(2) |
| FPGA | ASICs | Low(1) | Nil | Low(1) | Low(1) |

**Table 1.1 Comparison of layout styles. The number in parentheses indicates a rank to grade the layout style in comparison to others in the same category.**

to generate the product terms and an OR plane to generate the sums. PLAs are useful in constructing the control path of a digital circuit. They are also useful in implementing random logic. Table 1.1 assesses and compares several aspects of the various layout styles.

In order to reduce the complexity of a VLSI layout a step-wise solution approach is adopted where the layout process is broken into a sequence of physical design phases. The important design phases are *partitioning, floorplanning, placement, global routing,* and *detailed routing.* Each of these phases amounts to solving a combinatorial optimization problem. Unfortunately, these optimization problems are hard problems and no efficient algorithms are known to solve them exactly. As a result, heuristic techniques are employed and near-optimal solutions are obtained.

# 1.7 ORGANIZATION OF THE BOOK

This book is dedicated to problems and algorithms related to physical design. It is organized around the main physical design phases. For each design phase, relevant concepts and important algorithms are described in detail and illustrated with examples. The body of available literature on the subject is enormous. Therefore, it is impossible to describe and discuss every single reported work. Rather, we concentrate on describing those techniques that are most widely used (with satisfaction). Also, these techniques illustrate the complexity and the numerous difficult decisions that must be made to solve the particular design problem.

Chapters 2–9 are augmented with a section 'Other Approaches and Recent Work' where several other relevant techniques (recent or otherwise) are described and discussed. The book has a total of nine chapters.

The purpose of this introductory chapter is to motivate the student towards a study of Physical Design Automation of Integrated Circuits. It also introduces most of the basic terminology needed in the remaining chapters.

Chapter 2 concisely describes the circuit partitioning problem. Three of the most popular partitioning techniques are described. These are: Kernighan–Lin algorithm, Fiduccia–Mattheyses algorithm, and simulated annealing approach. Chapter 3 formally defines the floorplanning problem and describes the following techniques: cluster growth method, simulated annealing approach, mathematical programming approach, and graph dualization approach. Chapter 4 addresses the problem of module placement. The three most widely used placement techniques, that is, min-cut placement, simulated annealing approach, and force-directed approach, are described in detail. Also, genetic placement is fairly well described. Chapters 5, 6, and 7 are dedicated to the topic of routing. All aspects of routing are addressed (grid routing, global routing, channel routing, and switchbox routing). For each routing subproblem, popular algorithms are described and illustrated with examples. Chapter 8 considers the problem of silicon compilation and automatic generation of cells. Three different cell styles are examined, namely, standard-cell, gate matrix and PLA. Algorithms for automatic generation of layout in the above cell styles are covered. Finally, in Chapter 9, layout editors and compaction are considered. Techniques for hand-drawing of layouts are examined. The importance of *compaction* in hand layouts is explained, and the two main compaction approaches, that is, grid based and graph based, together with their related algorithms are described.

## EXERCISES

**Exercise 1.1** Design a control path for the serial adder discussed in Section 1.2.1. Use hard-wired control. There are three methods for designing a hard-wired control path—the state table method, the delay element method, and the sequence counter method. These methods are discussed in the text *Computer Organization and Architecture* by J.P. Hayes (1988).

**Exercise 1.2** Which of the following steps is (are) not part of physical design? Elaborate and explain.

1. Design rule verification.

2. Circuit extraction.
3. Transistor sizing for performance enhancement.
4. Maintenance of a standard-cell library.

**Exercise 1.3** What layout style is best suited for a high performance microprocessor? Justify your answer.

**Exercise 1.4** A chip is being designed for implementing a new speech processing algorithm. The chip must be released into the market at a short deadline. What layout method will you use? Why?

**Exercise 1.5** Visit the Computer Lab in your institution and familiarize yourself with the working of a *computer workstation*. How is a workstation different from other computers? Find out about the architecture of a workstation. In particular, find out about the disk capacity, the memory capacity, the graphics support, and networking capabilities of the workstation. Why is networking important in a design environment?

**Exercise 1.6** A number of computer vendors offer workstations for computer-aided design—Sun, Hewlett Packard, DEC, Apollo, and IBM to mention a few. Collect information about these machines and make a comparative evaluation. Keep in mind the cost, the hardware features, the operating system, and the software support offered by the vendor. What features would you look for if your main application is VLSI design?

**Exercise 1.7 Programming Exercise:** You have received a connectivity matrix $C$ of size $100 \times 100$ in a file of integers named cmat. The matrix describes the connectivity among 100 circuit elements. The element $C_{ij}$ of the matrix represents the number of connections between elements $i$ and $j$ of the circuit.

Write a program to verify the correctness of the matrix $C$. Your program must print out the row and column positions of any errors detected.

**Exercise 1.8 Programming Exercise:** An alternative way to describe the connectivity among circuit elements is the *netlist* description. A two-point net is a tuple of the form $(i, j)$ and describes a connection between elements $i$ and $j$.

An ASCII file named netlist.2 contains a list of two-point nets, one net per line. Write a program to read the file netlist.2 and generate the connectivity matrix of the circuit. The connectivity matrix must be written into a file of integers named cmat. Use the following netlist as a sample input to your program. The circuit elements are given names such as 'A', 'B', and so on. You may have to map these names to integers for convenience.

$$(A, a2s_1.in1)$$
$$(B, a2s_1.in2)$$
$$(A, a2s_2.in1)$$
$$(C, a2s_2.in2)$$
$$(C, a2s_3.in1)$$
$$(B, a2s_3.in2)$$
$$(a2s_1.out, o3s_1.in1)$$
$$(a2s_2.out, o3s_1.in2)$$
$$(a2s_3.out, o3s_1.in3)$$
$$(o3s_1.out, Z)$$

**Exercise 1.9 Programming Exercise:** An alternative form of a netlist is shown below.

A1 = AND2(A, B);
A2 = AND2(C, B);
A3 = AND2(C, A);
Z = OR3 (A1, A2, A3);

A1 = AND2(A, B); means the signal A1 is the output of a two-input AND gate whose inputs are A and B. The other statements are similarly interpreted. You are given such a netlist description in an ASCII file named `netlist`. Write a program to generate the connectivity matrix of the circuit.

**Exercise 1.10** (*) Find out about the `lex` and `yacc` utilities provided by the UNIX operating system. Use these to carry out the conversions stated in Exercises 1.8 and 1.9.

**Exercise 1.11** List and briefly describe the steps of designing a new integrated circuit and the CAD programs that may be used to carry out these steps.

## REFERENCES

Alpert, D. and D. Avnon, Architecture of the Pentium Microprocessor. *IEEE Micro,* pages 11–21, June 1993.

Brglez, F. A D&T Special Report on ACD/SIGDA Design Automation Benchmarks: Catalyst or Anathema? *IEEE Design & Test,* pages 87–91, September 1993.

Brown, S. D, R. J. Francis, J. Rose and Z. G. Vranesic, *Field-Programmable Gate Arrays*, Kluwer Academic Publishers, 1992.

Camposano, R, M. McFarland and A. C. Parker. The high-level sythesis of digital systems. *Proceedings of IEEE*, 78(2):301–317, February 1990.

Du, D. Z. and F. K. Hwang. Reducing the Steiner problem in a normed space. *SIAM Journal on Computing*, 21:1001-1007, December 1992.

Hayes, J. P. *Computer Organization and Architecture*. McGraw-Hill Book Company, 1988.

Mead C. and L. Conway. *Introduction to VLSI Systems*. Addison-Wesley, 1980.

OASIS. *Open Architecture Silicon Implementation Systems*. Version 2.0, Microelectronics Center of North Carolina, 1992.

Ousterhout, J. K, G. T. Hamachi, R. N. Mayo, W. S. Scott and G. S. Taylor. Magic: A VLSI layout system. *Proceedings of 21st Design Automation Conference*, pages 152–159, 1984.

# TWO

# CIRCUIT PARTITIONING

## 2.1 INTRODUCTION

With rapid advances in integration technology, it is possible to place a large number of logic gates on a single chip. Despite this fact, it may become necessary to partition a circuit into several subcircuits and implement the subcircuits as ICs. This is either because the circuit is too large to be placed on a single chip, or because of I/O pin limitations. The larger the gate count of the circuit, the larger the number of I/O pins associated with the circuit. The relationship between the number of gates and the number of I/O pins is estimated by Rent's rule,

$$IO = tG^r \qquad (2.1)$$

where $IO$ is the number of I/O pins, $t$ is the number of terminals per gate, $G$ the number of gates in the circuit, and $r$ is Rent's exponent which is a positive constant less than one. Unfortunately, a large pin count increases dramatically the cost of packaging the circuit. Further, the number of I/O pins must correspond to one of the standard packaging technologies—12, 40, 128, 256, and so on.

When it becomes necessary to split a circuit across packages, care must be exercised as to how this partition is carried out. Invariably some inter-connections will get 'cut' when a circuit is partitioned into subcircuits, while the interconnections within subcircuits can be implemented as 'on-chip' wiring. Off-chip wires are undesirable due to several reasons. Electrical signals travel slower along wires external to the chip; thus off-chip wires

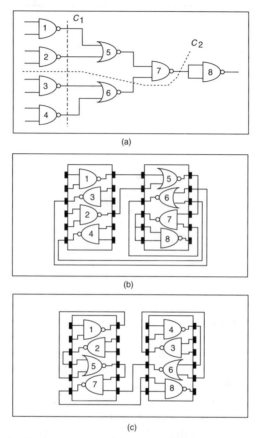

(a)

(b)

(c)

**Figure 2.1** (a) A circuit to be partitioned. (b) Partition using cutline $C_1$. There are four off-chip wires. (c) Partition using cutline $C_2$. There are only two off-chip wires.

cause performance degradation. Off-chip wires take up area on a printed circuit board. Off-chip wiring reduces the reliability of the system; printed wiring and plated-through holes are both likely sources of trouble in defective PCBs. Finally, since off-chip wires must originate and terminate into I/O pins, more off-chip wires essentially mean more I/O pins. These concepts are illustrated in Figure 2.1. In this chapter we shall discuss algorithms for the circuit partitioning problem. The general definition of the problem is given in Section 2.2. In Section 2.3, the problem is defined as a constrained optimization problem. Unfortunately, the partitioning problem is an NP-complete† problem; this means it is unlikely that a polynomial-time algorithm exists for solving the problem. Therefore, one must use heuristic techniques for generating approximate solutions. One such heuristic is the widely used Kernighan–Lin (KL) algorithm discussed

† See Appendix A for an explanation of NP-completeness.

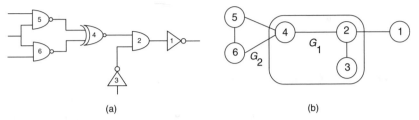

|  |  |
|---|---|
| (a) | (b) |

**Figure 2.2** (a) A circuit to be partitioned. (b) Its corresponding graph.

in Section 2.4.1 (Kernighan and Lin, 1970). As will be seen in Section 2.4.1 the Kernighan and Lin algorithm is applicable to a restricted version of the partitioning problem. Variations of the Kernighan–Lin algorithm are discussed in Section 2.4.2. Another algorithm that is a variation of the Kernighan–Lin heuristic but uses a different strategy and a different objective function is the Fiduccia–Mattheyses heuristic (Fiduccia and Mattheyses, 1982). Details of the Fiduccia–Mattheyses heuristic are discussed in Section 2.4.3. A more general purpose heuristic algorithm, known as simulated annealing, is discussed in Section 2.4.4. Simulated annealing is applicable to a variety of combinatorial optimization problems that arise in VLSI/CAD. In Section 2.5, we discuss other approaches and recent work in the area of circuit partitioning.

## 2.2 PROBLEM DEFINITION

The general case of the circuit partitioning problem is the $k$-way partitioning problem, which can be formulated as a graph partitioning problem. The idea is to model the circuit as a graph, whose vertices represent circuit elements, and edges represent the interconnects. For example, consider the circuit shown in Figure 2.2(a). The graph model of this circuit is shown in Figure 2.2(b). Observe that all the interconnects are 'two-pin' connections. For instance, the wire which connects the input pins of gates 5 and 6 is modelled as an edge between nodes 5 and 6 in Figure 2.2(b). A graph theoretic definition of the partitioning problem is as follows.

### Definition

Given a graph $G(V, E)$, where each vertex $v \in V$ has a *size* $s(v)$, and each edge $e \in E$ has a *weight* $w(e)$, the problem is to divide the set $V$ into $k$ subsets $V_1, V_2, \cdots, V_k$, such that an objective function is optimized, subject to certain constraints. The cost function and constraints are examined in the next section.

## 2.3 COST FUNCTION AND CONSTRAINTS

Before we formulate a cost function, let us reflect on the graph model described in the previous section. The size $s(v)$ of a node $v$ represents the area of the corresponding circuit element. Suppose that the circuit is partitioned into $k$ subcircuits. The partition divides the graph $G(V, E)$ into $k$ subgraphs $G_i(V_i, E_i)$, $i = 1, 2, \cdots, k$. In Figure 2.2(a), if we partition the circuit into two subcircuits, with gates 2,3,4 in one and gates 1,5,6 into another, the two subgraphs induced by the partition are shown in Figure 2.2(b); the subgraph $G_1$ consists of nodes 2,3,4 and edges (2,4) and (2,3). The subgraph $G_2$ consists of nodes 1,5,6 and edge (5,6). What about the edges (5,4), (1,2), and (4,6)? We say these edges are 'cut' by the partition. The name *cutset* is used to describe the set of these edges. The cutset of a partition is indicated by $\psi$ and is equal to the set of edges cut by the partition.

### 2.3.1 Bounded size partitions

As we mentioned in the beginning of the chapter, circuit partition arises due to size restriction on circuit packages. Again referring to the circuit of Figure 2.2(a), suppose that the six gates occupy too much area to be included on a single chip. (Of course this is just a supposition! VLSI technology allows tens of thousands of logic gates to be included on a single chip.) In order to package this circuit, it must be split into two or more subcircuits. In the general $k$-way partitioning problem, the size constraint is expressed by placing an upper bound on the size of each subcircuit. The size of the $i$th subcircuit is given by $\sum_{v \in V_i} s(v)$. If the upper bound on the size of this subcircuit is $A_i$, we have

$$\sum_{v \in V_i} s(v) \leq A_i \qquad (2.2)$$

It is desirable to divide the circuit into roughly equal sizes. This can be reflected by modifying Equation 2.2 as follows.

$$| V_i | = \sum_{v \in V_i} s(v) \leq \lceil \frac{1}{k} \sum_{v \in V} s(v) \rceil = \frac{1}{k} |V| \qquad (2.3)$$

where $|V_i|$ and $|V|$ are the sizes of sets $V_i$ and $V$ respectively. If all the circuit elements have the same size, then Equation 2.3 reduces to:

$$n_i \leq \frac{n}{k} \tag{2.4}$$

where $n_i$ and $n$ are the number of elements in $V_i$ and in $V$ respectively.


## 2.3.2 Minimize external wiring

If the subcircuits are implemented on separate packages, there is a need to connect these packages through external wires. In particular, the nets that belong to the cutset are to be implemented as external wiring. External wires are undesirable as explained earlier (see Section 2.1). It is highly desirable to minimize the external wiring. The weight $w(e)$ on an edge $e$ of the circuit graph represents the cost of wiring the corresponding connection as an external wire. Therefore, the cost function that must be minimized during partitioning is,

$$Cost = \sum_{e \in \psi} w(e) \tag{2.5}$$

Suppose that the partitions are numbered $1, 2, \cdots, k$. Let $p(u)$ indicate the partition number of node $u$. The condition $e \in \psi$ can then be written as $e = (u, v)$, and $p(u) \neq p(v)$. Thus Equation 2.5 can also be rewritten as

$$Cost = \sum_{\forall e = (u,v) \& p(u) \neq p(v)} w(e) \tag{2.6}$$


# 2.4 APROACHES TO PARTITIONING PROBLEM

The partitioning problem, as formulated in the previous section, is an *intractable* problem. Even the simplest case of the problem, namely two-way partitioning with identical node sizes and unit edge weights, is NP-complete. It is instructional to study this special case of the partitioning problem, which also finds wide use in practice. Suppose we have a circuit with $2n$ elements and we wish to generate a balanced two-way partition of the circuit into two subcircuits of $n$ elements each. The cost function is the size of the cutset. Convince yourself that this is indeed a special case of the $k$-way partitioning problem. If we do not place the constraint that the partitioning be *balanced*, the two-way partitioning problem (TWPP) is easy. One applies the celebrated max-flow mincut algorithm to get a minimum size cut. However, the balance criterion is extremely important in practice and cannot be overlooked.

To appreciate the complexity of TWPP, consider how many balanced partitions of a $2n$-node circuit exist (see Exercise 2.1). This is a number which grows exponentially with $n$. Even for moderate values of $n$, it is impractical to enumerate all the partitions and pick the best. The only way to deal with NP-complete problems such as TWPP is to 'satisfice', that is, be satisfied with an approximate solution to the problem. Such an approximate solution must satisfy the constraints, but *may not* necessarily possess the best cost. For instance, an approximate solution to the TWPP is a balanced partition which may not have a minimum-size cutset.

There are a number of 'heuristic' techniques to generate approximate solutions to the partitioning problem. These can be broadly classified into *deterministic* and *stochastic* algorithms. A deterministic algorithm progresses towards the solution by making deterministic decisions. On the other hand stochastic algorithms make random (coin tossing) decisions in their search for a solution. Therefore deterministic algorithms produce the same solution for a given input instance while this is not the case for stochastic algorithms.

Heuristic algorithms can also be classified as *constructive* and *iterative* algorithms. A constructive partitioning heuristic starts from a seed component (or several seeds). Then other components are selected and added to the partial solution until a complete solution is obtained. Once a component is selected to belong to a partition, it is never moved during future steps of the partitioning procedure. An iterative heuristic receives two things as inputs, one, the description of the problem instance, and two, an initial solution to the problem. The iterative heuristic attempts to modify the given solution so as to improve the cost function; if improvement cannot be attained by the algorithm, it returns a 'NO', otherwise it returns an improved solution. It is customary to apply the iterative procedure repeatedly until no cost improvement is possible. Frequently, one applies an iterative improvement algorithm to refine a solution generated by the constructive heuristic. Usually constructive algorithms are deterministic while iterative algorithms may be deterministic or stochastic.

Alternatively, one could generate an initial solution randomly and pass it as input to the iterative heuristic. Random solutions are, of course, generated quickly; but the iterative algorithm may take a large number of iterations to converge to either a local or global optimum solution. On the other hand, a constructive heuristic takes up time; nevertheless the iterative improvement phase converges rapidly if started off with a constructive solution.

Figure 2.3 gives the flow chart of applying a constructive heuristic followed by an iterative heuristic. The 'stopping criteria met' vary

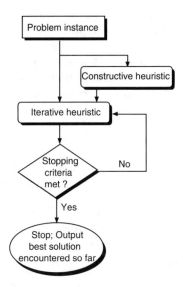

**Figure 2.3** General structure combining constructive and iterative heuristics.

depending on the type of heuristic applied. In case of deterministic heuristics, the stopping criterion could be the first failure in improving the present solution. Examples of these heuristics are the Kernighan–Lin and Fiduccia–Mattheyses heuristics discussed in Sections 2.4.1 and 2.4.3 respectively. While in the case of non-deterministic heuristics the stopping criterion could be the run time available or $k$ consecutive failures in improving the present solution. An example of this is the simulated annealing heuristic discussed in Section 2.4.4.

### 2.4.1 Kernighan–Lin algorithm

An iterative improvement algorithm due to Kernighan and Lin is one of the most popular algorithms for the two-way partitioning problem (Kernighan and Lin, 1990). The algorithm can also be extended to solve more general partitioning problems.

### Two-way uniform partition problem

We have introduced the two-way partitioning problem in the previous section. The problem is characterized by a *connectivity matrix C*. This is a square matrix with as many rows as there are nodes in the circuit graph.

The element $c_{ij}$ represents the sum of weights of the edges which connect elements $i$ and $j$. In the TWPP, since the edges have unit weights, $c_{ij}$ simply counts the number of edges which connect $i$ and $j$. The output of the partitioning algorithm is a pair of sets (or blocks) $A$ and $B$ such that $|A| = n = |B|$, and $A \cap B = \emptyset$, and such that the size of the cutset is as small as possible. The size of the cutset is measured by $T$,

$$T = \sum_{a \in A, b \in B} c_{ab} \qquad (2.7)$$

The Kernighan–Lin heuristic is an iterative improvement algorithm. It starts from an initial partition $(A, B)$ such that $|A| = n = |B|$, and $A \cap B = \emptyset$.

How can a given partition be improved? Let $P^* = \{A^*, B^*\}$ be the optimum partition, that is, the partition with a cutset of minimum cardinality. Further, let $P = \{A, B\}$ be the current partition. Then, in order to attain $P^*$ from $P$, one has to swap a subset $X \subseteq A$ with a subset $Y \subseteq B$ such that,

(1) $|X| = |Y|$
(2) $X = A \cap B^*$
(3) $Y = A^* \cap B$

Using basic set theory, the reader should be able to show that $A^* = (A - X) + Y$ and $B^* = (B - Y) + X$, where '+' and '−' are the union and difference operations on sets. This swap is illustrated in Figure 2.4.

However, the problem of identifying $X$ and $Y$ is as hard as that of finding $P^* = \{A^*, B^*\}$. Kernighan and Lin proposed an heuristic algorithm to approximate $X$ and $Y$.

Before we present the algorithm, it is necessary to develop two results which throw light on the effect of swapping a single node in block $A$ with another in block $B$.

Consider any node $a$ in block $A$. The contribution of node $a$ to the cutset is called the external cost of $a$, or $E_a$, and is simply the number of edges that emerge from $a \in A$ and terminate in $B$:

$$E_a = \sum_{v \in B} c_{av} \qquad (2.8)$$

Similarly, we can define the internal cost $I_a$ of node $a \in A$ as

$$I_a = \sum_{v \in A} c_{av} \qquad (2.9)$$

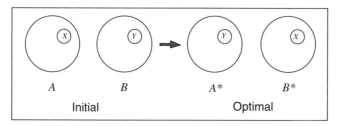

**Figure 2.4** Initial and optimal partitions.

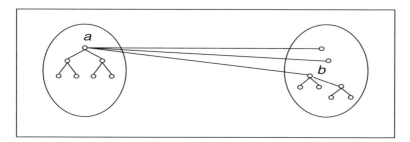

**Figure 2.5** Internal cost versus external cost.

What is the effect of moving node $a$ from block $A$ to block $B$? A moment's reflection shows that the size of the cutset would increase by a value $I_a$ and decrease by a value $E_a$. The benefit of moving $a$ from $A$ to $B$ is therefore $E_a - I_a$. This is known as the $D$-value of node $a$, or $D_a$:

$$D_a = E_a - I_a \qquad (2.10)$$

**Example 2.1** Referring to Figure 2.5, $I_a = 2$, $I_b = 3$, $E_a = 3$, $E_b = 1$, $D_a = 1$, and $D_b = -2$.

Since we want balanced partitions, we must move a node from $B$ to $A$ each time we move a node from $A$ to $B$. The following result characterizes the effect of swapping two modules among blocks $A$ and $B$.

**Lemma 1** If two elements $a \in A$ and $b \in B$ are interchanged, the reduction in the cost is given by

$$g_{ab} = D_a + D_b - 2c_{ab}$$

PROOF   From Equation 2.8 the external cost can be rewritten as

$$E_a = c_{ab} + \sum_{v \in B, v \neq b} c_{av} \tag{2.11}$$

Therefore,

$$D_a = E_a - I_a = c_{ab} + \sum_{v \in B, v \neq b} c_{av} - I_a \tag{2.12}$$

Similarly

$$D_b = E_b - I_b = c_{ab} + \sum_{u \in A, u \neq a} c_{bu} - I_b \tag{2.13}$$

Moving $a$ from $A$ to $B$ reduces the cost by

$$\sum_{v \in B, v \neq b} c_{av} - I_a = D_a - c_{ab} \tag{2.14}$$

Moving $b$ from $B$ to $A$ reduces the cost by

$$\sum_{u \in A, u \neq a} c_{bu} - I_b = D_b - c_{ab} \tag{2.15}$$

When both moves are carried out, the total cost reduction is given by the sum of Equations 2.14 and 2.15 and is equal to

$$g_{ab} = D_a + D_b - 2c_{ab} \tag{2.16}$$

The swapping of two nodes affects the $D$-values of all other nodes that are connected to either of the nodes swapped. The following lemma tells us how to update the $D$-values of the remaining nodes after two nodes $a$ and $b$ have been swapped.

**Lemma 2**  If two elements $a \in A$ and $b \in B$ are interchanged, then the new $D$-values, indicated by $D'$, are given by

$$D'_x = D_x + 2c_{xa} - 2c_{xb}, \quad \forall x \in A - \{a\} \tag{2.17}$$

$$D'_y = D_y + 2c_{yb} - 2c_{ya}, \quad \forall y \in B - \{b\} \tag{2.18}$$

PROOF  Refer to Figure 2.6. Consider a node $x \in A - \{a\}$. Since $b$ has entered block $A$, the internal cost of $x$ increases by $c_{xb}$. Similarly, since $a$ has entered the opposite block $B$, the internal cost of $x$ must be decreased by $c_{xa}$. The new internal cost of $x$ therefore is

$$I'_x = I_x - c_{xa} + c_{xb}$$

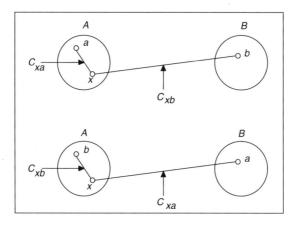

**Figure 2.6**  Updating $D$-values after an exchange.

One can similarly show that the new external cost of $x$ is

$$E'_x = E_x + c_{xa} - c_{xb}$$

Thus the new $D$-value of $x \in A - \{a\}$ is

$$D'_x = E'_x - I'_x = D_x + 2c_{xa} - 2c_{xb}$$

Similarly, the new $D$-value of $y \in B - \{b\}$ is

$$D'_y = E'_y - I'_y = D_y + 2c_{yb} - 2c_{ya}$$

Notice that if a module '$x$' is connected neither to '$a$' nor to '$b$' then $c_{xa} = c_{xb} = 0$, and, $D'_x = D_x$.

## Improving a partition

Assume we have an initial partition $\{A, B\}$ of $n$ elements each. Kernighan–Lin used Lemmas 1 and 2 and devised a greedy procedure to identify two subsets $X \subseteq A$, and $Y \subseteq B$, of equal cardinality, such that when interchanged, the partition cost is improved. $X$ and $Y$ may be empty, indicating in that case that the current partition can no longer be improved.

The procedure works as follows. The gains of interchanging any two modules $a \in A$ and $b \in B$ are computed. The pair $(a_1, b_1)$ leading to maximum gain $g_1$ is selected and the elements $a_1$ and $b_1$ are locked so as not to be considered for future interchanges. The $D$-values of remaining free cells are updated as in Lemma 2 and gains recomputed as indicated in Lemma 1. Then a second pair $(a_2, b_2)$ with maximum gain $g_2$ is selected and locked. Notice that $g_2$ is the gain of swapping $a_2 \in A$ with $b_2 \in B$ given that

$a_1 \in A$ has already been interchanged with $b_1 \in B$. Hence, the gain of swapping the pair $(a_1, b_1)$ followed by the $(a_2, b_2)$ swap is $G_2 = g_1 + g_2$. The process continues selecting $(a_1, b_1), (a_2, b_2), \cdots, (a_i, b_i) \cdots (a_n, b_n)$ with corresponding gains $g_1, g_2, \cdots, g_i, \cdots, g_n$. Obviously $G = \sum_{i=1}^{n} g_i = 0$ since this amounts to swapping all the elements of $A$ with those of $B$, which generates a partition identical to the initial one. In general, the gain of making the swap of the first $k$ pairs $(a_1, b_1), (a_2, b_2), \cdots, (a_k, b_k), 1 \le k \le n$ is $G_k = \sum_{i=1}^{k} g_i$. If there is no $k$ such that $G_k > 0$ then the current partition cannot be improved and remains as is; otherwise we choose the $k$ that maximizes $G_k$, and make the interchange of $\{a_1, a_2, \cdots, a_k\}$ with $\{b_1, b_2, \cdots, b_k\}$ permanent.

## Iterative improvement

As pointed out earlier, the Kernighan–Lin algorithm is an iterative algorithm. The above improvement procedure constitutes a single pass of the Kernighan–Lin procedure. The partition obtained after the $i$th pass constitutes the initial partition of the $i + 1$st pass. Iterations are terminated when $G_k \le 0$, that is, no further improvements can be obtained by pairwise swapping. The entire algorithm is shown in Figure 2.7.

**Example 2.2**   The circuit given in Figure 2.2(a) is to be partitioned into two subcircuits. Apply Kernighan–Lin heuristic to break the circuit into two equal size partitions, so as to minimize the number of interconnections between partitions. Assume that all gates are of the same size.

SOLUTION   The graph corresponding to the circuit is given in Figure 2.2(b).

Step 1: Initialization.

Let the initial partition be a random division of vertices into the partition $A = \{2,3,4\}$ and $B = \{1,5,6\}$.

$A' = A = \{2,3,4\}$,   and   $B' = B = \{1,5,6\}$.

Step 2: Compute $D$-values.

$D_1 = E_1 - I_1 = 1 - 0 = +1$
$D_2 = E_2 - I_2 = 1 - 2 = -1$
$D_3 = E_3 - I_3 = 0 - 1 = -1$
$D_4 = E_4 - I_4 = 2 - 1 = +1$
$D_5 = E_5 - I_5 = 1 - 1 = +0$
$D_6 = E_6 - I_6 = 1 - 1 = +0$

**Algorithm** KL_TWPP;
**Begin**
Step 1. $V$ = set of $2n$ elements;
   *{$A, B$} is initial partition such that
   $|A|=|B|$; $A \cap B = \emptyset$; and $A \cup B = V$;
Step 2. Compute $D_v$ for all $v \in V$;
   $queue \leftarrow \phi$; and $i \leftarrow 1$;
   $A' = A$; $B' = B$;
Step 3. Choose $a_i \in A', b_i \in B'$, which maximizes
   $g_i = D_{a_i} + D_{b_i} - 2c_{a_i b_i}$;
   add the pair $(a_i, b_i)$ to $queue$;
   $A' = A' - \{a_i\}$; $B' = B' - \{b_i\}$;
Step 4. **If** $A'$ and $B'$ are both empty **then Goto** Step 5
   Else
      recalculate $D$-values for $A' \cup B'$;
      $i \leftarrow i + 1$; **Goto** Step 3;
   **EndIf**
Step 5. Find $k$ to maximize the partial sum G= $\sum_{i=1}^{k} g_i$;
   **If** $G > 0$ then
      Move $X = \{a_1, \cdots, a_k\}$ to $B$, and $Y = \{b_1, \cdots, b_k\}$ to $A$;
      **Goto** Step 2
   Else STOP
   **EndIf**
**End**.

**Figure 2.7** Kernighan–Lin algorithm for TWPP.

## Step 3: Compute gains.

$$g_{21} = D_2 + D_1 - 2c_{21} = (-1) + (+1) - 2(1) = -2$$
$$g_{25} = D_2 + D_5 - 2c_{25} = (-1) + (+0) - 2(0) = -1$$
$$g_{26} = D_2 + D_6 - 2c_{26} = (-1) + (+0) - 2(0) = -1$$
$$g_{31} = D_3 + D_1 - 2c_{31} = (-1) + (+1) - 2(0) = +0$$
$$g_{35} = D_3 + D_5 - 2c_{35} = (-1) + (+0) - 2(0) = -1$$
$$g_{36} = D_3 + D_6 - 2c_{36} = (-1) + (+0) - 2(0) = -1$$
$$g_{41} = D_4 + D_1 - 2c_{41} = (+1) + (+1) - 2(0) = +2$$
$$g_{45} = D_4 + D_5 - 2c_{45} = (+1) + (+0) - 2(1) = -1$$
$$g_{46} = D_4 + D_6 - 2c_{46} = (+1) + (+0) - 2(1) = -1$$

In the above list, the largest $g$ value is $g_{41}$ and corresponds to a maximum gain which results in the interchange of 4 and 1. Thus the pair $(a_1, b_1)$ is $(4, 1)$, the gain $g_{41} = g_1 = 2$, and $A' = A' - \{4\} = \{2, 3\}$, $B' = B' - \{1\} = \{5, 6\}$.

Since $A'$ and $B'$ are both not empty, we update the $D$-values in the next step and repeat the procedure from Step 3.

Step 4: Update $D$-values.

Since $D$-values of only those nodes that are connected to vertices $(4,1)$ are changed, only these are updated. The vertices connected to $(4,1)$ are vertex $(2)$ in set $A'$ and vertices $(5,6)$ in set $B'$. The new $D$-values for vertices of $A'$ and $B'$ are given by

$$D'_2 = D_2 + 2c_{24} - 2c_{21} = -1 + 2(1-1) = -1$$
$$D'_5 = D_5 + 2c_{51} - 2c_{54} = +0 + 2(0-1) = -2$$
$$D'_6 = D_6 + 2c_{61} - 2c_{64} = +0 + 2(0-1) = -2$$

To repeat Step 3, we assign $D_i = D'_i$ and then recompute the gains:

$$g_{25} = D_2 + D_5 - 2c_{25} = (-1) + (-2) - 2(0) = -3$$
$$g_{26} = D_2 + D_6 - 2c_{26} = (-1) + (-2) - 2(0) = -3$$
$$g_{35} = D_3 + D_5 - 2c_{35} = (-1) + (-2) - 2(0) = -3$$
$$g_{36} = D_3 + D_6 - 2c_{36} = (-1) + (-2) - 2(0) = -3$$

In the above list, all the $g$ values are equal, so we arbitrarily choose $g_{36}$, and hence the pair $(a_2, b_2)$ is $(3,6)$, gain $g_{36} = g_2 = -3$, and $A' = A' - \{3\} = \{2\}$, $B' = B' - \{6\} = \{5\}$. The new $D$-values are:

$$D'_2 = D_2 + 2c_{23} - 2c_{26} = -1 + 2(1-0) = 1$$
$$D'_5 = D_5 + 2c_{56} - 2c_{53} = -2 + 2(1-0) = 0$$

The corresponding new gain is:

$$g_{25} = D_2 + D_5 - 2c_{52} = (+1) + (0) - 2(0) = +1$$

Therefore the last pair $(a_3, b_3)$ is $(2,5)$ and the corresponding gain is $g_{25} = g_3 = +1$.

Step 5: Determine $k$.

We see that $g_1 = +2$, $g_1 + g_2 = -1$, and $g_1 + g_2 + g_3 = 0$. The value of $k$ that results in maximum $G$ is 1. Therefore elements of set $X = \{a_1\} = \{4\}$ and set $Y = \{b_1\} = \{1\}$. The new partition that results from moving $X$ to $B$ and $Y$ to $A$ is, $A = \{1,2,3\}$ and $B = \{4,5,6\}$. The entire procedure is repeated again with this new partition as the initial partition.

We leave it to the reader to work through the rest of the example (see Exercise 2.5). The reader may verify that the second iteration of the algorithm is also the last, and that the best solution obtained is $A = \{1,2,3\}$ and $B = \{4,5,6\}$.

## Time complexity analysis

Computing the $D$-values of any single node requires $O(n)$ time. Since Step 2 computes $D$-values for all the nodes, the step takes $O(n^2)$ time. It takes constant time to update any $D$-value. We update as many as $(2n - 2i)$ $D$-values after swapping the pair $(a_i, b_i)$. Therefore the total time spent in updating the $D$-values can be

$$\sum_{i=1}^{n}(2n - 2i) = O(n^2) \qquad (2.19)$$

The pair selection procedure is the most expensive step in the Kernighan–Lin algorithm. If we want to pick $(a_i, b_i)$, there are as many as $(n - i + 1)^2$ pairs to choose from leading to an overall complexity of $O(n^3)$. Kernighan and Lin proposed a clever technique to avoid looking at all the pairs. Recall that, while selecting $(a_i, b_i)$, we want to maximize $g_i = D_{a_i} + D_{b_i} - 2c_{a_i b_i}$. Suppose that we sort the $D$-values in a non-increasing order of their magnitudes. Thus, in block $A$,

$$D_{a_1} \geq D_{a_2} \geq \cdots \geq D_{a_{(n-i+1)}}$$

Similarly, in block $B$,

$$D_{b_1} \geq D_{b_2} \geq \cdots \geq D_{b_{(n-i+1)}}$$

The sorting can be completed in $O(n \log n)$ time since only a linear number of items are to be sorted. Now suppose that we begin examining $D_{a_i}$ and $D_{b_j}$ pairwise. If we come across a pair $(D_{a_k}, D_{b_l})$ such that $(D_{a_k} + D_{b_l})$ is less than the gain seen so far in this improvement phase, then we do not have to examine any more pairs. In other words, if $D_{a_k} + D_{b_l} < g_{ij}$ for some $i, j$ then $g_{kl} < g_{ij}$. The proof is straightforward.

Since it is almost never required to examine all the pairs $(D_{a_i}, D_{b_j})$, the overall complexity of selecting a pair $(a_i, b_i)$ is $O(n \log n)$. Since $n$ exchange pairs are selected in one pass of the Kernighan–Lin algorithm, the complexity of Step 3 is $O(n^2 \log n)$. Step 5 takes only linear time. The complexity of the Kernighan–Lin algorithm is $O(pn^2 \log n)$, where $p$ is the number of iterations of the improvement procedure. Experiments on large practical circuits have indicated that $p$ does not increase with $n$. Therefore, the Kernighan–Lin algorithm has a time complexity of $O(n^2 \log n)$.

The time complexity of the pair selection step can be improved by scanning the unsorted list of $D$-values and selecting $a$ and $b$ which maximize $D_a$ and $D_b$. Since this can be done in linear time, the algorithm's time complexity reduces to $O(n^2)$. This scheme is suited for sparse matrices where the probability of $c_{ab} > 0$ is small. Of course, this is an approximation of the greedy selection procedure, and may generate a different solution as compared to greedy selection.

## 2.4.2 Variations of Kernighan–Lin algorithm

The Kernighan–Lin algorithm may be extended to solve several other cases of the partitioning problem. These are discussed below.

*Unequal sized blocks.* To partition a graph $G = (V, E)$ with $2n$ vertices into two subgraphs of unequal sizes $n_1$ and $n_2$, $n_1 + n_2 = 2n$, the procedure shown below may be employed:

1. Divide the set $V$ into two subsets $A$ and $B$, one containing $MIN(n_1, n_2)$ vertices and the other containing $MAX(n_1, n_2)$ vertices. This division may be done arbitrarily.
2. Apply the algorithm of Figure 2.7 starting from Step 2, but restrict the maximum number of vertices that can be interchanged in one pass to $MIN(n_1, n_2)$.

Another possible solution to the unequal sized blocks problem instance is the following. Without loss of generality, assume that $n_1 < n_2$. To divide $V$ such that there are *at least* $n_1$ vertices in block $A$ and *at most* $n_2$ vertices in block $B$, the procedure shown below may be used:

1. Divide the set $V$ into blocks $A$ and $B$; $A$ containing $n_1$ vertices and $B$ containing $n_2$ vertices.
2. Add $n_2 - n_1$ dummy vertices to block $A$. *Dummy* vertices have no connections to the original graph.
3. Apply the algorithm of Figure 2.7 starting from Step 2.
4. Remove all dummy vertices.

*Unequal sized elements.* To generate a two-way partition of a graph whose vertices have unequal sizes, we may proceed as follows:

1. Without loss of generality assume that the smallest element has unit size.
2 Replace each element of size $s$ with $s$ vertices which are fully connected with edges of infinite weight. (In practice, the weight is set to a very large number $M$.)
3. Apply the algorithm of Figure 2.7.

*k-way partition.* Assume that the graph has $k \cdot n$ vertices, $k > 2$, and it is required to generate a $k$-way partition, each with $n$ elements.

1. Begin with a random partition of $k$ sets of $n$ vertices each.
2. Apply the two-way partitioning procedure on each pair of partitions.

Pairwise optimality is only a necessary condition for optimality in the $k$-way partitioning problem. Sometimes a complex interchange of three or more items from three or more subsets will be required to reduce the pairwise optimal to the global optimal solution. Since there are $\binom{k}{2}$ pairs to consider, the time complexity for one pass through all pairs for the $O(n^2)$-*procedure* is $\binom{k}{2}n^2 = O(k^2 n^2)$.

In general, more passes than this will be actually required, because when a particular pair of partitions is optimized, the optimality of these partitions with respect to others may change.

The above heuristic works very well for partitioning graphs, but does not take into account a special property of electrical circuits, that is, a group of vertices connected by a single *net*† do not have to be pairwise interconnected but can be connected by a spanning or Steiner tree. Since pins belonging to the same net in each partition are interconnected, then, a *single wire* is sufficient to connect each net across two partitions. Such a situation arises in circuits which have gates with fan-outs greater than one. This did not arise in Example 2.2.

A simple way to overcome this problem is to define the cost in terms of nets rather than edges. The cost then is the number of nets that have at least one pin in each partition.

### 2.4.3 Fiduccia–Mattheyses heuristic

In the previous section we presented the Kernighan–Lin algorithm which partitions a circuit modelled as a graph into two blocks ($A$ and $B$) such that the cost of the edges cut by the partition is minimized. In case of two point nets, the number of *edges* cut by a partition is equal to the number of nets cut. In case of multipoint nets, however, this is not the case. Figure 2.8 illustrates a circuit and its equivalent graph representation. If we partition the graph corresponding to the circuit in Figure 2.8 into two blocks $A = \{1,2,3\}$ and $B = \{4,5,6\}$, then the number of edges cut is equal to four while only three wires are required to connect cells of block $A$ to cells in block $B$. Therefore reducing the number of nets cut is more realistic than reducing the number of edges cut.

---

† A net is a set of points in a circuit that are always at the same electrical potential.

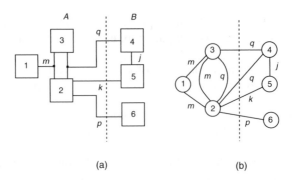

(a)              (b)

**Figure 2.8** Illustration of (a) Cut of nets. (b) Cut of edges.

Fiduccia–Mattheyses presented an iterative heuristic that takes into consideration multipin nets as well as sizes of circuit elements. Fiduccia–Mattheyses heuristic is a technique used to find a solution to the following bipartitioning problem: Given a circuit consisting of $C$ cells connected by a set of $N$ nets (where each net connects at least two cells), the problem is to partition circuit $C$ into two blocks $A$ and $B$ such that the number of nets which have cells in both the blocks is minimized and the balance factor $r$ is satisfied (Fiduccia and Mattheyses, 1982). Below we enumerate the principal differences and similarities between Kernighan–Lin and Fiduccia–Mattheyses heuristics.

1. Unlike Kernighan–Lin heuristic in which during each pass a *pair* of cells, one from each block, is selected for swapping, in the Fiduccia–Mattheyses heuristic a *single* cell at a time, from either block is selected and considered for movement to its complementary block.
2. The Kernighan–Lin heuristic partitions a graph into two blocks such that the cost of *edges* cut is minimum, whereas the Fiduccia–Mattheyses heuristic aims at reducing the cost of *nets* cut by the partition.
3. The Fiduccia–Mattheyses heuristic is similar to the Kernighan–Lin in the selection of cells. But the gain due to the movement of a single cell from one block to another is computed instead of the gain due to swap of two cells. Once a cell is selected for movement, it is locked for the remainder of that pass. The total number of cells that can change blocks is then given by the best sequence of moves $c_1, c_2, \cdots, c_k$. In contrast, in Kernighan–Lin the first best $k$ pairs in a pass are swapped.
4. The above modification can cause an imbalance arising from all cells wanting to migrate to a single partition. Therefore, the Fiduccia-Mattheyses heuristic is designed to handle imbalance, and it

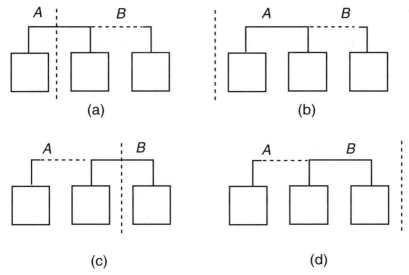

**Figure 2.9** Illustration of critical nets. Block to the left of partition is designated as '$A$' and to the right as '$B$'. (a) $A(n) = 1$. (b) $A(n) = 0$. (c) $B(n) = 1$. (d) $B(n) = 0$.

produces partitions balanced with respect to size. The balance factor $r$ (called *ratio*) is user specified and is defined as follows: $r = \frac{|A|}{|A|+|B|}$, where $|A|$ and $|B|$ are the sizes of partitioned blocks $A$ and $B$.

5. Some of the cells can be initially locked to one of the partitions.
6. The time complexity of Fiduccia–Mattheyses heuristic is linear. In practice only a very small number of passes are required leading to a fast approximate algorithm for min-cut partitioning.

Before we explain this heuristic we present some definitions and terms used in the explanation of the procedure.

## Definitions

Let $p(j)$ be the number of pins of cell '$j$', and $s(j)$ be the size of cell '$j$', for $j = 1, 2, \cdots, C$. If $V$ is the set of the $C$ cells, then $|V| = \sum_{i=1}^{C} s(i)$.

'*Cutstate of a net*': A net is said to be *cut* if it has cells in both blocks, and is *uncut* otherwise. A variable *cutstate* is used to denote the state of a net. That is, cutstate of a net is either cut or uncut.

'*Cutset of partition*': The *cutset* of a partition is the cardinality of the set of all nets with cutstate equal to cut.

'*Gain of cell*': The gain $g(i)$ of a cell '$i$' is the number of nets by which the cutset would decrease if cell '$i$' were to be moved. A cell is *moved* from its current block (the From_block) to its complementary block (the To_block).

'*Balance criterion*': To avoid having all cells migrate to one block a balancing criterion is maintained. A partition $(A, B)$ is balanced if

$$r \times |V| - s_{max} \leq |A| \leq r \times |V| + s_{max} \tag{2.20}$$

where $|A| + |B| = |V|$; and $s_{max} = Max[s(i)]$, $i \in A \cup B = V$.

'*Base cell*': The cell selected for movement from one block to another is called '*base cell*'. It is the cell with maximum gain and the one whose movement will not violate the balance criterion.

'*Distribution of a net*': Distribution of a net $n$ is a pair $(A(n), B(n))$ where $(A, B)$ is an arbitrary partition, and, $A(n)$ is the number of cells of net $n$ that are in $A$ and $B(n)$ is the number of cells of net $n$ that are in $B$.

'*Critical net*': A net is critical if it has a cell which if moved will change its cutstate. That is, if and only if $A(n)$ is either 0 or 1, or $B(n)$ is either 0 or 1, as illustrated in Figure 2.9.

**Algorithm** FM_TWPP
**Begin**
Step 1. Compute gains of all cells;
Step 2. $i = 1$;
    Select 'base cell' and call it $c_i$;
    **If** no base cell **Then Exit**; **EndIf**;
    A base cell is the one which
        (i) has maximum gain;
        (ii) satisfies balance criterion;
        **If** *tie* **Then** use   Size criterion or
                            Internal connections;
        **EndIf**;
Step 3. Lock cell $c_i$;
    Update gains of cells of those affected critical nets;
Step 4. **If** free_cells $\neq \phi$
        **Then** $i = i + 1$;
        select next base cell $c_i$;
        **If** $c_i \neq \phi$ then **Goto** Step 3;
Step 5. Select best sequence of moves $c_1, c_2, \cdots, c_k (1 \leq k \leq i)$
        such that $G = \sum_{j=1}^{k} g_j$ is maximum;
    **If** tie **then** choose subset that achieves a superior balance;†
    **If** $G \leq 0$ **Then Exit**;
Step 6. Make all $k$ moves permanent;
    Free all cells;
    **Goto**  Step 1
**End**.
**Figure 2.10** Fiduccia–Mattheyses bipartitioning algorithm.

† This is not mentioned in the original Fiduccia–Mattheyses algorithm.

# General idea

A general description of the heuristic is given in Figure 2.10. Next we give a step-by-step explanation of the algorithm (Fiduccia and Mattheyses, 1982).

**Step 1.** The first step consists of computing the gains of all *free* cells. Cells are considered to be free if they are not locked either initially by the user, or after they have been moved during this pass. Similar to the Kernighan–Lin algorithm, the effect of the movement of a cell on the cutset is quantified with a gain function. Let $F(i)$ and $T(i)$ be the From_block (current block) and To_block (destination block) of cell $i$ respectively, $1 \leq i \leq C$. The gain $g(i)$ resulting from the movement of cell $i$ from block $F(i)$ to block $T(i)$ is:

$$g(i) = FS(i) - TE(i) \qquad (2.21)$$

$FS(i) =$ the number of nets connected to cell $i$ and not
connected to any other cell in the From_block $F(i)$ of cell $i$.
$TE(i) =$ the number of nets that are connected to
cell $i$ and not crossing the cut.

It is easy to verify that $-p(i) \leq g(i) \leq p(i)$, where $p(i)$ is the number of pins on cell $i$.

Let us apply the above definitions to some cells of the circuit given in Figure 2.8(a). Consider cell 2, its From_block is $A$ and its To_block is $B$. Nets $k, m, p,$ and $q$ are connected to cell 2 of block $A$, of these only two nets $k$ and $p$ are *not* connected to any other cell in block $A$. Therefore, by definition, $FS(2) = 2$. And $TE(2) = 1$ since the only net connected and not crossing the cut is net $m$. Hence $g(2) = 2 - 1 = 1$. Which means that the number of nets cut will be reduced by 1 (from 3 to 2) if cell 2 were to be moved from $A$ to $B$.

Consider cell 4. In block $B$, cell 4 has only one net (net $j$) which is connected to it and also not crossing the cut, therefore $TE(4) = 1$. $FS(4) = 1$ and $g(4) = 1 - 1 = 0$, that is, no gain.

| Cell $i$ | F | T | $FS(i)$ | $TE(i)$ | $g(i)$ |
|---|---|---|---|---|---|
| 1 | A | B | 0 | 1 | -1 |
| 2 | A | B | 2 | 1 | +1 |
| 3 | A | B | 0 | 1 | -1 |
| 4 | B | A | 1 | 1 | 0 |
| 5 | B | A | 1 | 1 | 0 |
| 6 | B | A | 1 | 0 | +1 |

**Table 2.1 Gains of cells.**

Finally consider cell 5. Two nets $j$ and $k$ are connected to cell 5 in block $B$, but one of them, that is, net $k$ is crossing the cut, while net $j$ is not. Therefore, $TE(5)$ is also 1. The values of $F$, $T$, $FS$, $TE$ and $g$ for all cells are tabulated in Table 2.1. The above observation can be translated into an efficient procedure to compute the gains of all free cells. One such procedure is given in Figure 2.11.

**Algorithm** Compute_cell_gains;
**Begin**
    **For** each free cell '$i$' **Do**
       $g(i) \leftarrow 0$;
       $F \leftarrow From\_block$ of cell $i$;
       $T \leftarrow To\_block$ of cell $i$;
       **For** each net '$n$' on cell '$i$' **Do**
          **If** $F(n) = 1$ **Then** $g(i) \leftarrow g(i) + 1$;
          (*Cell $i$ is the only cell in the From_block connected to net $n$.*)
          **If** $T(n) = 0$ **Then** $g(i) \leftarrow g(i) - 1$
          (* All of the cells connected to net $n$ are in the From_block. *)
       **EndFor**
    **EndFor**
**End**.

**Figure 2.11** Procedure to compute gains of free cells.

**Example 2.3** Apply the procedure given in Figure 2.11 to the circuit of Figure 2.8(a) and compute the gains of all the free cells of the circuit.

SOLUTION We first compute the values of $A(n)$ and $B(n)$ (where $A(n)$ and $B(n)$ are the numbers of cells of net $n$ that are in block $A$ and block $B$ respectively). For the given circuit we have,

$$A(j) = 0, A(m) = 3, A(q) = 2, A(k) = 1, A(p) = 1,$$
$$B(j) = 2, B(m) = 0, B(q) = 1, B(k) = 1, B(p) = 1.$$

For cells in block $A$ we have, the From_block $A$ ($F = A$) and To_block is $B$ ($T = B$). For this configuration we get,

$$F(j) = 0, F(m) = 3, F(q) = 2, F(k) = 1, F(p) = 1,$$
$$T(j) = 2, T(m) = 0, T(q) = 1, T(k) = 1, T(p) = 1.$$

where $F(i)$ is the number of cells of net $i$ in From_block.

Since only critical nets affect the gains, we are interested only in those values which have, for cells of block $A$, $A(n) = 1$ and $B(n) = 0$, and for cells of block $B$, $B(n) = 1$ and $A(n) = 0$. Therefore, values of interest for block $A$ are $F(k) = 1$, $F(p) = 1$, and $T(m) = 0$. Now applying the procedure of Figure 2.11 we get:

$i = 1$; $F = A$; $T = B$; net on cell 1 is $m$. Values of interest are $T(m) = 0$; therefore, $g(1) = 0 - 1 = -1$.

$i = 2$; $F = A$; $T = B$; nets on cell 2 are $m$, $q$, $k$, and $p$. Values of interest are $F(k) = 1$; $F(p) = 1$; and $T(m) = 0$; therefore, $g(2) = 2 - 1 = 1$.

$i = 3$; $F = A$; $T = B$; nets on cell 3 are $m$ and $q$, but only $T(m) = 0$; therefore, $g(3) = 0 - 1 = -1$.

**Algorithm** Select_Base_Cell;
**Begin**
**ForEach** cell c with maximum gain **Do**
  **Begin**
    **If** moving the cell c creates imbalance
      **Then** discard this cell
      **Else Return** c
    **End If**;
  **End For**;
  **If** neither block has a qualifying cell **Then** Exit **EndIf**;
**End**.

**Figure 2.12** Procedure summarizing selection step for each candidate cell.

We leave it to the reader to complete the above example (see Exercise 2.14).

**Step 2. Selection of 'base cell':** Having computed the gains of each cell, we now choose the 'base cell'. The base cell is one that has a maximum gain and does not violate the balance criterion. If no base cell is found then the procedure stops. The procedure given in Figure 2.12 summarizes the selection step for each candidate cell.

When the balance criterion is satisfied then the cell with maximum gain is selected as the base cell. In some cases, the gain of the cell is non-positive. However, we still move the cell with the expectation that the move will allow the algorithm to 'escape out of a local minimum'. As mentioned before, to avoid migration of all cells to one block, during each move, the balance criterion is maintained. The notion of a tolerance factor is used in order to speed up convergence from an unbalanced situation to a balanced one. The balance criterion is therefore relaxed from Equation 2.20 to the inequality below:

$$r \times |V| - k \times s_{max} \le |A| \le r \times |V| + k \times s_{max} \qquad (2.22)$$

where $k$ is an increasing function of the number of free cells. Initially $k$ is large and is slowly decreased with each pass until it reduces to unity. If more than one cell of maximum gain exists, and all such cells satisfy the balance criterion, then ties may be broken depending on the size, internal connectivity, or any other criterion.

**Step 3. Lock cell and update gains:** After each move the selected cell is locked in its new block for the remainder of the pass. Then the gains of cells of affected critical net are updated using the procedure given in Figure 2.13.

**Algorithm** Update_Gains;
**Begin**
(*move base cell and update neighbours' gains*)
$F \leftarrow$ the From_block of base cell;
$T \leftarrow$ the To_block of base cell;
Lock the base cell and complement its blocks;
**For** each net $n$ on base cell **Do**
    (*check critical nets before the move*)
    **If** $T(n) = 0$ **Then** increment gains of all free cells on that net $n$
    **Else If** $T(n) = 1$ **Then** decrement gain of the only $T$ cell on net $n$, if it is free
    **End If;**
    (*change the net distribution $F(n)$ and $T(n)$ to reflect the move*)
    $F(n) \leftarrow F(n) - 1$ ; $T(n) \leftarrow T(n) + 1$;
    (*check for critical nets after the move*)
    **If** $F(n) = 0$ **Then** decrement gains of all free cells on net $n$
    **Else If** $F(n) = 1$ **Then** increment the gain of the only $F$ cell on net $n$, if it is free
    **End If**
**End For**
**End**.
**Figure 2.13** Algorithm to update gains after movement.

**Step 4. Select next base cell:** In this step, if more free cells exist then we search for the next base cell. If found then we go back to Step 3, lock the cell, and repeat the update. If no free cells are found then we move on to Step 5.

**Step 5. Select best sequence of moves:** After all the cells have been considered for movement, as in the case of Kernighan–Lin, the best partition encountered during the pass is taken as the output of the pass. The number of cells to move is given by the value of $k$ which yields maximum positive gain $G_k$, where $G_k = \sum_{i=1}^{k} g_i$.

**Step 6. Make moves permanent:** Only the cells given by the best sequence, that is, $c_1, c_2, \cdots, c_k$ are permanently moved to their complementary blocks. Then all cells are freed and the procedure is repeated from the beginning. We will now illustrate the above procedure with an example.

**Example 2.4** The procedure to compute gains due to the movement of cells was explained above. Apply the remaining steps of the Fiduccia–

|  | Before move | | After move | |
| --- | --- | --- | --- | --- |
| *Net* | *F* | *T* | *F'* | *T'* |
| *Net* | *F* | *T* | *F'* | *T'* |
| $k$ | 1 | 1 | 0 | 2 |
| $m$ | 3 | 0 | 2 | 1 |
| $q$ | 2 | 1 | 1 | 2 |
| $p$ | 1 | 1 | 0 | 2 |

**Table 2.2  Change in net distribution ($T(n)$, $F(n)$) to reflect the move.**

Mattheyses heuristic to the circuit of Figure 2.8(a) to complete one pass. Let the desired balance factor be 0.4 and the sizes of cells be as follows: $s(c_1) = 3$, $s(c_2) = 2$, $s(c_3) = 4$, $s(c_4) = 1$, $s(c_5) = 3$, and $s(c_6) = 5$.

SOLUTION  Earlier in this section we found that cell $c_2$ is the candidate with maximum gain. Verify that this candidate also satisfies the balance criterion (Equation 2.20).

Now, for each net $n$ on cell $c_2$ we find its distribution $F(n)$ and $T(n)$ (that is, the number of cells on net $n$ in the From_block and in the To_block respectively before the move). Similarly we find $F'(n)$ and $T'(n)$, the number of cells after the move. These values for cell $c_2$ are tabulated in Table 2.2. Observe in Figure 2.13 that the change in net distribution to reflect the move is a decrease in $F(n)$ and an increase in $T(n)$.

We now apply the procedure of Step 3 to update the gains of cells and determine the new gains. For each net $n$ on the base cell we check for the critical nets before the move. If $T(n)$ is zero then the gains of all free cells on the net $n$ are incremented. If $T(n)$ is one then the gains of the only $T$ cell on net $n$ is decremented (if the cell is free).

In our case, the selected base cell $c_2$ is connected to nets $k$, $m$, $p$, and $q$, and all of them are critical, with $T(m) = 0$, and $T(k) = T(q) = T(p) = 1$. Therefore, the gains of the free cells connected to net $m$ ($c_1$ and $c_3$) are incremented, while the gains of the free T_cells connected to nets $k$, $p$ and $q$ ($c_5$, $c_6$, and $c_4$) are decremented.

| | Gain due to $T(n)$ | | | | Gain due to $F(n)$ | | | | Gains | |
| --- | --- | --- | --- | --- | --- | --- | --- | --- | --- | --- |
| Cells | $k$ | $m$ | $q$ | $p$ | $k$ | $m$ | $q$ | $p$ | Old | New |
| $c_1$ | | +1 | | | | | | | -1 | 0 |
| $c_3$ | | +1 | | | | | +1 | | -1 | 1 |
| $c_4$ | | | -1 | | | | | | 0 | -1 |
| $c_5$ | -1 | | | | -1 | | | | 0 | -2 |
| $c_6$ | | | | -1 | | | | -1 | 1 | -1 |

**Table 2.3  Incremental values of gains of cells on critical nets before and after the move.**

These values are tabulated in the first four columns (Gain due to $T(n)$) of Table 2.3. We continue with the procedure of Figure 2.13 and check for the critical nets after the move. If $F(n)$ is zero then the gains of all free cells on net $n$ are decremented and if $F(n)$ is one then the gain of the only $F$ cell on net $n$ is incremented, if it is free. Since we are looking for the net distribution after the move, we look at the values of $F'$ in Table 2.2. Here we have $F'(k) = F'(p) = 0$ and $F'(q) = 1$. The contribution to gain due to cell 5 on net $k$ and cell 6 on net $p$ is $-1$, and since cell 3 is the only $F$ cell (cell on From_block), the gain due to it is $+1$. These values are tabulated in the next four columns (Gain due to $F(n)$) of Table 2.3.

From Table 2.3, the updated gains are obtained. The second candidate with maximum gain (say $g_2$) is cell $c_3$. This cell also satisfies the balance criterion and therefore is selected and locked.

We continue the above procedure of selecting the base cell (Step 2) for different values of $i$ (Figure 2.10). Initially $A_0 = \{1,2,3\}$, $B_0 = \{4,5,6\}$. The results are summarized below.

$i = 1$ : The cell with maximum gain is $c_2$. $|A| = 7$. This move satisfies the balance criterion. Maximum gain $g_1 = 1$. Lock cell $\{c_2\}$. $A_1 = \{1,3\}$, $B_1 = \{2,4,5,6\}$.

$i = 2$ : Cell with maximum gain is $c_3$. $|A| = 3$. The move satisfies the balance criterion. Maximum gain $g_2 = 1$. Locked cells are $\{c_2, c_3\}$. $A_2 = \{1\}$, $B_2 = \{2,3,4,5,6\}$.

$i = 3$ : Cell with maximum gain $(+1)$ is $c_1$. If $c_1$ is moved then $A = \{\}$, $B = \{1,2,3,4,5,6\}$. $|A| = 0$. This *does not* satisfy the balance criterion. Cell with next maximum gain is $c_6$. $|A| = 8$. This cell satisfies the balance criterion. Maximum gain $g_3 = -1$. Locked cells are $\{c_2, c_3, c_6\}$. $A_3 = \{1,6\}$, $B_3 = \{2,3,4,5\}$.

$i = 4$ : Cell with maximum gain is $c_1$. $|A| = 5$. This satisfies the balance criterion. Maximum gain $g_4 = 1$. Locked cells are $\{c_1, c_2, c_3, c_6\}$. $A_4 = \{6\}$, $B_4 = \{1,2,3,4,5\}$.

$i = 5$ : Cell with maximum gain is $c_5$. $|A| = 8$. This satisfies the balance criterion. Maximum gain $g_5 = -2$. Locked cells are $\{c_1, c_2, c_3, c_5, c_6\}$. $A_5 = \{5,6\}$, $B_5 = \{1,2,3,4\}$.

$i = 6$ : Cell with maximum gain is $c_4$. $|A| = 9$. This satisfies the balance criterion. Maximum gain $g_6 = 0$. All cells are locked. $A_6 = \{4,5,6\}$, $B_6 = \{1,2,3\}$.

Observe in the summary above that when $i = 3$, cell $c_1$ is the cell with maximum gain, but since it violates the balance criterion, it is discarded and the next cell $(c_6)$ is selected. When $i = 4$ cell $c_1$ again is the cell with maximum gain, but this time, since the balance criterion is satisfied, it is selected for movement.

We now look for $k$ that will maximize $G = \sum_{j=1}^{k} g_j$; $1 \le k \le i$. We have a tie with two candidates for $k$, $k = 2$ and $k = 4$, giving a gain of +2. Since the value of $k = 4$ results in a better balance between partitions, we choose $k = 4$. Therefore we move across partitions the first four cells selected, which are cells $c_2$, $c_3$, $c_6$, and $c_1$. The final partition is $A = \{6\}$, and $B = \{1,2,3,4,5\}$. The cost of nets cut is reduced from 3 to 1.

We leave it to the reader to work through the rest of the example (see Exercise 2.15).

## 2.4.4 Simulated annealing

Simulated annealing is perhaps the most well developed and widely used iterative technique for solving several combinatorial optimization problems (Kirkpatrick *et al.*, 1983). It has been applied to almost all known CAD problems, including partitioning. It is an *adaptive*† heuristic and belongs to the class of *non-deterministic* algorithms. This heuristic was first introduced by Kirkpatrick, Gelatt and Vecchi in 1983.

The simulated annealing heuristic, as the name suggests, derives inspiration from the process of carefully cooling molten metals in order to obtain a good crystal structure. During annealing, a metal is heated to a very high temperature (whereby the atoms gain enough energy to break the chemical bonds and become free to move), and then slowly cooled. By cooling the metal at a proper rate, atoms will have an increased chance to regain proper crystal structure. If we compare optimization to the annealing process, the attainment of global optimum is analogous to the attainment of a good crystal structure.

## Background

Every combinatorial optimization problem may be discussed in terms of a *state space*. A *state* is simply a configuration of the combinatorial objects involved. For example, in the two-way partitioning problem, any division of $2n$ nodes into two equal sized blocks is a configuration. There are a large number of such configurations in any combinatorial optimization. Only some of these correspond to global optima, i.e., states with optimum cost.

An iterative improvement scheme starts with some given state, and examines a *local neighbourhood* of the state for better solutions. A local

---

† In adaptive heuristics some (or all) parameters of the algorithm are changed during the execution.

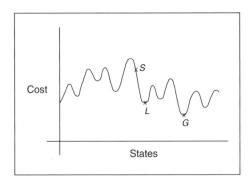

**Figure 2.14** Local versus global optima.

neighbourhood of a state $S$ is the set of all states which can be reached from $S$ by making a small change to $S$. For instance, if $S$ represents a two-way partition of a graph, the set of all partitions which are generated by swapping two nodes across the partition represents a local neighbourhood. The iterative improvement algorithm moves from the current state to a state in the local neighbourhood, if the latter has a better cost. If all the local neighbours have inferior costs, the algorithm is said to have *converged* to a *local optimum*. This is illustrated in Figure 2.14. Here, the states are shown along the $x$-axis, and it is assumed that two consecutive states are local neighbours. It is further assumed that we are discussing a *minimization* problem. The cost curve is *non-convex*, i.e., it has multiple minima. A greedy iterative improvement algorithm started off with an initial solution such as $S$ in Figure 2.14, can slide along the curve and find a local minimum such as $L$. There is no way such an algorithm can find the global minimum $G$ of Figure 2.14, unless it 'climbs the hill' at the local minimum $L$. In other words, an algorithm which occasionally accepts an inferior solution can escape from getting stuck at a local optimum solution. Simulated annealing is such a hill-climbing algorithm.

During annealing, a metal is maintained at a certain temperature $T$ for a precomputed amount of time, before reducing the temperature by a precomputed amount. The molecules have a greater degree of freedom to move at higher temperatures than at lower temperatures. *The movement of molecules is analogous to the generation of new (neighbourhood) states in an optimization process.* In order to simulate the annealing process, much flexibility is allowed in neighbourhood generation at higher 'temperatures', i.e., many uphill moves are permitted at higher temperatures. The temperature parameter is lowered gradually as the algorithm proceeds. At temperatures close to absolute zero, very few uphill moves are permitted. In fact, at the absolute zero temperature, the simulated annealing algorithm turns greedy, allowing only downhill moves.

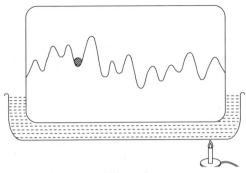

**Figure 2.15** Design space analogous to a hilly terain.

**Example 2.5** We can understand simulated annealing by considering the analogy of a ball placed in a hilly terrain, as shown in Figure 2.15. The hilly terrain is nothing but the variation of the cost function over the configuration space, as shown by Figure 2.14. If a ball is placed at point $S$, it will roll down into a pit such as $L$, which represents a local minimum. In order to get the ball out of the local minimum (our intention is to get the ball into $G$, the global minimum), we do the following. We enclose the hilly terrain in a box and place the box in a water bath. When the water bath is heated, the box begins to shake, and the ball has a chance to climb out of the local minimum $L$ (Kirkpatrick *et al.*, 1983; Nahar *et al.*, 1989).

If we are to apply simulated annealing to this problem, we would initially heat the water bath to a high temperature, making the box wobble violently. At such high temperatures, the ball moves rapidly into and out of local minima. As time proceeds, we cool the water bath gradually. The lower the temperature, the gentler the movement of the box, and the less likelihood of the ball jumping out of a minimum. The search for a local minimum is more or less random at high temperatures; the search becomes more greedy as temperature falls. At absolute zero, the box is perfectly still, and the ball rolls down into a minimum, which, hopefully, is the global minimum $G$.

## The algorithm

The simulated annealing algorithm is shown in Figure 2.16. The core of the algorithm is the *Metropolis* procedure (Figure 2.17), which simulates the annealing process at a given temperature $T$ (Metropolis *et al*; 1953). The procedure is named after a scientist who devised a similar scheme to simulate a collection of atoms in equilibrium at a given temperature. *Metropolis* also receives as input the current solution $S$ which it improves through local search. Finally, *Metropolis* must also be provided with the value $M$, which is the amount of time for which annealing must be applied at temperature $T$. The procedure *Simulated_annealing* simply invokes *Metropolis* at various (decreasing) temperatures. Temperature is initialized to a value $T_0$ at the beginning of the procedure, and is slowly reduced in a geometric progression; the parameter $\alpha$ is used to achieve this cooling. The amount of time spent in annealing at a temperature is gradually *increased* as temperature is lowered. This is done using the parameter $\beta > 1$. The variable *Time* keeps track of the time being expended in each call to the *Metropolis*. The annealing procedure halts when *Time* exceeds the allowed time.

The *Metropolis* procedure is shown in Figure 2.17. It uses the procedure *neighbour* to generate a local neighbour *NewS* of any given solution $S$. The function *Cost* returns the cost of a given solution $S$. If the cost of the new solution *NewS* is better than the cost of the current solution $S$, then certainly the new solution is acceptable, and we do so by setting $S = NewS$. If the new solution has an inferior cost in comparison to the original solution $S$, *Metropolis* will accept the new solution on a *probabilistic* basis. A random number is generated in the range 0 to 1. If this random number is smaller than $e^{-\Delta h/T}$, where $\Delta h$ is the difference in costs, and $T$ is the temperature, the inferior solution is accepted. This criterion for accepting the new solution is known as the *Metropolis criterion* named after its inventor. The *Metropolis* procedure generates and examines $M$ solutions.

**Algorithm** Simulated_annealing $(S_0, T_0, \alpha, \beta, M, Maxtime)$;
    $(*S_0$ is the initial solution $*)$
    $(*T_0$ is the initial temperature $*)$
    $(*alpha$ is the cooling rate $*)$
    $(*beta$ a constant $*)$
    $(*Maxtime$ is the total allowed time for the annealing process$*)$
    $(*M$ represents the time until the next parameter update $*)$
**begin**
    $T = T_0$;
    $S = S_0$;
    $Time = 0$;
      **repeat**
        Call $Metropolis(S, T, M)$;
        $Time = Time + M$;
        $T = \alpha \times T$;
        $M = \beta \times M$
      **until** $(Time \geq MaxTime)$;
      Output Best solution found
**End.** $(*of\,Simulated\_annealing*)$

**Figure 2.16** Procedure for simulated annealing algorithm.

**Algorithm** Metropolis($S, T, M$);
**begin**
   **repeat**
     $NewS = neighbour(S)$;
     $\Delta h = (Cost(NewS) - Cost(S))$;
     **If** $((\Delta h < 0)$ **or** $(random < e^{-\Delta h/T}))$ **then** $S = NewS$;
     {accept the solution}
     $M = M - 1$
   **until** $(M = 0)$
**End.** (*of Metropolis*).

**Figure 2.17** The Metropolis procedure.

The probability that an inferior solution is accepted by the *Metropolis* is given by $P(random < e^{-\Delta h/T})$. The random number generation is assumed to follow a *uniform distribution*, i.e., all numbers in the range 0 to 1 are equally likely to be generated by $random()$. In that case, the probability reduces to $e^{-\Delta h/T}$. Remember that $\Delta h > 0$ since we have assumed that *NewS* is inferior compared to $S$. At very high temperatures, say $T \to \infty$, the above probability approaches 1. On the contrary, when $T \to 0$, the probability $e^{-\Delta h/T}$ falls to 0.

## Partitioning using simulated annealing

Consider the two-way partitioning problem, and assume that nodes as well as nets have weights. It is required to generate an almost balanced partition with a minimum weighted cutset.

In order to use simulated annealing to solve the problem, the first step is to formulate a cost function which reflects both the balance criterion as well as the weight of the cutset. For any given partition $(A, B)$ of the circuit, we define

$$Imbalance(A, B) = Size\ of\ A - Size\ of\ B \qquad (2.23)$$

$$= \sum_{v \in A} s(v) - \sum_{v \in B} s(v) \qquad (2.24)$$

$$Cutset\ weight(A, B) = \sum_{n \in \psi} w_n \qquad (2.25)$$

As before, $s(v)$ is the size of vertex $v$, and $w_n$ is the weight of net $n$, and $\psi$ is the set of nets with terminals in both $A$ and $B$.

$$Cost(A, B) = W_c * Cutset\ weight(A, B) + W_s * Imbalance(A, B) \quad (2.26)$$

$W_s$ and $W_c$ are constants in the range of [0,1] which indicate the relative importance of balance and minimization of cutset, respectively. These are usually selected by the user. Notice that, unlike the Kernighan–Lin algorithm, the balance constraint is part of the cost function. Thus if the simulated annealing algorithm halts at a local minimum, it may yield an unbalanced partition. If balance is of significant importance, the user must specify a $W_s$ close to 1. On the other hand, if the imbalance is tolerable and minimizing the cutset is of paramount importance, the user has the option to lower $W_s$ and specify a large value of $W_c$.

## Neighbour function

The simplest neighbour function is the pairwise exchange. Here, two elements, one from each partition are swapped to disturb the current solution. A subset of elements may also be randomly selected from each partition and exchanged. Another possibility is to select those elements whose contribution to the external cost is high, or those that are internally connected to the least number of vertices. Neighbour functions used by simulated annealing to solve other physical design problems are further discussed in later chapters. The actions of simulated annealing can be best explained using an example.

**Example 2.6** The circuit in Figure 2.18 contains 10 cells and 10 nets. Assume that all cells are of the same size. The nets of the circuit are indicated along with their weights in Table 2.4.
*Initial solution:* Randomly assign nodes $C_1, \cdots, C_5$ to block $A$;
*Neighbour function:* Pairwise exchange, i.e., exchange a node $a \in A$ with a node $b \in B$;
*Initial temperature:* $T_0 = 10$;
*Constants:* $M = 10$; $\alpha = 0.9$; $\beta = 1.0$;
*Termination criterion:* $T$ reduces to 30 per cent of its initial value.

SOLUTION Pairwise exchange will not disturb the balance of partition, since all elements are of equal size. Therefore we may set $W_s = 0$ and $W_c = 1$. The objective function reduces to:

$$Cost(A, B) = \sum_{n \in \psi} w_n$$

Table 2.5 shows results for some iterations, where $(c_i, c_j)$ represents the two cells selected for swapping, one from each partition. Only the accepted moves are listed. For example, when $count = 9$, the temperature is equal to 10, the cells selected for pairwise interchange

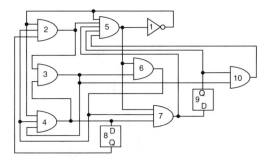

**Figure 2.18** Circuit for Example 2.6.

| Net | Weight |
|---|---|
| $N_1 = \{C_1, C_2, C_4, C_5\}$ | $w_1 = 1$ |
| $N_2 = \{C_2, C_3, C_5\}$ | $w_2 = 1$ |
| $N_3 = \{C_3, C_6, C_{10}, C_4\}$ | $w_3 = 2$ |
| $N_4 = \{C_4, C_8, C_3, C_7\}$ | $w_4 = 1$ |
| $N_5 = \{C_5, C_7, C_1, C_6\}$ | $w_5 = 3$ |
| $N_6 = \{C_6, C_4, C_7, C_2\}$ | $w_6 = 3$ |
| $N_7 = \{C_7, C_9, C_5\}$ | $w_7 = 2$ |
| $N_8 = \{C_8, C_2\}$ | $w_8 = 3$ |
| $N_9 = \{C_9, C_{10}, C_5\}$ | $w_9 = 2$ |
| $N_{10} = \{C_{10}, C_5\}$ | $w_{10} = 4$ |

**Table 2.4  Netlist for Example 2.6.**

are (3,8), the cost of current solution $S$ is $Cost(S) = 13$, and the cost of the new solution is $Cost(NewS) = 10$. Therefore the interchange is automatically accepted. For the next iteration at the same temperature, $Count = 10$, the cells selected are (1,4), the cost of new solution $Cost(NewS) = 13$, which is larger than the cost of the current solution $(Cost(S) = 10)$. In this case a random number is generated. Since the value of this number is 0.11618 which is less than $e^{-\Delta h/T} = e^{-3/10} = 0.74082$, the move is accepted.

Figure 2.19 plots the *cost* for both a greedy algorithm and for simulated annealing. Note that in simulated annealing, the cost reduces to 10 in the 9th iteration. This is due to the acceptance of some previous bad moves. The plot of greedy pairwise exchange shows plateaux or a decrease and converges to 10 after 50 iterations. For this example, the final partition obtained by both deterministic pairwise exchange and simulated annealing procedures is the same, with $A = \{C_2, C_4, C_6, C_7, C_8\}$ and $B = \{C_1, C_3, C_5, C_9, C_{10}\}$. The cost of this partition is 10.

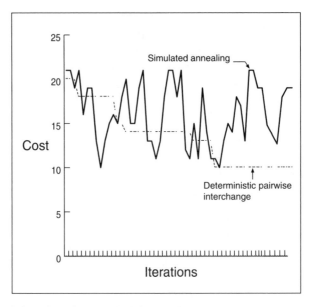

**Figure 2.19** Variation of cost in Example 2.6.

## 2.5 OTHER APPROACHES AND RECENT WORK

In this chapter, we studied two deterministic algorithms, one developed by Kernighan and Lin, and the other by Fiduccia and Mattheyses. A non-deterministic algorithm that uses simulated annealing to solve the two-way partition problem was also discussed. The exercises at the end of this chapter shed more light on some of the finer details of these algorithms. We now discuss some of the recent developments in the area of circuit partitioning. We look at some of the significant reported work to improve the quality of solution obtained by the Kernigan–Lin heuristic, and use of other objective functions and optimization tools to solve the bipartitioning problem.

The classical two-way partitioning technique that was proposed by Kernighan–Lin is used to partition graphs. This heuristic was modified by Schweikert and Kernighan (1972) who proposed a net-cut model to handle multipin nets. Fiduccia and Mattheyses (Section 2.4.3) improved their algorithm by reducing the time complexity. Another multipin net model proposed by Sechen and Chen, (1988) also produced excellent results.

The Kernighan–Lin algorithm has been found to be highly sensitive to the choice of initial partition. The initial partition chosen is generally random, and the final results vary significantly as a consequence of different starting points (Krishnamurthy, 1984). Therefore many runs† on

| Count | $\alpha * T$ | $(c_i, c_j)$ | Random | $Cost(S)$ | $Cost(NewS)$ | $e^{-\Delta h}/T$ |
|---|---|---|---|---|---|---|
| 1 | 10.000 | (1,6) | 0.73222 | 20 | 21 | 0.90484 |
| 2 | | (1,2) | | 21 | 19 | |
| 3 | | (2,3) | 0.13798 | 19 | 21 | 0.81873 |
| 4 | | (2,7) | | 21 | 16 | |
| 5 | | (1,3) | 0.64633 | 16 | 19 | 0.74082 |
| 6 | | (2,3) | 0.46965 | 19 | 19 | 1.00000 |
| 8 | | (3,5) | | 19 | 13 | |
| 9 | | (3,8) | | 13 | 10̲ | |
| 10 | | (1,4) | 0.11618 | 10 | 13 | 0.74082 |
| | | | | | | |
| 11 | 9.000 | (2,4) | 0.47759 | 13 | 15 | 0.80074 |
| 12 | | (3,4) | 0.19017 | 15 | 16 | 0.89484 |
| 13 | | (4,6) | | 16 | 15 | |
| 14 | | (4,9) | 0.26558 | 15 | 18 | 0.71653 |
| 15 | | (1,5) | 0.19988 | 18 | 20 | 0.80074 |
| 16 | | (2,5) | | 20 | 15 | |
| 17 | | (3,5) | 0.28000 | 15 | 15 | 1.00000 |
| 18 | | (4,5) | 0.90985 | 15 | 15 | 1.00000 |
| 19 | | (5,7) | 0.06332 | 15 | 19 | 0.64118 |
| 20 | | (5,10) | | 19 | 16 | |
| | | | | | | |
| 21 | 8.100 | (2,6) | 0.15599 | 16 | 21 | 0.53941 |
| 22 | | (3,6) | | 21 | 19 | |
| 23 | | (4,6) | 0.36448 | 19 | 21 | 0.78121 |
| 24 | | (5,6) | | 21 | 13 | |
| 25 | | (6,8) | 0.53256 | 13 | 13 | 1.00000 |
| 26 | | (3,7) | | 13 | 11 | |
| 27 | | (4,7) | 0.18104 | 11 | 13 | 0.78121 |
| 28 | | (5,7) | 0.51371 | 13 | 18 | 0.53941 |
| 29 | | (7,8) | 0.37249 | 18 | 21 | 0.69048 |
| 30 | | (7,9) | 0.57933 | 21 | 21 | 1.00000 |
| | | | | | | |
| 31 | 7.290 | (1,8) | | 21 | 18 | |
| 32 | | (4,8) | 0.10814 | 18 | 21 | 0.66264 |
| 33 | | (5,8) | | 21 | 12 | |
| 37 | | (1,9) | | 12 | 11 | |
| 39 | | (3,9) | 0.14080 | 11 | 15 | 0.57770 |
| | | | | | | |
| 41 | 6.561 | (6,9) | | 15 | 11 | |
| 44 | | (2,10) | 0.21728 | 11 | 19 | 0.29543 |
| 45 | | (3,10) | | 19 | 14 | |
| 49 | | (1,2) | | 14 | 11 | |
| | | | | | | |
| 50 | | (1,3) | 0.84921 | 11 | 11 | 1.00000 |
| | | | | | | |
| 52 | 5.905 | (1,7) | | 11 | 10 | |
| 53 | | (1,8) | 0.54613 | 10 | 13 | 0.60167 |

Table 2.5 Table of execution run of Example 2.6.

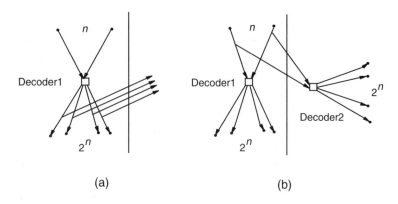

(a)                                    (b)

**Figure 2.20** (a) Graph of an $n$ to $2^n$ decoder. (b) Replication of decoder reduces the size of cut from $2^n$ to $n$.

randomly generated initial partitions are needed to avoid being trapped at local minima (Krishnamurthy, 1984). This process is very time consuming. Furthermore, the probability of finding the optimal solution in a single trial drops exponentially as the size of the circuit increases. This problem of unpredictable performance or '*instability*' of the quality of solution was addressed by Cheng and Wei (1991). Cheng and Wei use a recursive top-down partitioning technique to divide the circuit into small highly connected *groups*. The groups generated by each partition run get progressively smaller and the size of the problem is reduced at every single run. The final step is to rearrange the small groups into two subsets that meet the size constraint. Because the number of groups is relatively small in comparison with the number of modules in the circuit, many trials of rearrangements can be attempted. Using this method, the probability of getting a near optimum solution is high since the number of groups generated is small. A ratio-cut approach is used that removes the constraint on predefined subset size, and tends to identify natural clusters in the circuit. The ratio-cut objective function that is minimized is as follows:

$$\frac{Cost(A, B)}{|A| \times |B|} \tag{2.27}$$

where $A$ and $B$ are two disjoint blocks. This ratio-cut metric of Wei and Cheng has been proved to be a highly successful objective function for many applications ranging from layout to hardware simulation (Wei and Cheng, 1991). The technique intuitively allows freedom to find natural

---

† A run is one trial of Kernighan–Lin algorithm on a given initial partition.

partitions. The numerator captures the minimum cut criterion while the denominator favours an even partition since $|A| \times |B|$ is largest when $|A| = |B|$.

The minimum ratio-cut metric can be used not only to assign *modules* to the two sides of the partition but, equivalently, to assign nets to the two sides of the partition. This metric is used by Cong *et al.* (1992) to find partitioning of nets rather than partition of modules with the objective of maximizing the number of nets that are not cut by the partition.

A novel technique that dramatically reduces the size of cut by replicating hardware was presented by Hwang and El Gamal (1992). They defined *vertex replication* as a transformation on graphs and presented an algorithm for determining optimal min-cut replication sets in a *k*-partitioned directed graph. Vertex replication transforms a circuit into a larger equivalent one. Vertex replication is illustrated in Figure 2.20. If the cost of the cut is more than the cost of hardware, the technique becomes extremely useful. Furthermore, in design methodologies that use FPGAs, the redundant or unused logic on the chip can be used to help in reducing the size of cut. During the mapping of large logic networks into multiple FPGAs, replication can be used to reduce the number of FPGAs, the number of wires interconnecting FPGAs, and the number of inter-chips wires. The technique is also helpful in improving the performance of digital systems.

Cost of cut and imbalance are not the only objective functions to be minimized in partitioning for physical design. Completion of designs to achieve predictable performance such as avoidance of timing problems, etc., fall into a category known as *performance directed physical design*. The problem of assigning functional blocks into slots on multichip modules during high level design was addressed by Shih and others (1992). Their method minimizes nets cut while satisfying timing and capacity constraints.

Finally, we refer to the work of Yih and Mazumder (1989) who used *neural networks* for circuit bipartitioning. They represent bipartitioning in the form of states of the neural network. They describe the criterion for selecting components for changing locations in terms of neuron connections and the corresponding weights between neurons. The problem of balancing partitions while minimizing the cut was also addressed. Their work exploits the massive parallelism of neural networks to obtain results comparable to those obtained by Fiduccia–Mattheyses heuristic. In later chapters we will touch upon attempts to apply neural networks to solve other physical design problems such as VLSI cell placement.

## 2.6 CONCLUSION

In this chapter, we have examined an important subproblem that arises in electronic circuit layout, namely, circuit partitioning. The two-way balanced partitioning problem was examined in greater detail, due to its practical importance. The two-way partitioning problem is used in a circuit placement procedure known as min-cut placement, which is discussed in Chapter 4. Furthermore, a two-way partitioning algorithm can form the basis for a $k$-way partitioning procedure. One recursively applies the two-way partitioning algorithm $\log_2 k$ times. However, this scheme has some drawbacks; first, it assumes that $k$ is a power of 2, and second, there is no reason to believe that it will generate a better partition than another $k$-way partitioning algorithm. In general, the $k$-way partitioning problem is more challenging than the two-way partitioning problem; researchers are developing heuristic algorithms for generating good $k$-way partitions.

## 2.7 BIBLIOGRAPHIC NOTES

The most popular algorithm for two-way graph partition was given by Kernighan and Lin in 1970. The algorithm has been widely used in CAD programs and has been extended by numerous authors. Particularly notable among such extensions is the work of Fiduccia–Mattheyses, who gave an efficient implementation of Kernighan–Lin algorithm for sparse circuits. If it can be asserted that each circuit element is connected to no more than $p$ other elements, where $p$ is a positive integer constant, the Fiduccia–Mattheyses algorithm generates a two-way partition in $O(n)$ time. Simulated annealing, a general technique for solving combinatorial optimization problems, was first proposed in 1983 (Kirkpatrick *et al.*, 1983). It is inspired by the physical process of annealing metals and uses randomization as a powerful tool. Simulated annealing has been applied to numerous problems such as travelling salesman, circuit partitioning, floorplanning, circuit placement, routing, and so on. Some of these applications are treated in later chapters.

## EXERCISES

### Exercise 2.1 Programming Exercise:

1. Given a set of $2 \cdot n$ elements, show that the number of *balanced* two-way partition of the set is $P(2n) = \frac{2n!}{2 \cdot n! \cdot n!}$.

2. Use Stirling's approximation for $n!$ and simplify the expression for $P(2n)$ in Exercise 2.1.1 above. Express $P(2n)$ using the Big-Oh notation.

3. A *brute force* algorithm for the two-way partition problem enumerates all the $P(2n)$ solutions and selects the best. Write a computer program which implements such a brute force algorithm. What is the time complexity of your program? (*Programming Hint:* You may represent a set of $2n$ elements using an array of $2n$ integers. In this representation, the first $n$ elements of the array belong to partition $A$, and the remaining belong to partition $B$.)

4. Plot the running time of the *brute force* partition program for $n = 1, \cdots, 10$. If the maximum permitted execution time for the program is 24 hours, what is the maximum value of $n$ for which your program can run to completion?

**Exercise 2.2 Programming Exercise:**
1. Given a set of $k \cdot n$ elements, write an expression for $P(k \cdot n)$, the number of balanced, $k$-part partitions of the set.

2. Plot $P(k \cdot n)$ as a function of $k$ for $k \cdot n = 24$. Use $k = 1, 2, \cdots, 24$. What is your conclusion?

**Exercise 2.3 Programming Exercise:** Suppose we are given a circuit with $2n$ identical elements. All the nets in the circuit are two-pin nets, and a matrix $C$ specifies the connectivity information between elements; for example, $c_{ij}$ gives the number of connections between elements $i$ and $j$. Let $A$ and $B$ represent the two blocks of a balanced partition of the circuit.

Define a variable $x_i$ for each element $i$ as follows. $x_i$ is set to 0 if the element $i$ belongs to block $A$, and $x_i$ is set to 1 if $i$ belongs to $B$.

1. Define $p_{ij} = x_i + x_j - 2 \cdot x_i \cdot x_j$. Show that $p_{ij}$ takes on the value 1 if and only if $x_i \neq x_j$.

2. Formulate an expression for the external cost of the partition (size of the cutset) in terms of $x_i$. (*Hint* : Use the result 1 above.)

3. Express the balance constraint as a linear equation in $x_i$. (*Hint* : What should the sum of $x_i$ be, if the partition is balanced?)

4. (*) Using Exercises 2.3.2 and 2.3.3 above, formulate the partitioning problem as an optimization problem in terms of $x_i$. For the moment, pretend that $x_i$ are *continuous* variables in the range 0 to 1. In that case, the partition problem may be solved using numerical optimization. Construct an heuristic algorithm which applies such a numerical technique to solve for $x_i$; if $x_i \leq 0.5$,

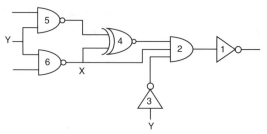

**Figure. 2.21** Circuit for Exercise 2.6.

the element $i$ is assigned to $A$. If $x_i > 0.5$, it is assigned to $B$. Write a program and test your heuristic against the examples in Chapter 2. Does the algorithm retain balance?

5. **(*)** The cost function in Exercise 2.3.2 is *quadratic* because of the product term involving $x_i$ and $x_j$. Is it possible to derive a cost function that is *linear* in the optimization variables? If yes, you have a 0–1 *integer program formulation* of the partitioning problem.

**Exercise 2.4** Give an example of a circuit with four nodes, and an initial partition, such that on the application of Kernighan–Lin algorithm, the $g$ value corresponding to the first selected pair is

(1) Zero;
(2) Negative.

**Exercise 2.5** Complete Example 2.2.

**Exercise 2.6** Suppose that the circuit given in Figure 2.2(a) is modified as shown in Figure 2.21; Nets marked $X$ and $Y$ are both 3-pin nets. Net $X$ can be modelled by including three edges in the graph of Figure 2.2(b)—between 6 and 4, between 4 and 2, and between 6 and 2. How will you modify the graph to model net Y? In general, how many edges are required to model a $k$-pin net?

**Exercise 2.7** Draw the graph to represent the circuit given in Figure

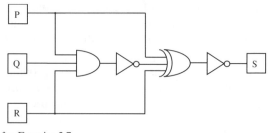

**Figure. 2.22** Circuit for Exercise 2.7.

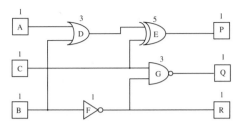

**Figure 2.23** Circuit for Exercise 2.8.

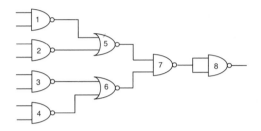

**Figure 2.24** Circuit for Exercise 2.11.

2.22. Using the method of Kernighan–Lin divide the graph into two subgraphs so that the number of edges crossing the partitions is minimized. Assume that all gates of the circuits are of the same size.

**Exercise 2.8**
1. In the circuit given in Figure 2.23, the number adjacent to the gate is its size. Divide the circuit into two partitions of equal *sizes* such that the number of edges crossing the partitions is minimum.
2. Repeat the above problem, such that the number of *nets* crossing the partitions is reduced.

**Exercise 2.9** Refer to Section 2.4.2 for a two-way partitioning algorithm that generates unequal sized blocks $A$ and $B$. Show that the algorithm works, i.e., prove that the algorithm terminates with $|A| = n_1$ and $|B| = n_2$.

**Exercise 2.10** If $g_1, g_2, \cdots, g_n$ are the gains of swapping node pairs in the improvement procedure of the Kernighan–Lin algorithm, show that

$$g_1 + g_2 + \cdots + g_n = 0.$$

**Exercise 2.11** For the circuit shown in Figure 2.24, apply the Kernighan–Lin algorithm to generate a two-way balanced partition. The initial partition is $A = \{1, 2, 3, 4\}$ and $B = \{5, 6, 7, 8\}$. You can guess the optimum partition for this example; does the Kernighan–Lin algorithm generate the optimum? What is the balanced partition of the circuit that has the *maximum* value of cutset?

**Exercise 2.12** Apply the Kernighan–Lin algorithm to the following circuit. Use the improvement suggested by Kernighan–Lin concerning the sorting of the $D$-values before computing/updating the cell gains.

**Nets:**

$N_1 = \{C_4, C_5, C_6\}$  $\qquad N_6 = \{C_4, C_7, C_9\}$  $\qquad N_{11} = \{C_2, C_6 C_7\}$

$N_2 = \{C_4, C_3, C_{12}\}$  $\qquad N_7 = \{C_2, C_8, C_{10}\}$  $\qquad N_{12} = \{C_{10}, C_{12}\}$

$N_3 = \{C_2, C_4\}$  $\qquad N_8 = \{C_1, C_7\}$  $\qquad N_{13} = C_4, C_7, C_9\}$

$N_4 = \{C_3, C_7, C_8\}$  $\qquad N_9 = \{C_3, C_5, C_9\}$  $\qquad N_{14} = \{C_3, C_9, C_{11}\}$

$N_5 = \{C_2, C_3, C_6\}$  $\qquad N_{10} = \{C_6, C_8, C_{11}\}$

**Exercise 2.13** Apply the Fiduccia–Mattheyses heuristic on the circuit of Exercise 2.12 and compare both the solutions.

**Exercise 2.14** Complete Example 2.3.

**Exercise 2.15** Complete Example 2.4.

**Exercise 2.16 (\*) Programming Exercise:** In a graph $G = (V, E)$, the *degree $d_i$* of an node $i \in V$ is defined as the number of (other) nodes $i$ is connected to. Construct an example of a graph with 10 nodes, such that the nodes have a large degree, say 5 to 10. You may assume that all the nodes have unit sizes. Apply the Kernighan–Lin algorithm to generate a two-way balanced partition of the graph. Also apply the simulated annealing algorithm on the same example. Which algorithm gives you better results?

**Exercise 2.17 (\*) Programming Exercise:**

1. Construct a graph with 10 nodes and 50 nets. Starting from a random partition apply both the Kernighan–Lin and simulated annealing algorithms to this graph and generate balanced two-way partitions. Which algorithm gave you a better cutset?
2. Starting from the solution obtained from the first pass of Kernighan–Lin heuristic, apply simulated annealing. Comment on any noticeable improvement in quality of solution and runtime?

**Exercise 2.18** Modify the terminating condition of the simulated annealing algorithm so that the final annealing temperature is $T_f$. Estimate the time complexity of the simulated annealing procedure in terms of $M$, $T_0$, $\alpha$, $\beta$ and $T_f$. (*Hint* : First estimate the number of temperatures during the annealing process.)

# REFERENCES

Cheng, C. and Y. A. Wei. An improved two-way partitioning algorithm with stable performance. *IEEE Transactions on Computer-Aided Design*, 10(12):1502–1511, 1991.

Cong, J., L. Hagen and A. Kahng. Net partitions yield better module partitions. In *29th Design Automation Conference*, pages 47–52, 1992.

Fiduccia, C. M. and R. M. Mattheyses. A linear time heuristic for improving network partitions. In *19th Design Automation Conference*, pages 175–181, 1982.

Hwang, J. and A. El Gamal. Optimal replication for min-cut partitioning. In *IEEE International Conference on Computer-Aided Design*, pages 432–435, 1992.

Kernighan, B. W. and S. Lin. An efficient heuristic procedure to partition graphs. *Bell System Technical Journal*, 49(2):291–307, February 1970.

Kirkpatrick, S. Jr., C. Gelatt, and M. Vecchi. Optimization by simulated annealing. *Science*, 220(4598):498–516, May 1983.

Krishnamurthy, B. An improved mincut algorithm for partitioning VLSI Networks. *IEEE Transactions on Computers*, C-33:438–446, 1984.

Metropolis, N. *et al.* Equation of state calculations by fast computing machines. *Journal of Chem. Physics*, 21:1087–1092, 1953.

Nahar, S., S. Sahni, and E. Shragowitz. Simulated annealing and combinatorial optimization. *Journal on Computer Aided Design*, 1:1–23, 1989.

Schweikert, D. G. and B. W. Kernighan. A proper model for the partitioning of electrical circuits. In *9th Design Automation Workshop*, pages 57–62, 1972.

Sechen, C. and D. Chen. An improved objective function for mincut circuit partitioning. In *IEEE International Conference on Computer-Aided Design*, pages 502–505, 1988.

Shih, M., E. S. Kuh, and R. Tsay. Performance-driven system partitioning on multi-chip modules. In *29th Design Automation Conference*, pages 53–56, 1992.

Wei, Y. A. and C. Cheng. Ratio cut partitioning for hierarchical designs. *IEEE Transactions on Computer-Aided Design*, 10(7):911–921, 1991.

Yih, J. and P. Mazumder. A neural network design for circuit partitioning. In *26 Design Automation Conference*, pages 406–411, 1989.

# THREE

# FLOORPLANNING

## 3.1 INTRODUCTION

As discussed in Chapter 1, the increasing complexity of VLSI circuits led to the breaking of the design process into steps, as well as the introduction of several semi-custom ASIC design methodologies such as general cell, standard-cell, and gate-array. As opposed to the full-custom design approaches, semi-custom approaches impose some structure on the layout of circuit elements, thus restricting the design space and, by the same token, reducing the complexity of solving the problem. Since the dawn of mankind, men (and engineers in particular) have also been using the divide-and-conquer technique to solve large problems. For the divide-and-conquer solution technique, the input problem is recursively subdivided into subproblems yielding problems of manageable size and complexity (Horowitz and Sahni, 1984). It is usually the case that the subproblems are of the same kind as the original problem. The solution of the individual subproblems when combined yields a solution to the original problem. Partitioning, which has been explained in the previous chapter, is used to divide the original circuit into subcircuits which, somehow, can be dealt with separately.

Each design step of the design process represents (and usually requires) a different level of abstraction. At the floorplanning as well as placement steps, the VLSI circuit is seen as a set of rectangular blocks interconnected by signal nets. Placement consists of placing these blocks on a two dimensional surface such that no two blocks overlap, while optimizing certain objectives (area of the surface, interconnection length, performance).

Floorplanning is an essential design step when a hierarchical/building block design methodology is used. For such methodologies, the following sequence of tasks must be executed in order to carry a design from specification to layout:

1. Define the layout hierarchy.
2. Estimate the overall required area.
3. Determine the aspect ratios for each module.
4. Assign the pin and pad locations.
5. Perform placement.
6. Route the design.

Floorplanning helps solve some of the above problems. It is closely related to placement. It can be seen as a feasibility study of the layout (placement). Sometimes it is referred to as topological layout. Where for placement shape and pin positions on the periphery of circuit components are fixed, in floorplanning these have some specified flexibility. The flexibility in the shape of the component represents the designer's freedom to select one among several implementations of the element.

## 3.2 PROBLEM DEFINITION

As we said, floorplanning is a generalization of the placement problem. During floorplanning, designers have additional flexibility and freedom in terms of chip and component geometries (block orientations, shapes, and maybe sizes). This added flexibility must be captured by the floorplan model. Obviously, the aspects that need to be modelled should consist of the components, the interconnections, the flexible interfaces (blocks and chip), the chip carrier (layout surface), any designer stated constraints, and the objective to optimize.

### 3.2.1 Floorplanning model

A formal representation of the floorplanning problem is as follows (Wong and Liu, 1986; Sutanthavibul et al, 1990).

Given the following:

1. a set $S$ of $n$ rectangular modules $S = \{1, 2, \cdots, i, \cdots, n\}$;
2. $S_1$ and $S_2$, a partition of $S$, where $S_1$ and $S_2$ are the set of the modules with fixed and free orientation respectively;
3. an interconnection matrix $C_{n \times n} = [c_{ij}]$, $1 \leq i, j \leq n$, where $c_{ij}$ indicates the connectivity between modules $i$ and $j$;
4. a list of $n$ triplets $(A_1, r_1, s_1), \cdots (A_i, r_i, s_i), \cdots (A_n, r_n, s_n)$, where $A_i$ is the area of block $i$ (i.e., $A_i = w_i \times h_i$, with $w_i$ and $h_i$ the width and height of block $i$), $r_i$ and $s_i$ are lower and upper bound constraints on the shape of block $i$ ($r_i \neq s_i$ if the block is flexible, and $r_i = s_i$ if the block is rigid);
5. two additional positive numbers $p$ and $q$ ($p \leq q$), which are lower and upper bound constraints on the shape of the rectangle enveloping the $n$ blocks.

The required output is:

A feasible floorplan solution, i.e., an enveloping rectangle $R$ subdivided by horizontal and vertical line segments into $n$ non-overlapping rectangles labelled $1, 2, \cdots, i, \cdots, n$, such that the following constraints are satisfied:

1. $w_i \times h_i = A_i$, $1 \leq i \leq n$;
2. $r_i \leq \frac{h_i}{w_i} \leq s_i$ for all modules $i$ with fixed orientation ($i$ is an element of $S_1$);
3. $r_i \leq \frac{h_i}{w_i} \leq s_i$ or $\frac{1}{s_i} \leq \frac{h_i}{w_i} \leq \frac{1}{r_i}$ for all modules $i$ with free orientation ($i$ is an element of $S_2$);
4. $x_i \geq w_i$ and $y_i \geq h_i$ $1 \leq i \leq n$, where $x_i$ and $y_i$ are the dimensions of basic rectangle $i$, (every rectangle $i$ is large enough to accommodate module $i$);
5. $p \leq \frac{H}{W} \leq q$, where $H$ and $W$ are the height and width of the enveloping rectangle $R$.

A feasible floorplan optimizing the desired cost function is an optimum floorplan. For example, if the cost function is the area of $R$, then an optimum feasible floorplan is one with minimum area.

**Example 3.1** Assume we have five rigid blocks with dimensions as indicated in Table 3.1. Assume further that all blocks have free orientations. Figure 3.1 gives several feasible floorplans. Notice that all of these floorplans have the same area. If area is the cost function used to measure the quality of distinct floorplans, then all of the floorplans of Figure 3.1 will be equally good. In the following section, we will discuss many of the cost measures that are used to evaluate and compare floorplan solutions.

| | Module | Width | Height |
|---|---|---|---|
| **Table 3.1 Dimensions of modules for floorplans of Figure 3.1** | 1 | 1 | 1 |
| | 2 | 1 | 1 |
| | 3 | 2 | 1 |
| | 4 | 1 | 2 |
| | 5 | 1 | 3 |

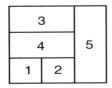 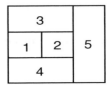

**Figure 3.1** Some of the feasible floorplans for modules of Table 3.1

## 3.2.2 Cost functions

The number of feasible solutions for a given instance of a floorplanning problem is very large (if $R$ is unbounded the number is infinity). Besides tremendously reducing the search of a floorplanning solution, the introduction of an objective function allows us to select superior floorplan(s). However, this will change the problem into an optimization problem that is much harder.

The floorplanning problem is an NP-hard problem. This is an obvious statement since floorplanning is a generalization of the placement problem, a generalization of the quadratic assignment problem, which is NP-hard (Garey and Johnson, 1979). There are no universally accepted criteria for measuring the quality of floorplans. Possible criteria can be

1. minimize area,
2. minimize wirelength,
3. maximize routability,
4. minimize delays, or
5. a combination of two or more of the above criteria.

Next we discuss some of these objective functions when used as floorplan cost functions.

## Area of the enveloping rectangle

Here the objective is to find a feasible floorplan with the smallest overall area. In this case, floorplanning can be seen as a generalized two-dimensional bin-packing problem. In most general terms, bin-packing is the problem of determining an optimal packing of an arbitrary list of rectangles. Even this simplified version of floorplanning problem has been shown to be NP-hard (Baker *et al.*, 1980(b)). However, this problem had been extensively studied and several polynomial-time approximation algorithms have been reported (Baker *et al.*, 1980(a), 1981, 1983). For these algorithms, the rectangles are kept in a sorted queue (usually sorted on height or width). Then the rectangles are packed in the specified order. The order in the queue is dynamic, and new rectangles may join the queue as packing progresses.

Two-dimensional bin-packing will not be described in this chapter as this is an oversimplified model of the floorplanning problem. However, a generalization of this approach which will take into consideration other objectives and constraints is worth further study.

## Overall interconnection length

In this case, the objective is to find a feasible floorplan with minimum overall interconnection length. For floorplanning, only a coarse measure of wirelength is used, which is based on a rough global routing step where all I/O pins of the block are merged and assumed to reside at its centre. In this case, a possible estimate of the overall wiring length is $L = \sum_{i,j} c_{ij} \cdot d_{ij}$, where $c_{ij}$ is the connectivity between modules $i$ and $j$, and $d_{ij}$ is the Manhattan distance between the centres of rectangles to which modules $i$ and $j$ have been assigned. Another possible estimate is to determine for each net the length of the minimum spanning tree which covers all the modules of the net. Then, a measure of $L$ will be the sum of the length of all these minimum spanning trees.

## Area and interconnection length

Here, the cost function is a weighted sum of the area $A$ of the bounding rectangle of the floorplan and the overall interconnection length $L$, i.e.,

$$Cost = \alpha \times A + \beta \times L \tag{3.1}$$

where $\alpha$ and $\beta$ are usually user specified parameters. The objective is to find a feasible floorplan which minimizes the above cost function.

## Performance (speed)

Since the early 1980s, circuit speed has become a major issue (Burstein and Youssef, 1985; Dunlop *et al.*, 1984; Lin and Shragowitz, 1992; Sutanthavibul and Shragowitz, 1991; Youssef and Shragowitz, 1990; Youssef *et al.*, 1992). Usually, a feasible solution which satisfies the timing requirements of the circuit is sought. This is achieved by identifying the critical paths and nets of the circuit such that modules belonging to such paths are assigned locations in topological proximity.

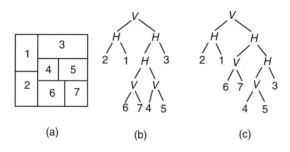

    (a)          (b)          (c)

**Figure 3.2** (a) Slicing floorplan. (b) Slicing tree. (c) Another possible slicing tree.

## 3.2.3 Terminology

Before we go any further, let us first clarify some of the terminology that will be used throughout this chapter.

*Rectangular dissection:* It is a subdivision of a given rectangle by a finite number of horizontal and vertical line segments into a finite number of non-overlapping rectangles. These rectangles are named basic rectangles. The floorplans of Figure 3.1 are rectangular dissections.

*Slicing structure:* A rectangular dissection that can be obtained by repetitively subdividing rectangles horizontally or vertically into smaller rectangles is called a slicing structure. The floorplans of Figure 3.1 are also slicing structures.

*Slicing tree:* A slicing structure can be modelled by a binary tree with $n$ leaves and $n - 1$ nodes, where each node represents a vertical cut line or horizontal cut line, and each leaf a basic rectangle. A slicing tree is also known as slicing floorplan tree. Figure 3.2 gives a slicing structure and two possible slicing trees corresponding to that structure. We shall be using the letters $H$ and $V$ to refer to horizontal and vertical cut operators respectively.

A skewed slicing tree is one in which no node and its right son are the same (Wong and Liu, 1986). For example, the slicing tree of Figure 3.2(b) is skewed, while the tree of Figure 3.2(c) is not skewed.

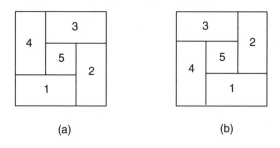

(a)        (b)

**Figure 3.3** Non-slicing floorplans. The two possible wheels.

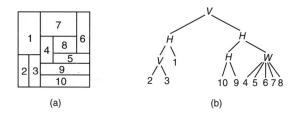

(a)        (b)

**Figure 3.4** (a) A hierarchical floorplan of order five. (b) Corresponding floorplan tree.

## Slicing and non-slicing floorplans

A floorplan that corresponds to a slicing structure is called a slicing floorplan, otherwise, it is called a non-slicing floorplan. For example, the floorplans of Figure 3.1 and Figure 3.2(a) are slicing floorplans. The floorplans of Figure 3.3 are non-slicing floorplans. These are known as *wheels*. A wheel is the smallest non-slicing floorplan. There are only two possible wheels and these are given in Figure 3.3. A generalization of slicing floorplans is hierarchical floorplans of order 5. A floorplan is said to be hierarchical of order 5 if it can be obtained by recursively subdividing each rectangle into either two parts by a horizontal or a vertical line segment, or into five parts by a wheel (Wang and Wong, 1992). Therefore, for trees of such floorplans, each internal node represents a vertical cut operator, a horizontal cut operator, or a wheel. Figure 3.4 gives an example of a hierarchical floorplan of order 5.

*Floorplan tree:* A tree representing the hierarchy of partitioning is called a floorplan tree. Each leaf represents a basic rectangle and each node a composite rectangle. If each node has two sons (each composite rectangle is cut either horizontally or vertically in two rectangles that are basic or composite) then the floorplan is a slicing floorplan. Otherwise, it is a non-slicing floorplan. For example, the floorplan of Figure 3.4 is a non-slicing floorplan.

## Graph representations of floorplans

There have been several graph models suggested for the representation of floorplans. These graphs have different names and are slightly different. Like floorplan trees, they constitute a concise abstract representation of the floorplan topology. As we will see, they serve other purposes such as global routing. In this section, we will briefly describe the four main graph models:

(1) Polar graphs
(2) Adjacency graphs
(3) Channel intersection graphs
(4) Channel position graphs.

## Polar graphs

Any floorplan (or topological placement in general), slicing or non-slicing, can be modelled by a pair of directed acyclic graphs: a horizontal polar graph, and a vertical polar graph (Ohtsuki *et al.*, 1970). For the horizontal (vertical) polar graph, each vertical (horizontal) channel is represented by a vertex. A directed edge $(u, v)$ indicates that $u$ is the left (top) side of a block, and $v$ is the right (bottom) side of the same block (refer to Figure 3.5). Each arc has a positive weight representing the width (height) of the block. Polar graphs can be used to determine the area of the smallest enveloping rectangle of the corresponding floorplan. The longest path in the horizontal polar graph is equal to the minimum required chip width. Similarly, the length of the longest path in the vertical polar graph is the minimum required chip height. Therefore, the area of the smallest enveloping rectangle is equal to the product of these two quantities.

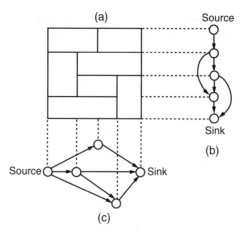

**Figure 3.5** Polar graphs. (a) A rectangular structure. (b) Vertical polar graph. (c) Horizontal polar graph.

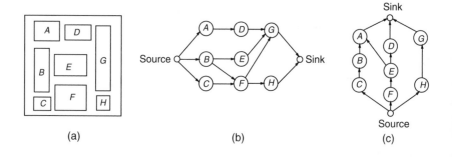

**Figure 3.6** Adjacency graphs. (a) A rectangular structure. (b) Horizontal adjacency graph. (c) Vertical adjacency graph.

## Adjacency graphs

Adjacency graphs constitute another graph representation of topological placements. Any floorplan (or topological placement in general), slicing or non-slicing, can be modelled by a pair of directed acyclic graphs: a horizontal adjacency graph, and a vertical adjacency graph. For both the horizontal and vertical adjacency graphs, each block is modelled by a vertex. However, for the horizontal adjacency graph, arcs model vertical channels, while for the vertical adjacency graph, arcs model horizontal

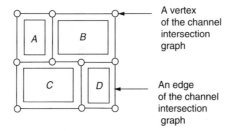

Figure 3.7 Channel intersection graph (also called floorplan graph).

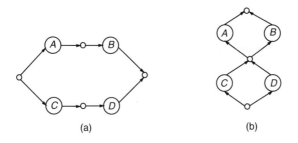

**Figure 3.8** Position graphs corresponding to the rectangular structure of Figure 3.7. (a) Horizontal channel position graph. (b) Vertical channel position graph.

channels. Figure 3.6 illustrates these graphs. As the reader can see, this representation is equivalent to polar graphs. Similar to polar graphs the adjacency graphs can be used to determine the minimum required width, height, and area of the corresponding rectangular floorplan.

## Channel intersection graphs

A rectangular floorplan consists of basic rectangles that can be represented as a rectangular dissection. This dissection can be modelled by an undirected planar graph (graph that can be drawn with no arc crossings). This graph is called the channel intersection graph or floorplan graph (Preas and Lorenzetti, 1988). In this graph each channel intersection is represented by a vertex. Arcs model intersections adjacencies (refer to Figure 3.7).

## Channel position graphs

Channel position graphs constitute another topological abstract representation of floorplans. A floorplan can be represented by a pair of directed acyclic bipartite graphs, one horizontal and one vertical. For the horizontal graph, blocks and vertical channels are represented by vertices. Arcs extend from a block vertex to a channel vertex or vice versa. An arc $(b, c)$ extending from block $b$ to channel $c$ indicates that vertical channel $c$ is bordering the right side of block $b$. The vertical channel position graph is similarly constructed. Figure 3.8 illustrates these graphs. The reader can see that channel position graphs and polar graphs are equivalent representations.

## 3.3 APPROACHES TO FLOORPLANNING

Several approaches have been reported to tackle the floorplanning problem. The reported approaches belong to three general classes: (1) constructive, (2) iterative, and (3) knowledge-based. As explained in Chapter 2, constructive algorithms attempt to build a feasible solution (feasible floorplan) by starting from a seed module; then other modules are selected one (or a group) at a time and added to the partial floorplan. This process continues until all modules have been selected. Among the approaches that fall into this class are cluster growth, partitioning and slicing, connectivity clustering, mathematical programming, and rectangular dualization.

Iterative techniques start from an initial floorplan. Then this floorplan undergoes a series of perturbations until a feasible floorplan is obtained or no more improvements can be achieved. Typical iterative techniques which have been successfully applied to floorplanning are simulated annealing, force directed interchange/relaxation, and genetic algorithm.

The knowledge-based approach has been applied to several design automation problems including cell generation and layout, circuit extraction, routing, and floorplanning. In this approach, a knowledge expert system is implemented which consists of three basic elements: (a) a knowledge base that contains data describing the floorplan problem and its current state, (b) rules stating how to manipulate the data in the knowledge base in order to progress toward a solution, and (c) an inference engine controlling the application of the rules to the knowledge base.

It is extremely difficult to describe all these techniques in a single chapter. However, we will limit ourselves to those techniques that attracted the most interest among designers and researchers.

Floor plan
growth

**Figure 3.9** Cluster growth floorplanning.

### 3.3.1 Cluster growth

In this approach, the floorplan is constructed in a greedy fashion one module at a time until each module is assigned to a location of the floorplan. A seed module is selected and placed into a corner of the floorplan (lower left corner). Then, the remaining modules are selected one at a time and added to the partial floorplan, while trying to grow evenly on upper, diagonal, and right sides simultaneously (see Figure 3.9), and

**Algorithm** Linear_Ordering;
$S$: Set of all modules;
Order: Sequence of ordered modules; (*initially empty*)
**Begin**
    Seed:= Select Seed module;
    Order:=[Seed];
    $S$:= $S-$ {Seed};

    **Repeat**
        **ForEach** module $m \in S$ **Do**
            Compute the gain for selecting module $m$;
            $gain_m$:= number of nets terminated by $m$–number of new nets started by $m$;
        **End ForEach** ;
        Select the module $m*$ with maximum gain;
        **If** there is a tie **Then**
            Select the module that terminates the largest number of nets
        **ElseIf**  there is a tie **Then**
            Select the module that has the largest number of continuing nets
        **ElseIf**  there is a tie **Then**
            Select the module with the least number of connections;
        **Else** break remaining ties as desired;
        **EndIf**
        Order:= [!Order,$m*$]; (*append $m*$ to the ordered sequence*)
        $S$:= $S - $ {$m*$}
    **Until** $S = \emptyset$
**End**.

**Figure 3.10** Linear ordering algorithm.

maintaining any stated aspect ratio constraint on the chip. To determine the order in which modules should be selected, the modules are initially organized into a linear order. Linear ordering algorithms order the given module netlist into a linear list so as to minimize the number of nets that will be cut by any vertical line drawn between any consecutive modules in the linear order (Goto *et al.*, 1977; Kang, 1983). Linear ordering is one of the most widely used techniques for constructively building an initial placement configuration. A general description of a linear ordering algorithm is given in Figure 3.10. The algorithm is based on the linear ordering heuristic reported by Kang (1983). First a seed module is selected. The seed selection could be random or based on the module connectivity with the I/O pads and/or the remaining modules. Then the algorithm enters a **Repeat** loop. At each iteration of this loop, a gain function is computed for each module in the set of the remaining unordered modules. The module with the maximum gain is selected, removed from the set of unordered modules, and added to the sequence of ordered modules. In case of a tie between several modules, the module which terminates the largest number of started nets is selected. In case of another tie, the module that is connected to the largest number of continuing nets is preferred. If we have one more tie, the most lightly connected module is selected. Remaining ties are broken as desired. The concepts of net termination, starting of new nets, and continuing nets are illustrated in Figure 3.11.

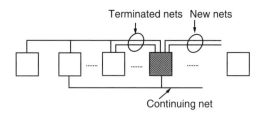

**Figure 3.11** Classification of nets during linear ordering.

**Algorithm** Cluster_Growth;
$S$ : Set of all modules;
**Begin**
   Order:= Linear_Ordering ($S$);
   **Repeat**
      nextmodule:= $b$ where Order = [$b$, !rest];
      Order:= rest;
      Select a location for $b$ that will result in minimum increase in cost function;
      (*cost may be function of the contour of the partial
      floorplan, size and shape of $b$, and wiring length*)
   **Until** Order = $\emptyset$
**End.**

**Figure 3.12** Cluster growth algorithm.

In the description of Figure 3.10, the notation $!L$ is used to mean the elements of sequence $L$. Curly braces ( { } ) are used with sets and square brackets ( [ ] ) are employed with sequences. A general description of the cluster growth algorithm is given in Figure 3.12.

**Example 3.2** Given the following netlist with six cells $[C_1, C_2, C_3, C_4, C_5, C_6]$ and six nets $N_1 = \{C_1, C_3, C_4, C_6\}$, $N_2 = \{C_1, C_3, C_5\}$, $N_3 = \{C_1, C_2, C_5\}$, $N_4 = \{C_1, C_2, C_4, C_5\}$, $N_5 = \{C_5, C_2, C_6\}$, $N_6 = \{C_3, C_6\}$. Assume that all the cells are rigid but have free orientations. In case we start from cell $C_1$ as a seed, then the linear ordering heuristic of Figure 3.10 will produce the following sequence: $[C_1, C_4, C_5, C_2, C_3, C_6]$. A step-by-step execution of the algorithm showing how cells are selected one at a time is given in Table 3.2. At the first step, cell $C_4$ has maximum gain and is selected to hold second position. At the second step, we have a tie between cells $C_2, C_3$, and $C_5$ which all have a gain equal to $-1$. All three cells do not terminate any net. However cell $C_5$ has the largest number of continuing nets and is therefore selected to occupy third position in the order. At the third step, cell $C_2$ has the maximum gain equal to $+1$. Therefore $C_2$ is given position 4. Finally, at the fourth step, we have a tie between the remaining cells $C_3$ and $C_6$. Both cells terminate one net each, have the same number of continuing nets, and are connected to the same number of nets. Therefore, we arbitrarily choose to select the cell with the smaller index, i.e., cell $C_3$. Hence the final linear order is $[C_1, C_4, C_5, C_2, C_3, C_6]$.

| Step # | Cell | New nets | Terminated nets | Gain | Continuing nets |
|---|---|---|---|---|---|
| 0 | $C_1^*$ | $N_1, N_2, N_3, N_4$ | – | $-4$ | – |
| 1 | $C_2$ | $N_5$ | – | $-1$ | $N_3, N_4$ |
|   | $C_3$ | $N_6$ | – | $-1$ | $N_1, N_2$ |
|   | $C_4^*$ | – | – | $0$ | $N_1, N_4$ |
|   | $C_5$ | $N_5$ | – | $-1$ | $N_2, N_3, N_4$ |
|   | $C_6$ | $N_5, N_6$ | – | $-2$ | $N_1$ |
| 2 | $C_2$ | $N_5$ | – | $-1$ | $N_3, N_4$ |
|   | $C_3$ | $N_6$ | – | $-1$ | $N_1, N_2$ |
|   | $C_5^*$ | $N_5$ | – | $-1$ | $N_2, N_3, N_4$ |
|   | $C_6$ | $N_5, N_6$ | – | $-2$ | $N_1$ |
| 3 | $C_2^*$ | – | $N_4$ | $+1$ | $N_3, N_5$ |
|   | $C_3$ | $N_6$ | $N_2$ | $0$ | $N_1$ |
|   | $C_6$ | $N_6$ | – | $-1$ | $N_1, N_5$ |
| 4 | $C_3^*$ | $N_6$ | $N_2$ | $0$ | $N_1$ |
|   | $C_6$ | $N_6$ | $N_5$ | $0$ | $N_1$ |

Table 3.2 Linear ordering example. Selected cells are marked with asterisk (*).

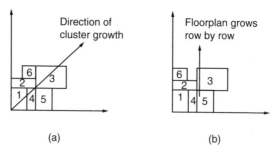

(a)                    (b)

**Figure 3.13** Cluster growth example. (a) Growth along the diagonal. (b) Row-based growth.

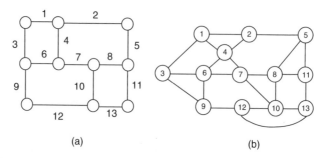

(a)                    (b)

**Figure 3.14** (a) Channel intersection graph. (b) Corresponding channel connectivity graph.

Once the modules are ordered, topological layout can proceed. Here, several approaches can be adopted. As mentioned earlier, one approach is to start from one corner of the floorplan. Then modules are selected in the order suggested by linear ordering. For each selected module a location is chosen so as to make the floorplan grow evenly on the upper and right sides, while satisfying shape constraints on the modules as well as the chip itself, and optimizing other criteria. Criteria might include: minimization of wiring length, minimization of dead space, or both. This approach is illustrated in Figure 3.13.

Another approach may consist of folding the linear order in a row structure while satisfying also shape constraints on the chip as well as on all the modules (Figure 3.13).

Notice that routability issue has been ignored so far. The quality of a floorplan solution cannot be properly assessed without performing some routability analysis (however crude it might be). Therefore, the next natural step is to predict the required routing space and estimate the pins and pads locations. It is extremely difficult to approximate the best pin and pad locations or predict routability without actually performing routing. Therefore, it is usually the case that floorplanning is followed by a global routing step. Global routing is executed in order to appraise the net routes, therefore leading to a fairly accurate measure of the required routing space.

A common approach to global routing for building block design style (which is closely related to floorplanning) is to build a global routing graph which models the regions of the floorplan, as well as relationships (the routing regions also called routing channels) between these regions. This graph is also called the channel connectivity graph. This graph is the dual of the channel intersection graph illustrated in Figure 3.7. Let $CIG$ be a channel intersection graph, and $CCG$ be the corresponding channel connectivity graph. Each vertex in $V(CCG)$ corresponds to an edge in $E(CIG)$. There is an edge $e = (u, v) \in E(CCG)$ if and only if the edges corresponding to $u$ and $v$ in $E(CIG)$ are incident to a same vertex in $V(CIG)$ (i.e., the channels touch at a common intersection point). Figure 3.14 gives a channel intersection graph and its corresponding channel connectivity graph. The vertices in the channel connectivity graph are usually assigned weights specifying the cost of assigning a net to the channels.

Global routing consists of performing a routing plan for each net, thus, determining for each net the set of channels through which the net will be routed. This amounts to performing the following tasks for each net:

1. marking of the channel vertices in which the particular net has pins; and
2. finding a minimum cost Steiner tree connecting the marked vertices.

Because the Steiner tree problem is computationally intensive, usually it is approximated by a minimum spanning tree. Global routing will be described in detail in Chapter 6.

## 3.3.2 Simulated annealing

The first application of simulated annealing to placement was reported by Jepsen and Gelatt (1983). Since then, there have been several successful applications of simulated annealing to the floorplanning problem (Otten and Ginneken, 1984; Wong and Liu, 1986; Sechen, 1988; Wong and The, 1989). Recall that in simulated annealing, first an initial solution is selected. Then a controlled walk through the search space is performed until no sizeable improvement can be made or we run out of time. In this section, we shall explain the application of this general combinatorial optimization technique to the floorplanning problem.

As pointed out by Rutenbar (1989), two approaches can be used to perform floorplanning by simulated annealing: the direct approach and the indirect approach. In the direct approach the annealing algorithm is applied directly to the physical layout, manipulating actual physical coordinates, sizes, and shapes of modules. Because it is very difficult to guarantee that every solution perturbation leads to a feasible solution with no module overlaps, such overlaps are allowed to exist in intermediate solutions. However, a penalty measure is included in the cost function to penalize any module overlap. Obviously, the final solution must be free from any overlaps. In the second approach, the annealing algorithm will be working on an abstract representation of the floorplan describing topological proximity between modules. The abstract representation usually consists of a graph representation similar to one of those graph models introduced at the beginning of this chapter, or a floorplan tree. Then a subsequent mapping process is required to generate a real floorplan from its corresponding abstract representation. The advantage of this approach is that all intermediate floorplan solutions are feasible.

The work reported by Sechen (1988) belongs to the first approach. The work reported by Wong and Liu (1986) and Wong and The (1989) adopted the second approach. We believe that the second approach is a more elegant formulation. Therefore, in this section, we will focus on this approach only. The description is based on the work reported by Wong and Liu (1986) in which a concise representation of floorplans called normalized Polish expressions, was introduced.

Recall that to apply the simulated annealing technique we need to be able to: (1) generate an initial solution, (2) disturb a feasible solution to create another feasible solution, and (3) evaluate the objective function for these solutions.

## Terminology and solution representation

In the following discussion, for the purpose of illustrating this technique, we will restrict ourselves to slicing floorplans. Slicing floorplans present some disadvantages, most important of which is the generation of extra dead space. However, on the positive side, slicing floorplans are computationally much easier to manipulate (Otten, 1982, 1984).

**Definition** An expression $E = e_1 e_2 \cdots e_{2n-1}$, where each $e_i \in \{1, 2, \cdots, n, H, V\}, 1 \leq i \leq 2n - 1$, is a Polish expression of length $2n - 1$ if and only if:

1. every operand $j$, $1 \leq j \leq n$, appears exactly once in the expression; and
2. the expression $E$ has the balloting property, i.e., for every sub-expression $E_i = e_1 \cdots e_i$, $1 \leq i \leq 2n - 1$, the number of operands is greater than the number of operators (Wong and Liu, 1986).

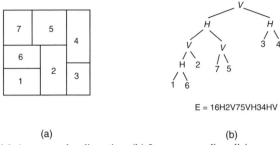

E = 16H2V75VH34HV

(a)           (b)

**Figure 3.15** (a) A rectangular dissection. (b) Its corresponding slicing tree.

As illustrated in Figure 3.2 the hierarchical structure of a slicing structure can be represented by a binary tree with $n$ leaves representing the $n$ basic rectangles, and $n - 1$ nodes representing the dissection operators ($H$ for horizontal and $V$ for vertical dissection). A postorder traversal of a slicing tree will produce a Polish expression with operators $H$ and $V$, and with operands the basic rectangles $1, 2, \cdots, n$. Figure 3.15 gives a rectangular dissection, its corresponding slicing tree, and its Polish expression representation. In a postorder traversal of a binary tree, the tree is traversed by visiting at each node the left subtree, the right subtree, and then the node itself. The general algorithm for performing a postorder traversal of a binary tree is given in Figure 3.1. In that description, the functions Root, LeftSubtree, and RightSubtree return respectively the root, the left subtree, and the right subtree of the given binary tree. Since there is only one way of performing a postorder traversal of a binary tree, then there is one to one correspondence between floorplan trees and their corresponding Polish expressions.

**Algorithm** PostorderTraversal($T$: Binary Tree)
**Begin**
  **If** Root($T$) $\neq$ nil **Then**
  **Begin**
    PostorderTraversal(LeftSubtree($T$));
    PostorderTraversal(RightSubtree(T));
    Visit(root($T$));
  **End**;
**End.**

**Figure 3.16** Postorder traversal of a binary tree.

The operators $H$ and $V$ carry the following meanings:

$ijH$ means rectangle $j$ on-top-of rectangle $i$;
$ijV$ means rectangle $i$ on-the-left-of rectangle $j$.

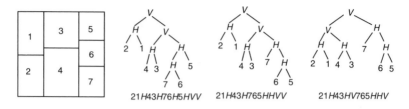

**Figure 3.17** A rectangular dissection with several slicing tree representations.

**Definition** A Polish expression $E = e_1 e_2 ... e_{2n-1}$ is called normalized if and only if $E$ has no consecutive $H$'s or $V$'s (Wong and Liu, 1986).

For example, $E_1 = 12H43VH$ is a normalized Polish expression, while $E_2 = 12V43HH$ is not normalized.

The classification of Polish expressions into normalized versus non-normalized Polish expressions is for the purpose of removing redundant solutions from the solution space. As we have stated earlier, there is a one-to-one correspondence between the set of Polish expressions of length $2n - 1$ and the set of slicing trees with $n$ leaves. However, in general, there may be several Polish expressions that correspond to the same slicing floorplan. This is an undesirable property because of the following:

1. the search space will be enlarged with several duplicate solutions, since several Polish expressions may represent the same slicing floorplan; and
2. the number of Polish expressions corresponding to a given slicing floorplan can vary from structure to structure; this will bias the search for floorplans with a larger number of corresponding slicing trees.

Figure 3.17 shows a floorplan example which has several slicing tree representations. Notice the one-to-one correspondence between the slicing trees and Polish expressions. Also, notice that all the slicing trees are not skewed.

A clever observation, embodied in the following lemmas, has been made by Wong and Liu (1986) which avoids these problems:

**Lemma 1** There is a one-to-one correspondence between the set of skewed slicing trees with $n$ leaves and the set of normalized Polish expressions of length $2n - 1$.

**Lemma 2** There is a one-to-one correspondence between the set of normalized Polish expressions of length $2n - 1$ and the set of slicing structures with $n$ basic rectangles.

Lemma 2 says, that given a normalized Polish expression, we can construct a unique rectangular slicing structure (i.e., a floorplan). Figure 3.15 gives a slicing floorplan, its corresponding skewed slicing tree, and its normalized Polish expression representation.

**Definition** A sequence $C = op_1 op_2 ... op_k$ of $k$ *operators* is called a chain of length $k$ if and only if $op_i \neq op_{i+1}$, $1 \leq i \leq k - 1$.

Let $E = e_1 e_2 ... e_{2n-1}$ be a normalized Polish expression that can be expressed as $E = P_1 C_1 P_2 C_2 ... P_n C_n$, where the $C_i$'s are chains (possibly of zero length), and $P_1 P_2 \cdots P_n$ is a permutation of the operands $1, 2, \cdots, n$.

**Definition** Two operands in $E$ are called adjacent if and only if they are consecutive elements in $P_1 P_2 ... P_n$. An operand and an operator are adjacent if and only if they are consecutive elements in $e_1 e_2 ... e_{2n-1}$ (Wong and Lui, 1986).

**Example 3.3**
$E = 123VH54HV = P_1 P_2 P_3 C_3 P_4 P_5 C_5$

$C_1 = C_2 = C_4$ are empty chains
$C_3 = VH$, $C_5 = HV$

$P_1 = 1$, $P_2 = 2$, $P_3 = 3$, $P_4 = 5$, $P_5 = 4$
1 and 2 are adjacent operands;
3 and 5 are also adjacent operands;
3 and $V$ are adjacent operand and operator.

## Solution perturbation — the move set

Floorplan solutions are represented by normalized Polish expressions. The perturbation of a given floorplan solution amounts to incurring some change to its corresponding normalized Polish expression. Three types of moves are suggested to perturb a given normalized Polish expression:

$M_1$: swap two adjacent operands;
$M_2$: complement some chain of non-zero length; (where $\overline{V} = H$ and $\overline{H} = V$);
$M_3$: swap two adjacent operand and operator.

Two normalized Polish expressions are called neighbours if one can be obtained from the other using one of the above three moves. Care must be taken to make sure that neighbours of normalized expressions are also

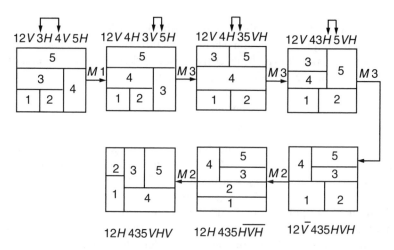

**Figure 3.18** An example of a walk through a floorplan solution space with five modules.

normalized. It is obvious that the first two moves always produce a normalized Polish expression from a normalized expression. However, the third move may at times result in a non-normalized Polish expression. Therefore, whenever an $M_3$ move is made, we must check that the resulting expression is a normalized Polish expression, i.e., (a) it does not contain two identical consecutive operators, and (b) it does not violate the balloting property. In case an $M_3$ move violates either (a) or (b), the move is rejected.

Checking that the new expression $E$ does not contain two identical consecutive operators is straightforward and achievable in $O(1)$ time. Furthermore, the following quick test is sufficient to know whether an $M_3$ move will violate the balloting property or not.

**Lemma 3** Let $N_k$ be the number of operators in the Polish expression $E = e_1 e_2 \ldots e_k, 1 \le k \le 2n - 1$. Assume that the $M_3$ move swaps the operand $e_i$ with the operator $e_{i+1}, 1 \le i \le k - 1$. Then, the swap will not violate the balloting property if and only if $2N_{i+1} < i$ (Wong and Liu, 1986).

Note that '$i$' is the number of operands in $e_1 e_2 \ldots e_{2i-1}$, and that $e_{2i-1}$ is an operator. Note also that the aforementioned three moves are complete in the sense that it is possible to generate from a given normalized Polish expression any other normalized Polish expression through a sequence of moves. Figure 3.18 is an example which illustrates the completeness of the three moves (Wong and Liu, 1986). It gives a walk through the floorplan solution space of a floorplanning problem with five modules, using the $M_1$ to $M_3$ moves.

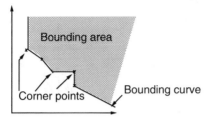

**Figure 3.19** A piecewise linear bounding curve.

## Solution evaluation

To measure how good a given floorplan solution is, a cost function is evaluated for each solution generated. Usually the principal goal is to achieve the floorplan with minimum area and overall interconnection length. A possible cost function is a linear combination of these two measures (Wong and Liu, 1986), that is,

$$Cost(F) = \alpha A + \lambda W \qquad (3.2)$$

where $A$ is the area of the smallest rectangle enveloping the $n$ basic rectangles, and $W$ is a measure of the overall wiring length. The parameters $\alpha$ and $\lambda$ control the relative importance of area versus wirelength. A possible estimate of $W$ may be defined as follows,

$$W = \sum_{ij} c_{ij} \cdot d_{ij} \qquad (3.3)$$

where $c_{ij}$ is equal to the number of connections between blocks $i$ and $j$, and $d_{ij}$ is the centre to centre distance between basic rectangles $i$ and $j$.

## Area evaluation

**Definition** Let $\Gamma$ be a continuous curve on the plane. $\Gamma$ is called a bounding curve if it satisfies the following conditions:

1. it is decreasing, i.e., for any two points $(x_1, y_1)$ and $(x_2, y_2)$ on $\Gamma$, if $x_1 \leq x_2$ then $y_2 \leq y_1$;
2. $\Gamma$ lies completely in the first quadrant, i.e., $\forall (x, y) \in \Gamma$, $x > 0$ and $y > 0$; and
3. it partitions the first quadrant into two connected regions. The connected region containing all the points $(x, x)$ for very large $x$ is

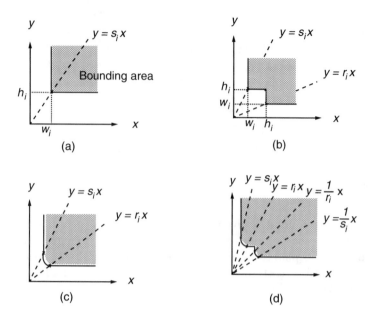

**Figure 3.20** Bounding curves for various classes of modules. Module $i$: (a) is rigid and has fixed orientation; (b) is rigid and has free orientation; (c) is flexible and has fixed orientation; (d) is flexible and has free orientation.

called the bounded area with respect to the bounding curve $\Gamma$ (Figure 3.19).

**Definition** Let $\Gamma$ and $\Lambda$ be two bounding curves. Two arithmetic operations on bounding curves are defined as follows:

1. the bounding curve corresponding to $\Gamma H \Lambda$ is obtained by summing the two curves along the $y$-axis, i.e., $\Gamma H \Lambda = \{(u, v + w) | (u, v) \in \Gamma$ and $(u, w) \in \Lambda\}$;
2. the bounding curve corresponding to $\Gamma V \Lambda$ is obtained by summing the two curves along the $x$-axis, i.e., $\Gamma V \Lambda = \{(u + v, w) | (u, w) \in \Gamma$ and $(v, w) \in \Lambda\}$.

Note that, a piecewise linear bounding curve is completely characterized by an ordered list of its corner points. Moreover, to add two piecewise linear curves along either direction, it is sufficient to sum up the two curves at their corner points.

Coming back to our floorplan problem, recall that each module $i$, $1 \leq i \leq n$ is constrained as follows:

1. height $= h_i$, width $= w_i$, and area $A_i = w_i h_i$;

2. $r_i \leq \frac{h_i}{w_i} \leq s_i$, if module $i$ has fixed orientation;

3. $r_i \leq \frac{h_i}{w_i} \leq s_i$ or $\frac{1}{s_i} \leq \frac{h_i}{w_i} \leq \frac{1}{r_i}$, if module $i$ has free orientation; and

4. $r_i = s_i$, if module $i$ is rigid, and $r_i \neq s_i$ if module $i$ is flexible.

Each basic rectangle $i$, $1 \leq i \leq n$ must be large enough to accommodate module $i$. Hence, $x_i \geq w_i$ and $y_i \geq h_i$, where $x_i$ and $y_i$ are the width and height of basic rectangle $i$.

The bounding curves corresponding to the various kinds of shape and area constraints for a given module $i$ are illustrated in Figure 3.20.

Let $T_E$ be the floorplan tree corresponding to the normalized Polish expression $E$. Let $R_E$ be the rectangular slicing structure corresponding to $T_E$, and $D_E$ be the set of all possible dimensions of $R_E$. The set of points in $D_E$ constitute a bounding curve $\Gamma_E$ corresponding to the rectangular dissection $R_E$. Next, we give a general description as to how $\Gamma_E$ is computed.

Every leaf node $i$, $1 \leq i \leq n$ of $T_E$ has associated with it a bounding curve $\Gamma_i$ consistent with the shape, size, flexibility, and orientation of the corresponding module $i$. Then the slicing tree is traversed from the leaves upwards, towards the root, computing on the way the bounding curves $\Gamma_v$ corresponding to each internal node $v$. Note that each internal node $v$ is labelled either with a horizontal cut operator $H$ or a vertical cut operator $V$. Hence, $\Gamma_v = \Gamma_l H \Gamma_r$ or $\Gamma_v = \Gamma_l V \Gamma_r$, where $l$ and $r$ are the left and right sons of $v$. This process continues until we get to the root of the tree. For efficiency reasons, all bounding curves are approximated by piecewise staircase linear curves. The accuracy of area estimation is a function of this staircase approximation.

Once all $\Gamma$'s are computed, the bounding curve $\Gamma_E$ of $R_E$ is as follows:

1. let $(a_1, b_1)$ and $(a_{k+1}, b_{k+1})$ be the points of intersection between $\Gamma_E$ and the lines $y = px$ and $y = qx$ respectively (consequence of the shape constraint on $R_E$, which states that $p \leq \frac{H}{W} \leq q$);

2. let $(a_1, b_1), (a_2, b_2), ..., (a_k, b_k)$ be all the corner points of the bounding curve $\Gamma_E$ which lie between the lines $y = px$ and $y = qx$.

Then, the dimensions of a minimum area realization of the floorplan tree $T_E$ are given by the corner point $(a_i, b_i)$ such that $a_i b_i = \min_j (a_j b_j)$. Hence, the minimum area enveloping rectangle has width $a_i$, height $b_i$, and area $A = a_i b_i$.

Finally, once the area and dimensions of the minimum enveloping rectangle have been found, we traverse the tree from the root down to the leaves, tracing back the shapes and orientations of the rectangles (composite or basic) that were selected in the upward traversal of the tree.

We should point out that, when dealing with rigid blocks, we might have width or height mismatch. In that case, the summation of the corresponding two bounding curves along the $x$ or $y$ directions should be changed to the following:

**Definition** Let $\Gamma$ and $\Lambda$ be two bounding curves.

1. The bounding curve corresponding to $\Gamma H \Lambda$ is obtained by summing the two curves along the $y$-axis, i.e., $\Gamma H \Lambda = \{(u, v + w) | (u_1, v)\} \in \Gamma$ and $(u_2, w) \in \Lambda$ and $\{u = \max(u_1, u_2)\}$.

2. The bounding curve corresponding to $\Gamma V \Lambda$ is obtained by summing the two curves along the $x$-axis, i.e., $\Gamma V \Lambda = \{(u + v, w) | (u, w_1)\} \in \Gamma$ and $(v, w_2) \in \Lambda$ and $\{w = \max(w_1, w_2)\}$.

**Definition** Let $(x_1, y_1)$ and $(x_2, y_2)$ be two possible implementations of a given rectangle. $(x_2, y_2)$ is a redundant implementation of $(x_1, y_1)$ if and only if $x_2 \geq x_1$ and $y_2 > y_1$, or $x_2 > x_1$ and $y_2 \geq y_1$.

Redundant implementations should be identified during the summation of the bounding curves and eliminated, since a minimum area enveloping rectangle cannot possibly include such redundant rectangles. Only corner points are non-redundant implementations, therefore we should only consider corner point implementations.

**Example 3.4** The purpose of this example is to illustrate how a minimum area enveloping rectangle is found for a particular slicing tree. For simplicity, we will assume that all modules are rigid and can be rotated 90° with respect to their original orientation.

For the normalized Polish expression $E = 12H34V56VHV$, the floorplan tree is given in Figure 3.21. Assume that the sizes of the modules are as indicated in Table 3.3. We would like to determine a minimum area enveloping rectangle $R_E$ corresponding to the normalized Polish expression $E$.

SOLUTION In the slicing tree of Figure 3.21, the set of points enclosed between curly braces and appearing next to each node (leaf or internal) is the bounding curve of that node. For example, the leaf node labelled '1' corresponds to module 1 whose width and height are $w_1 = 2$ and $h_1 = 3$. Since the module can be rotated, then for basic rectangle 1 to enclose module 1, its height and width must satisfy the following inequalities (refer to Figure 3.21(b)):

Table 3.3 Dimensions of modules for Example 3.4.

| Module | Width | Height |
|--------|-------|--------|
| 1 | 2 | 3 |
| 2 | 2 | 2 |
| 3 | 1 | 3 |
| 4 | 2 | 3 |
| 5 | 1 | 2 |
| 6 | 2 | 2 |

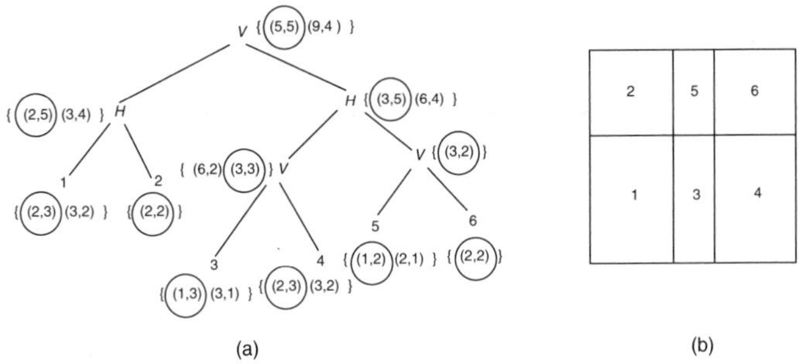

(a)

(b)

**Figure 3.21** Example of floorplan area computation.

$x_1 \geq 2$ and $y_1 \geq 3$ (normal orientation of module 1) or $x_1 \geq 3$ and $y_1 \geq 2$ (module 1 rotated 90°).

The bounding curves for the remaining leaf nodes are obtained in a similar fashion. Let $\Gamma_{56}$ be the bounding curve corresponding to the subtree '56V'. Then, $\Gamma_{56}$ is computed as follows: $\Gamma_{56} = \Gamma_5 V \Gamma_6$, where the operation $V$ is the summation of the two curves along the $x$-axis; $\Gamma_5 = \{(1,2);(2,1)\}$ and $\Gamma_6 = \{(2,2)\}$, therefore $\Gamma_{56} = \{(3,2);(4,2)\}$. Since the point (4,2) is not a corner point, it is eliminated. Hence $\Gamma_{56} = \{(3,2)\}$. The points (1,2) and (2,3) that were used in the computation of the unique corner point (3,2) of $\Gamma_{56}$ are marked (encircled in the figure). This marking is needed during the downward traversal of the tree to identify the module sizes, orientations, and shapes that are consistent with the computed minimum enveloping rectangle $R_E$.

The bounding curves of the remaining nodes are determined in a similar manner, until we reach the root. The bounding curve associated with the root gives the set of points whose coordinates are the sizes of possible enveloping rectangles of the rectangular slicing structure. For this example the bounding curve of the root is $\Gamma_E = \{(5,5);(9,4)\}$. Since $5 \times 5 = 25$ is less than $9 \times 4 = 36$, then a minimum area enveloping rectangle corresponding to the given Polish expression is a $5 \times 5$ rectangle. Now, tracing back from the root to the leaves the size and

| | Module | (w,h) | Orientation |
|---|---|---|---|
| | 1 | (2,3) | Original |
| **Table 3.4 Module sizes and orienation** | 2 | (2,2) | Original |
| **for the floorplan solution.** | 3 | (1,3) | Original |
| | 4 | (2,3) | Original |
| | 5 | (1,2) | Original |
| | 6 | (2,2) | Original |

orientation choices that led to the minimum area, we find that the modules must have the sizes and orientations indicated in Table 3.4. The rectangular slicing structure corresponding to this solution is given in Figure 3.21(b).

## The algorithm

When using the simulated annealing technique, there are several important decisions that must be made, which consist of the following:

1. a choice of the initial solution;
2. a choice of a cooling schedule, that is, (a) choice of the initial temperature, (b) how long before we reduce the temperature, and, (c) the temperature reduction rate;
3. a perturbation function;
4. a termination condition of the algorithm.

The general floorplanning algorithm using the simulated annealing technique is given in Figure 3.22. The algorithm starts from the initial Polish expression $E_0 = 12V3V4V \cdots nV$, which corresponds to the slicing of the initial chip into $n$ vertical slices.

To determine a value for the initial temperature $T_0$, a sequence of random moves are performed and the average cost change for all uphill moves $\Delta_{avg}$ is computed. Then $T_0$ is chosen such that $e^{\Delta_{avg}}/ T_0 = P$, where $P$ is the initial probability of accepting uphill moves. $P$ is initially set very close to 1.

The perturbation function used consists of the following. A neighbour of a given normalized Polish expression is selected as follows. First, the type of move is randomly selected. Then a pair of adjacent elements are chosen. In case the move is of type $M_3$, we should make sure that the perturbation does not lead to a non-normalized Polish expression. In case it does, another pair of elements is selected. This is repeated until the swapping of the two elements does not violate the normality property of the Polish

Algorithm Simulated_Annealing_Floorplanning;
$E_0 = 12V3V4V\ldots nV;$      (*Initial solution*)
$E = E_0;$      (*Assume that $E = e_1 e_2 \ldots e_i, \ldots e_n$*)
$Best = E_0$
$T_0 = \frac{\Delta_{avg}}{\ln(P)}$      (*Initial temperature*)
uphill = 0;      (*Number of uphill moves made at a given temperature*)
$MT = 0;$      (*Total number of moves at a given temperature*)
$M = 0;$      (*Overall number of moves at all temperatures*)
**Repeat**
     $MT = $ uphill $= $ Reject $= 0;$
     **Repeat**
        Select_Move($M$);
        Case $M$ of
           $M_1$ : Select two adjacent operands $e_i$ and $e_j$;
              $NewE = $ Swap$(E, e_i, e_j);$
           $M_2$ : Select a nonzero length chain $C$ of operators;
              $NewE = $ Complement$(E, C);$
           $M_3$ : Done = False
              **While** NOT(Done) **Do**
              **Begin**
                 Select two adjacent operand $e_i$ and operator $e_{i+1}$;
                 **If** $(e_{i-1} \neq e_{i+1})$ and $(2N_{i+1} < i)$ **Then** Done = TRUE;
              **End**;
              $NewE = $ Swap$(E, e_i, e_{i+1});$
        **EndCase**;
        $MT = MT + 1;$
        $\Delta Cost = Cost(NewE) - Cost(E);$
        **If** $(\Delta Cost < 0)$ OR $($RANDOM $< e^{-\Delta Cost/T})$ **Then**
           **Begin**
              **If** $(\Delta Cost > 0)$ **Then** uphill = uphill+1;
              $E = NewE;$ (*Accept $NewE$*);
              **If** $Cost(E) < Cost(Best)$ Then Best = $E$;
           **End**
           **Else** Reject = Reject + 1; (*reject the move*)
        **EndIf**
     **Until** (uphill $> N$) OR $(MT > 2N);$
     $T = \lambda T$
**Until** (Reject$/MT < .05$) OR $(T \leq \epsilon)$ OR Out_of_Time;
**End.**

**Figure 3.22** General description of the floorplanning algorithm using simulated annealing.

expression. Each generated normalized Polish expression is evaluated with respect to its cost (i.e., area of enveloping rectangle and overall wiring length). If this expression has an improved cost, then it is accepted. Otherwise, if it has a higher cost (worse solution) then it is accepted with a probability that is a decreasing function of the annealing temperature. At each temperature, a number of trials are attempted until either we make $N$ uphill moves (bad moves), or the total number of moves exceeds $2N$, where $N$ is an increasing function of $n$, the number of basic rectangles. When we exit from the inner **Repeat** loop, the temperature is reduced by a fixed ratio $\lambda$. A recommended value for $\lambda$ is $\lambda = 0.85$.

The algorithm terminates when the number of good moves becomes too small ($\leq 5\%$ of all moves made), or when the temperature becomes too low.

Recently, the simulated annealing formulation described in this section has been extended to include L-shaped blocks (Wang and Wong, 1992).

### 3.3.3 Analytical technique

This technique adopts an equation solving approach. The constraints specifying a feasible floorplan are described by a set of mathematical equations, and solved using mathematical programming techniques. This approach has two major problems:

1. The floorplanning problem is a non-linear problem. Therefore, the mathematical formulation leads to a non-linear program. To overcome this obstacle, linear approximation is adopted. However, this affects the optimality of the global solution.

2. The number of equations describing a feasible floorplan is very large leading to a very large mathematical program. To overcome this obstacle, a divide and conquer strategy is adopted.

In this section, rather than attempting to give a survey of different reported formulations, we will illustrate this analytical approach using a recently reported mixed integer linear program formulation (Sutanthavibul et al., 1990). Other mathematical formulations of the floorplanning problem have also been reported in literature (Markov *et al.*, 1984; Ying and Wong, 1989). A very good overview of the application of analytical techniques to the placement problem is given in a survey paper by Shahookar and Mazumder (1991).

For the mixed integer linear program formulation, the objective function minimized is the overall area of the rectangle enveloping all the basic rectangles (Sutanthavibul et al., 1990). Areas of the basic rectangles are inflated by an estimate of the routing space of the corresponding module. This is with the intention of including routability issues of the produced floorplan. However, for the purpose of our description, we will ignore this issue for the sake of simplicity.

| $x_{ij}$ | $y_{ij}$ | Meaning |
|------|------|---------|
| 0 | 0 | Constraint in Equation 3.4 is enforced |
| 0 | 1 | Constraint in Equation 3.5 is enforced |
| 1 | 0 | Constraint in Equation 3.6 is enforced |
| 1 | 1 | Constraint in Equation 3.7 is enforced |

**Table 3.5 Interpretation of the 0-1 integer variables.**

## Notation and problem definition

We are given a set of $n$ modules $S = \{1, 2, \cdots, n\}$. A subset $S_1$ of these modules have fixed orientation and the remaining subset $S_2$ consists of modules with free orientation. Each module $i$ has width $w_i$ and height $h_i$. Let $(x_i, y_i)$ be the $x$-$y$ coordinates of the lower left corner of module $i$, $1 \leq i \leq n$. Then, for two modules $i$ and $j$ not to overlap, $1 \leq i < j \leq n$, at least one of the following linear constraints must be satisfied (Sutanthavibul et al., 1990):

$$\text{if } i \text{ is to the left of } j: \quad x_i + w_i \leq x_j \qquad (3.4)$$

$$\text{if } i \text{ is below } j: \quad y_i + h_i \leq y_j \qquad (3.5)$$

$$\text{if } i \text{ is to the right of } j: \quad x_i - w_j \geq x_j \qquad (3.6)$$

$$\text{if } i \text{ is above } j: \quad y_i - h_j \geq y_j \qquad (3.7)$$

Notice that for two modules $i$ and $j$ not to overlap in either the $x$-direction or the $y$-direction, it is sufficient that one and only one of constraints given by Equations 3.4 to 3.7 is satisfied. Therefore, to avoid that modules $i$ and $j$ overlap we must enforce either Equation 3.4, or 3.5, or 3.6, or 3.7, not all. In order to state that in equations form, two additional 0-1 integer variables, $x_{ij}$ and $y_{ij}$, are introduced for each $(i, j)$ pair. These 0-1 variables have the interpretation indicated in Table 3.5.

Let $W$ and $H$ be upper bounds on the floorplan width and height. Hence, $|x_i - x_j| \leq W$ and $|y_i - y_j| \leq H$. If $W$ and $H$ are not given, then possible estimates of these quantities could be $W = \sum_i w_i$ and $H = \sum_i h_i$.

Therefore, to enforce that no two modules overlap, Equations 3.4 to 3.7 are rewritten as follows:

$$x_i + w_i \leq x_j + W(x_{ij} + y_{ij}) \qquad (3.8)$$

$$y_i + h_i \leq y_j + H(1 + x_{ij} - y_{ij}) \qquad (3.9)$$

$$x_i - w_j \geq x_j - W(1 - x_{ij} + y_{ij}) \qquad (3.10)$$

$$y_i - h_j \geq y_j - H(2 - x_{ij} - y_{ij}) \qquad (3.11)$$

These are four mixed linear constraints with four unknown real variables, $x_i, x_j, y_i, y_j$, and two unknown 0-1 integer variables $x_{ij}$ and $y_{ij}$. The reader can easily verify that there is always one and only one constraint enforced, consistent with the interpretation given on Table 3.3. Notice that in the above equations, whenever the multiplicative factor of either $W$ or $H$ is non-zero, the corresponding constraints are obviated.

## Linear programming formulation

In this formulation, we shall assume that one dimension of the chip, say $W$, is fixed. We will start by addressing the simplest case when all modules are rigid and have fixed orientation. Then the case when module rotation is allowed is solved. Finally, the general case when some of the modules have flexible shapes is addressed.

**Case 1: All modules are rigid and have fixed orientation.** A feasible floorplan is one which satisfies the following conditions:

(1) no two modules overlap (Equations 3.8 to 3.11 $\forall i, j: 1 \leq i < j \leq n$);
(2) each module is enclosed within the floorplan enveloping rectangle of width $W$ and height $Y$, i.e., $x_i + w_i \leq W$ and $y_i + h_i \leq Y, 1 \leq i \leq n$;
(3) all module coordinates are positive, $x_i \geq 0$ and $y_i \geq 0, 1 \leq i \leq n$.

Since the width $W$ is fixed, a possible objective to minimize would be $Y$, the height of the floorplan. To summarize, we end up with the following 0-1 integer linear program:

$$
\begin{cases}
Y \leftarrow minimize & \\
\text{Subject to :} & \\
x_i + w_i \leq W, & 1 \leq i \leq n \\
y_i + h_i \leq Y, & 1 \leq i \leq n \\
x_i + w_i \leq x_j + W(x_{ij} + y_{ij}), & 1 \leq i < j \leq n \\
x_i - w_j \geq x_j - W(1 - x_{ij} + y_{ij}), & 1 \leq i < j \leq n \\
y_i + h_i \leq y_j + H(1 + x_{ij} - y_{ij}), & 1 \leq i < j \leq n \\
y_i - h_j \leq y_j - H(2 - x_{ij} - y_{ij}), & 1 \leq i < j \leq n \\
x_i \geq 0, y_i \geq 0, & 1 \leq i \leq n
\end{cases}
$$

For an instance of the floorplan problem with $n$ modules, the above mixed integer linear program formulation requires overall $2 \times n$ continuous variables, $n(n-1)$ integer variables, and $2n^2$ linear constraints. For large $n$, this will lead to unacceptably large programs. Later on, we will see how a divide-and-conquer strategy can be used to reduce the size complexity of the problem.

**Case 2: All modules are rigid and rotation is allowed.** When a module is rotated by $90°$, its width becomes its height and vice versa. In order to include in the model this extra freedom for those blocks that are allowed to rotate, one 0–1 integer variable is introduced for each module belonging to such class. Therefore, for each module $i$ with free orientation, we associate a 0–1 variable $z_i$ with the following meaning:

$z_i = 0$ if module $i$ is placed in its original orientation ($0°$ rotation); and
$z_i = 1$ if module $i$ is rotated $90°$ with respect to its original orientation.

Hence the previous 0–1 mixed integer linear programming formulation has to be rewritten as follows:

$$
\begin{cases}
Y \leftarrow minimize \\
Subject\ to: \\
x_i + z_i h_i + (1 - z_i)w_i \leq W, & 1 \leq i \leq n \\
y_i + z_i w_i + (1 - z_i)h_i \leq Y, & 1 \leq i \leq n \\
x_i + z_i h_i + (1 - z_i)w_i \leq x_j + M(x_{ij} + y_{ij}), & 1 \leq i < j \leq n \\
x_i - z_j h_j + (1 - z_j)w_j \geq x_j - M(1 - x_{ij} + y_{ij}), & 1 \leq i < j \leq n \\
y_i + z_i w_i - (1 - z_i)h_i \leq y_j + M(1 + x_{ij} - y_{ij}), & 1 \leq i < j \leq n \\
y_i - z_j w_j - (1 - z_j)h_j \leq y_j - M(2 - x_{ij} - y_{ij}), & 1 \leq i < j \leq n \\
x_i \geq 0, y_i \geq 0, & 1 \leq i \leq n
\end{cases}
$$

In the above mixed integer linear program, $M$ could be set equal to $\max(W, H)$ or $W + H$. The number of equations did not change from the first formulation. However, the number of 0–1 integer variables have increased by $n$, which is equal to the number of modules.

**Case 3: Some of the modules are flexible.** Now, some of the modules are allowed to vary in shape as long as they keep a fixed area $A_i = w_i h_i$. This complicates the matter a bit as the equality $A_i = w_i h_i$ is a non-linear relationship. To maintain a linear program, we must linearize this relationship. Let $w_{i,\max}$ and $h_{i,\max}$ be the maximum width and height of module $i$, $1 \leq i \leq n$. Then, a possible linearization approach is to make a Taylor's series expansion of $A_i$ about the point $w_{i,\max}$, and use the first two terms of the series as an approximation of $A_i$ (Sutanthavibul et al., 1990). Recall that the Taylor's series expansion of a function $f(x)$ about the point $x_0$ is defined as follows:

$$f(x) = \sum_{k=0}^{\infty} \frac{(x - x_0)^k}{k!} \times f^{(k)}(x_0) \tag{3.12}$$

By evaluating the above Taylor's series expansion for $h_i = \frac{A_i}{w_i} = f(w_i)$ and $x_0 = w_{i,\max}$, and taking the first two terms, we get the following:

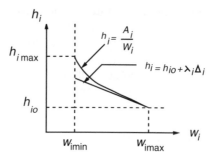

**Figure 3.23** Linear approximation of the relationship $A_i = w_i \times h_i$.

$$f(w_i) = h_i = \frac{A_i}{w_{i,\max}} + A_i \frac{(w_{i,\max} - w_i)}{w^2_{i,\max}} + O(w_i - w_{i,\max}) \qquad (3.13)$$

Let $h_{i,0} = \frac{A_i}{w_{i,\max}}$, $\Delta_i = w_{i,\max} - w_i$, and $\lambda_i = \frac{A_i}{w^2_{i,\max}}$. If we drop the error term, then the above equation can be rewritten as follows:

$$h_i = h_{i,o} + \lambda_i \Delta_i \qquad (3.14)$$

This linear approximation of the area of a module is illustrated in Figure 3.23. In Equation 3.14, $h_{i,0}$ and $\lambda_i$ are known constant parameters. Hence, this approximation will require the addition of only one continuous variable $\Delta_i$ for each module $i, 1 \leq i \leq n$.

Equations 3.4 to 3.7, which state the conditions of no overlapping between modules $i$ and $j$, must be rewritten to take into account the flexibility of some of the modules. Three cases can be distinguished:

1. **Both modules are rigid:** In this case, Equations 3.8 to 3.11 remain as they are. Here, we recall them for convenience.

$$x_i + w_i \leq x_j + W(x_{ij} + y_{ij}) \qquad (3.15)$$
$$x_i - w_j \geq x_j - W(1 - x_{ij} + y_{ij}) \qquad (3.16)$$
$$y_i + h_i \leq y_j + H(1 + x_{ij} - y_{ij}) \qquad (3.17)$$
$$y_i - h_j \geq y_j - H(2 - x_{ij} - y_{ij}) \qquad (3.18)$$

2. **Module i is flexible and module j is rigid:** In this case, the height of module $i$ should be replaced with its linear approximation in terms of $w_i$, i.e., $h_i = h_{i,o} + \lambda_i \Delta_i$. Hence, the constraints for no overlapping between flexible module $i$ and rigid module $j$ become,

$$x_i + w_{i,\max} - \Delta_i \le x_j + W(x_{ij} + y_{ij}) \tag{3.19}$$

$$y_i + h_{i,0} + \lambda_i \Delta_i \le y_j + H(1 + x_{ij} - y_{ij}) \tag{3.20}$$

$$x_i - w_j \ge x_j - W(1 - x_{ij} + y_{ij}) \tag{3.21}$$

$$y_i - h_j \ge y_j - H(2 - x_{ij} - y_{ij}) \tag{3.22}$$

Note that in  Equation 3.19, $w_i$ is replaced by $w_{i,\max} - \Delta_i$. This is for the purpose of reducing the number of variables in the overall mixed integer program.

3. **Both modules i and j are flexible**: In this case, both $h_i$ as well as $h_j$ must be replaced with their linear approximations. Here again we express $w_i = w_{i,\max} - \Delta_i$. Also, $w_j$ is expressed the same way.

$$x_i + w_{i,\max} - \Delta_i \le x_j + W(x_{ij} + y_{ij}) \tag{3.23}$$

$$y_i + h_{i,0} + \lambda_i \Delta_i \le y_j + H(1 + x_{ij} - y_{ij}) \tag{3.24}$$

$$x_i - w_{j,\max} + \Delta_j \ge x_j - W(1 - x_{ij} + y_{ij}) \tag{3.25}$$

$$y_i - h_{j,0} - \lambda_j \Delta_j \ge y_j - H(2 - x_{ij} - y_{ij}) \tag{3.26}$$

We leave it as an exercise to completely formulate the mixed integer program corresponding to the last two cases, as well as to determine the sizes of the programs (see Exercise 3.17).

## Successive augmentation

The major problem with analytical approaches to floorplan design is the size of the resulting problem. For example, for the mixed integer formulation described in this section, the smallest program (when all modules are rigid and have fixed orientation) will have $2 \times n$ continuous variables, $n(n-1)$ integer variables, and $2n^2$ linear constraints. For a value of $n = 100$ modules (medium size problem), the linear program will have 200 continuous variables, 990 integer variables, and 20,000 linear constraints. Moreover, the time complexity of a mixed integer linear programming problem grows exponentially with the size of the program. Therefore, for this approach to be realistic, the number of modules must be kept very small (around 10). Here, the classic cluster-growth greedy approach comes to the rescue. Instead of solving the original problem with the $n$ modules, a linear program is formulated using a subset $S_1$ of $n_1$ modules. Then a second subset $S_2$ of $n_2$ modules is selected and the corresponding linear program is formulated, with the additional constraints that the previously selected $n_1$ modules have fixed locations, shapes, and orientations. The floorplanning problem is solved when we solve problems corresponding to remaining subsets $S_2, ..., S_k$ such that, $\sum_{i=1}^{k} n_i = n$. This

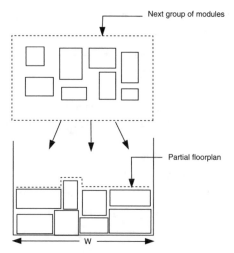

**Figure 3.24** Successive augmentation approach.

greedy approach is called successive augmentation (Sutanthavibul et al., 1990). At each step, the partial problem is solved optimally. However, the final solution will most likely not be globally optimal. This approach is graphically illustrated in Figure 3.24. As can be seen, the width of the floorplan is assumed fixed and the floorplan grows in height until all modules have been assigned. This greedy approach raises two new problems:

1. how to select the next subgroup of modules; and
2. how to formulate the successive mixed integer programs while minimizing the number of required integer variables.

For problem (1), a possible strategy is to use the linear ordering algorithm given in Figure 3.10 to order the modules into a linear list based on their connectivity. The second problem is the subject of the next subsection.

## Formulation of the successive problems

The size of each successive mixed integer program depends on the cardinality of the next group of modules as well as the partially constructed floorplan. Therefore, in order to keep the size of these mixed integer programs small, we must describe the partial floorplan using the smallest possible number of constraints and variables. One way of achieving this is as follows (Sutanthavibul et al., 1990). The main idea consists of replacing

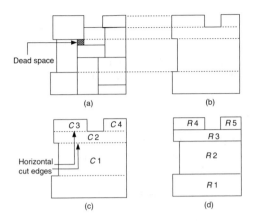

**Figure 3.25** Steps for determining a set of covering rectangles of the partial floorplan.

the already placed modules by a set of covering rectangles. The number of covering rectangles is guaranteed to be always less than the number of original modules (usually much less). The proof is left as an exercise. The idea of this covering algorithm is illustrated in Figure 3.25. As illustrated in that figure, first the covering polygon of the partial floorplan is identified. Internal holes are ignored, since new modules are always added to the top side of the partial floorplan. Then starting from the bottom of the floorplan, horizontal edges of the polygon are identified one at a time. At each such horizontal edge, a horizontal cut edge is drawn, thus delimiting a new rectangular slice. This process is repeated until we reach the top of the partial floorplan. Notice that only at the top row we could have collinear horizontal cut edges.

One final comment about this approach is routability. Although, as we have said earlier, module sizes are inflated proportionally to their interconnection requirements, a final global routing step is necessary to ensure routability of the produced floorplan. Global routing will be addressed in detail in Chapter 6.

A general description of the floorplanning algorithm is outlined in Figure 3.26 (Sutanthavibul et al., 1990). The mixed integer linear programs formulated at each step can be solved using standard mathematical programming software such as the LINDO package (Schrage, 1982).

### 3.3.4 Dual graph technique

The graph dualization technique seeks to find a topological layout (relative placement) of the modules which is consistent with the overall topological

**Algorithm** Greedy_Floorplanning;
**Begin**
  Order the $n$ modules;
  Select a first subset $S_1$ of $n_1$ modules;
  Formulate a mixed integer linear program for this first subset;
  Call the linear programming procedure to solve this first problem;
  $k = n_1$;
  $i = 1$;
  **While** $k < n$ **Do**
    **Begin**
      Select the next subset $S_i$ of $n_i$ modules;
      Find a set of $d_i$ covering rectangles of the partial floorplan;
      Formulate a mixed integer program with $n_i$ free modules and
      $d_i$ fixed basic rectangles;
      Call the linear programming procedure to solve this $i^{th}$ problem;
      $k = k + n_i$;
    **End**;
  Perform Global routing and adjust floorplan accordingly;
**End**.

**Figure 3.26** General greedy analytical floorplanning algorithm.

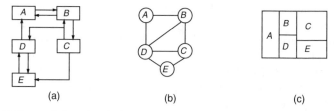

(a)                    (b)                    (c)

**Figure 3.27** Steps of floorplan design by rectangular dualization. (a) A circuit. (b) Graph model for (a). (c) Rectangular dual for (b).

relations of the blocks, as well as the sizes and shapes of these blocks. This approach requires a sound graph theoretical background.

In most general terms, this approach consists of modelling the original circuit by a graph $G = (V, E)$. The set of vertices $V$ model the modules and the set of edges $E$ model module interconnections. This graph is then planarized. Next, a rectangular dual of this planar graph is found, where faces of the dual correspond to modules, and edges correspond to interfaces between the modules (module adjacencies). The edges of the dual model the routing channels through which signal nets will be routed. Finally a drawing of the dual graph is sought such that the rectangular area assigned to each module is large enough to accommodate the module. A final adjustment step is usually necessary to provide sufficient routing space for the interconnections. Figure 3.27 illustrates the steps of this approach.

Before we proceed any further, let us introduce the necessary terminology required for the description of this approach.

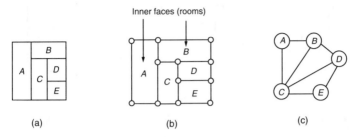

**Figure 3.28** (a) Rectangular floorplan. (b) Its channel intersection planar graph. (c) Its inner dual graph.

## Terminology

A plane graph is a graph that can be embedded in the plane with no two edges crossing each other. As we have seen earlier, a rectangular floorplan $R$ can be represented by a channel intersection graph $G = (V, E)$. The graph $G$ is a planar graph. Each vertex in $V(G)$ represents a line intersection point of $R$. There is an edge $(u, v) \in E(G)$ if and only if the intersection points modelled by $u$ and $v$ are adjacent (see Figure 3.28). $V(G)$ and $E(G)$ are the vertex set and edge set of graph $G$. The inner faces of $G$ are called rooms.

**Definition** Let $G$ and $H$ be two graphs. Let $f$ be a one-to-one mapping of $V(G)$ on to $V(H)$, and $g$ a one-to-one mapping of $E(G)$ on to $E(H)$. Let $\Theta$ be the ordered pair $(f, g)$. $\Theta$ is an isomorphism of $G$ on to $H$ if the following condition holds:

The vertex $v$ is incident with the edge $e \in G$ if and only if the vertex $f(v)$ is incident with the edge $g(e)$ in $H$. When such an isomorphism exists, we say that graphs $G$ and $H$ are isomorphic.

Clearly, for any two isomorphic graphs $G$ and $H$, we have $|V(G)| = |V(H)|$ and $|E(G)| = |E(H)|$.

**Definition** The inner dual graph of a graph $G$ is a graph $G^* = (V^*, E^*)$ modelling the adjacencies of the rooms of $G$ (see Figure 3.28). That is,

$V^* = \{v^* | v^* \text{ corresponds to an inner face of } G\}$
$E^* = \{(u^*, v^*) | u^* \text{ and } v^* \text{ have a common wall in } G\}$

**Definition** A rectangular dual of a graph $G$ is any rectangular dissection $D$ where the inner dual graph of $D$, $G^*$, and $G$ are isomorphic.

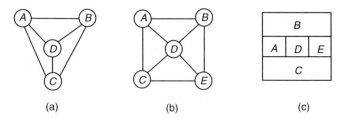

(a)                     (b)                     (c)

**Figure 3.29** (a) A triangular planar graph. (b) Properly triangulated graph. (c) Rectangular dual of (b).

The inner dual graph of a floorplan is a plane triangular graph, that is, a graph that can be embedded in a plane with no edge crossings, and each one of its faces is delimited by exactly three edges (also three vertices). Kozminski and Kinnen (1984), and Bhasker and Sahni (1985c) have given a formal definition of a properly triangulated graph.

A plane triangular graph has a rectangular dual if it is the inner dual of a rectangular dissection floorplan. Furthermore, the inner dual graph of a rectangular dissection is a properly triangulated graph, that is, a graph with no faces enclosing vertices. Hence, a plane triangulated graph has a rectangular dual if and only if it is properly triangulated.

The rectangular dualization approach to floorplan design is based on this last observation. Therefore, to apply this technique to floorplan design, the following steps must be carried out:

1. model the original circuit netlist by a graph;

2. convert the graph of Step 1 into a planar graph;

3. transform the planar graph of Step 2 into a planar triangulated graph;

4. check that the graph of Step 3 is a properly triangulated graph, i.e., it does not contain faces enclosing vertices;

5. find a rectangular dual floorplan graph (this graph is not unique).

Kozminski and Kinnen (1984) have given necessary and sufficient conditions for the existence of a properly triangulated graph. Leiwand and Lai (1984) have stated that a plane triangulated graph admits a rectangular dual if and only if it contains no complex triangular faces (a triangle enclosing vertices). Also they gave a quadratic algorithm to check a graph for that condition. Figure 3.29(a) gives a planar graph that is not properly triangulated. The reader can see that the graph is a $K_4$ (complete graph on four vertices) and cannot possibly have a rectangular dual (no way of drawing a rectangular dissection of four rectangles where every rectangle is adjacent to every other rectangle). Figure 3.29(b) gives a properly triangulated planar graph of the graph in Figure 3.29(a). An extra dummy vertex $E$ is added together with some extra edges. Now the graph of Figure 3.29(b) has a rectangular dual as illustrated in Figure 3.29(c).

In the remainder of this section, we shall assume that the graph is a properly triangulated graph. Suffice to say, that to planarize a graph, we can proceed in two ways;

1. identify the minimum number of edge crossings and add artificial vertices at these points of intersection, or
2. identify a minimum number of edges which, if removed, the resulting graph becomes planar.

Similar techniques are applied to transform a planar graph into a properly triangulated graph. Efficient algorithms for graph planarization and triangulation have recently been reported (Lokanathan and Kinnen, 1989).

Next, we focus on the problems of Steps 4 and 5, that is, (a) the problem of checking that the triangular planar graph is properly triangulated, and (b) the problem of obtaining a rectangular dual of a properly triangulated graph.

The remainder of this section is based on the work of Kozminski and Kinnen (1984) and Bhasker and Sahni (1985 a, b, c).

## Properly triangulated planar (PTP) graph

The outermost cycle of a planar graph is the cycle with the property that all edges of the graph lie either on this cycle or on its interior.

The vertices of the outermost cycle are called *outer*; all other vertices are internal vertices. Referring to Figure 3.29(b), the outermost cycle is $A \rightarrow C \rightarrow E \rightarrow B$. The outer vertices are also $A, C, E, B$. The only internal vertex is $D$. Next we formally characterize PTP graphs (Kozminski and Kinnen, 1984; Bhaskar and Sahni, 1985c).

**Definition** A graph is said to be a properly triangulated graph if it is a connected planar graph with the following properties.

P1: Every face except the exterior is a triangle.

P2: All internal vertices have degree greater than or equal to four.

P3: All cycles that are not faces have length greater than or equal to four.

The graph of Figure 3.29(a) violates properties P2 and P3 since the degree of the internal vertex $D$ is less than four. Also the cycle $A \rightarrow B \rightarrow C$ is not a

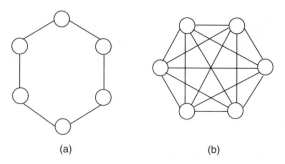

(a)                                (b)

**Figure 3.30**  Regular graphs. (a) A 2-connected graph. (b) A 6-connected graph.

face and has length less than four. On the other hand the graph of Figure 3.29(b) is a PTP graph.

**Lemma 4**   Let $G$ be a planar graph satisfying properties P1 and P3. Then $G$ must also satisfy property P2.

PROOF   If $G$ violates property P2, then it must have an internal vertex $v$ of degree 1, 2 or 3. If $v$ is of degree 1 or 2, then obviously $G$ cannot be a triangulated graph. If $v$ is of degree 3, then if $G$ is triangulated, then it must contain the subgraph $K_4$ (a complete graph on four vertices, such as the graph of Figure 3.29(a)). However, this cannot be drawn without violating P3.

**Definition**   A graph in which each vertex has the same degree (number of incident vertices) is called a regular graph (see Figure 3.30(a) and (b)).

**Definition**   A subgraph $G_1 = (V_1, E_1)$ of a graph $G(V, E)$ is a connected component of $G$ if and only if $G_1$ is a connected graph and $G$ contains no connected subgraph which properly contains $G_1$.

The connectivity of a graph is equal to the size of its disconnecting vertex set. If the size of this set is $n$ then the graph is called an $n$-connected graph. For example the graph of Figure 3.30(a) is a biconnected graph while that of Figure 3.30(b) is 6-connected. The graph of Figure 3.31 is 1-connected; each of the nodes $k$ and $l$, when removed disconnects the graph into two components. Disconnecting vertices such as $k$ and $l$ are called articulation points.

**Lemma 5**   A planar graph $G$ satisfies P1 and P3 if and only if it has a planar drawing that in addition to satisfying P1 and P3 also has all the articulation points of $G$ on the outer boundary.

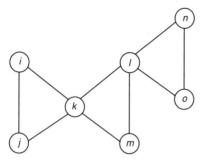

**Figure 3.31** A 1-connected graph with all articulation points ($k$ and $l$) on the outer boundary.

**Algorithm** Draw_PTP_Graph ($G$);
**Begin**
1. Identify all biconnected components of $G$;
2. Perform initial checks on each component of $G$ and discard those which fail the checks;
3. **For** each component **Do**
   3.1   Determine a drawing satisfying P1 and P3 and such that
         articulation points are on the outer boundary;
   3.2   **If** such drawing does not exit **Then** report failure;
4. Place all drawings obtained such that common vertices of the biconnected components abut.
**End.**

**Figure 3.32** Algorithm to draw a PTP graph.

PROOF   If an articulation point $a$ is not on the outer boundary, then at least one of the internal faces is not triangular.

The above lemma and definitions suggest that only biconnected components should be of concern to us. Hence an algorithm that seeks to construct a PTP drawing of a graph may be outlined as in Figure 3.32. Linear time algorithms for Steps 1 and 4 can be found in textbooks on computer algorithms (Horowitz and Sahni, 1984).

We shall now briefly describe a linear time algorithm for Step 3 due to Bhasker and Sahni (1985a). This algorithm to construct a PTP graph proceeds in a greedy fashion starting from any triangle of the given biconnected graphs (or component). Then remaining vertices are added one at a time while maintaining properties P1 and P3 satisfied. Before proceeding with the desired drawing, initial checks are performed on each component so that components for which no such drawing exists are eliminated right from the outset (Step 2 of Figure 3.32). Two checks performed are:

Check 1:   For each edge $(i, j)$, determine the number of common vertices of $(i, j)$, $cv_{ij}$. A vertex $k$ is common to $(i, j)$ if it is connected to both $i$ and $j$. In that case $i, j$ and $k$ form a triangle. If $cv_{ij} > 2$, then every

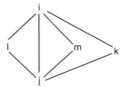

**Figure 3.33** A graph which fails Check 1; $(i, j)$ has more than two common vertices $(k, l, m)$. Therefore it has a cycle '$i - j - k$' that is not a face and is a triangle (length = 3 < 4).

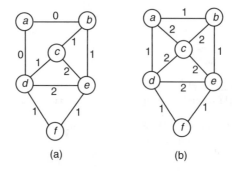

(a)                              (b)

**Figure 3.34** Example of biconnected graphs. (a) Graph which fails both checks. $f = 9 - 6 + 1 = 4$ and $t = \frac{9}{3} = 3$. (b) A graph which passes both checks. $f = 10 - 6 + 1 = 5$ and $t = \frac{15}{3} = 5$.

drawing of the component will have at least one cycle of length 3 that is not a face, which is a violation of property P3 (see Figure 3.33). If on the other hand $cv_{ij} = 0$ then every drawing will contain a face that is not a triangle.

Check 2: The number of interior faces must equal the number of triangles of the biconnected component. Let $f$ and $t$ be respectively the number of interior faces and number of triangles of a given biconnected component $H$. Then,

$$f = |E(H)| - |V(H)| + 1 \qquad (3.27)$$

$$t = \frac{1}{3} \times \sum_{\forall (i,j) \in E(H)} cv_{ij} \qquad (3.28)$$

If $f \neq t$ then $H$ cannot have a drawing that satisfies P1 and P3.

For example the graph of Figure 3.34(a) will fail Check 1 as well as Check 2 because $cv_{ad} = 0$ and $f \neq t$ ($f = 4$ and $t = 3$). However the graph of Figure 3.34(b) will pass both checks. A semi-formal description of the PTP construction algorithm is given in Figure 3.35 (Bhasker and Sahni, 1985a).

**Algorithm** Construct_PTP($H$);
**Begin**
0. Initialization.
   Mark all vertices as new and all edges as not covered;
   Choose arbitrarily an edge $(i,\ j)$ and one of its common vertices $k$;
   $(i - j - k)$ is the starting triangle;
   Mark the vertices $(i, j, k,)$ as old and the edges
   $(i, j),\ (i, k)$ and $(j, k)$ as covered;
   Set the outer-boundary *outer* to comprise vertices $i, j, k$;
   Arbitrarily choose a vertex among $i, j, k$ to be '*start-vertex*';
   Triangles_drawn $\leftarrow$ 1.
1. **Repeat** (*Successive greedy expansion.*)
   *next_vertex* $\leftarrow$ vertex that is anti-clockwise from *start_vertex* on *outer*;
   **If** (*start_vertex, next_vertex*) has a common vertex $i$ AND
   (one or both edges connecting $i$ to this edge are not covered)
   **Then** *common* $\leftarrow i$
   **Else** *common* $\leftarrow$ 0
   **EndIf**
   (*Attempt expansion. See Figure 3.36*).

**Figure 3.35** Algorithm to construct PTP graph.

**Example 3.5** Assume that we are given the following biconnected graph $G = (V, E)$ where $V(G) = \{1, 2, 3, 4, 5\}$ and $E(G) = \{(1, 2), (1, 3),\ (1, 4), (1, 5), (2, 3), (2, 4), (3, 4), (4, 5)\}$. We would like to obtain a PTP drawing of $G$ if one exists.

SOLUTION We can see right away that $G$ has no such drawing. The number of interior faces $f = |E(G)| - |V(G)| + 1 = 8 - 5 + 1 = 4$, while the number of triangles $t = \frac{15}{3} = 5$. Therefore we conclude that $G$ has no PTP drawing ($t \neq f$). However, to illustrate the working of the algorithm of Figure 3.35, we apply it on $G$. The stepwise expansion resulting from the application of this algorithm on $G$ is summarized in Figure 3.37.

Initially all vertices are new and all edges are not covered. Suppose that the triangle $2 \rightarrow 1 \rightarrow 3$ is arbitrarily chosen as the first triangle. The outer boundary is then $2 \rightarrow 1 \rightarrow 3$. Vertices $\{1,2,3\}$ are made old, edges $\{(1,2),(1,3),(2,3)\}$ are marked covered, and triangles_drawn is set to 1. Assume that we arbitrarily chose vertex 3 as the *start_vertex*. Since vertex 2 is one position anticlockwise from 3 on *outer* then *next_vertex* is equal to 2. We then attempt to expand the current drawing at the edge (*start_vertex, next_vertex*), that is, edge (3,2). Next, find a common vertex $i$ such that either edge (3,$i$) or (2,$i$) is not covered. In our case $i = 4$ is a new vertex, and both (2,4) and (3,4) are not covered. Therefore, the outer boundary *outer* becomes $2 \rightarrow 1 \rightarrow 3 \rightarrow 4$, vertices $\{1,2,3,4\}$ are old, and edges $\{(1,2),(1,3),(2,3),(2,4),(3,4)\}$ are covered,

(*__Algorithm__ Construct_PTP (*H*) continued.*)
    (*Attempt expansion*).
**Case** *common* **OF**
    *common* = 0: (*no *common* vertex*)
        **If** *next_vertex* has already been a *start_vertex*
            **Then Goto** 2
            **Else** (*advance *start_vertex* anticlockwise*)
                *start_vertex* ← *next_vertex*;
        **EndIf**;
    *common* is a new vertex:
        Draw the triangle (*common*, *next_vertex*, *start_vertex*);
        Add *common* to outer boundary between *start_vertex* and *next_vertex*;
        Mark *common* as old;
        Mark (*common*, *start_vertex*) and (*common*, *next_vertex*) as covered;
        Triangles_drawn ← Triangles_drawn +1;
    *common* is old: (*there are three sub-cases*)
        **Cases** *common* of:
            *common* is one position anticlockwise from *next_vertex* on *outer*;
                **If** (*next_vertex* is on *outer*) OR;
                (*next_vertex* has additional not covered incident edges)
                    **Then Exit** (*drawing is not possible*)
                    **Else** Draw the triangle (*common*, *next_vertex*, *start_vertex*);
                        Remove *next_vertex* from *outer*;
                        Mark (*common*, *start_vertex*) as covered;
                        Triangles_drawn ← Triangles_drawn +1
                **EndIf**;
            *common* is one position clockwise from *start_vertex* on *outer*;
                **If** *start_vertex* is on *outer* OR;
                *start_vertex* has additional not covered incident edges
                    **Then Exit** (*drawing is not possible*)
                    **Else** Draw the triangle (*common*, *start_vertex*, *next_vertex*);
                        Remove *start_vertex* from *outer*;
                        Mark (*common*, *next_vertex*) as covered;
                        Triangles_drawn ← Triangles_drawn +1
                **EndIf**;
            otherwise : (*continue, triangles involving *common* may be drawn later*)
                **If** *next_vertex* has already been a *start_vertex* once
                    **Then Goto** 2
                    **Else** *start_vertex* ← *next_vertex*
                **EndIf**;
        **EndCase**
    **EndCase**
    **Until** doomsday;
2. (*no further expansion is possible; test for success*)
    **If** (new vertices or not covered edges remain) OR (Triangles_drawn < *t*)
        **Then** no drawing is possible
        **Else** the desired drawing has been obtained
**End**. (*end of Construct_PTP*)


**Figure 3.36** Algorithm to construct PTP graph (continued).

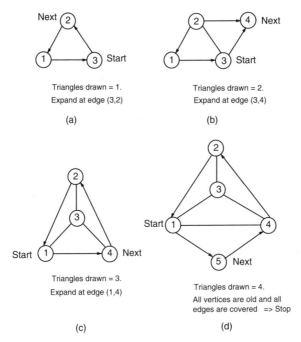

Triangles drawn = 1.
Expand at edge (3,2)

(a)

Triangles drawn = 2.
Expand at edge (3,4)

(b)

Triangles drawn = 3.
Expand at edge (1,4)

(c)

Triangles drawn = 4.
All vertices are old and all
edges are covered   => Stop

(d)

**Figure 3.37** Example of execution of algorithm of Figure 3.35 on the graph of Example 3.5.

and triangles drawn equal to 2. The *start_vertex* remains 3, but the new *next_vertex* becomes 4 (see Figure 3.37(b)). The two remaining expansion steps of the algorithm are as illustrated in Figure 3.37(c) and (d). At this point all vertices are old, and all edges are covered, hence the final check is made to see whether triangles drawn is equal to $t$. In our case triangles drawn is equal to 4 while $t$ equals 5. Therefore, we conclude that the graph has no PTP drawing.

In the following paragraphs we describe how to construct a rectangular dual of a given properly triangulated planar graph.

## Construction of a rectangular dual

Recall that a rectangular dual of an $n$-vertex graph $G(V, E)$ consists of $n$ non-overlapping rectangles such that to each vertex $i \in V$ corresponds a distinct rectangle $i$. Furthermore, rectangles $i$ and $j$ are adjacent if $(i, j) \in E(G)$. Not all graphs have a rectangular dual. But for a PTP graph the existence of a rectangular dual is guaranteed. However, such a dual is not unique.

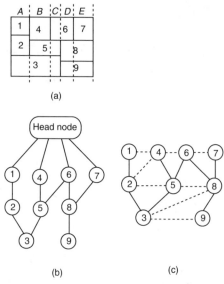

(a)

(b)                    (c)

**Figure 3.38** PDG construction from rectangular dual. (a) A rectangular dual. (b) Its PDG. (c) A corresponding PTP graph.

We will now present a linear time greedy algorithm which takes a PTP graph and constructs a rectangular dual (Bhasker and Sahni, 1985a).

## Algorithm overview

To obtain a rectangular dual from a PTP graph, the algorithm proceeds in two steps. First the PTP graph is transformed into what is referred to as a path digraph (PDG). The PDG is similar to a vertical adjacency graph, where we have a node $i$ for each rectangle $i$ and edges model the on-top relationship between corresponding rectangles. In the second step the PDG is used to construct a rectangular dual.

## Construction of a PDG

To understand the motivation of constructing the PDG and better illustrate the strategy adopted by the construction algorithm, we first examine the easier process of constructing the PDG of a given rectangular floorplan.

Any rectangular floorplan can be partitioned into columns along the vertical edges of its basic rectangles (see Figure 3.38). Now from this partitioning a directed graph is constructed where each basic rectangle $i$ is represented by a node $i$. There is a directed edge $(i, j)$ in the PDG if and only if a rectangle $i$ is on top of rectangle $j$ in the rectangular floorplan (see Figure 3.38(a) and (b)).

The headnode in Figure 3.38(b) is by definition on top of all rectangles. To each column in the partitioned floorplan corresponds a directed path from the headnode to a leafnode. For example, the path headnode$\rightarrow 4 \rightarrow 5$ $\rightarrow 3$ corresponds to column $B$. The immediate ancestors of a node are its parents. Similarly the immediate successors of a node constitute its children. For example, the parents of node 5 in Figure 3.38(b) are 4 and 6. Node 5 has 3 as its only child. The children of any node are ordered left to right, corresponding to the order in which the corresponding basic rectangles appear left-to-right in the rectangular floorplan. This leads to a similar ordering of the paths. Therefore, the first path corresponds to the leftmost column in the rectangular floorplan, the second path to the second column, and so on.

A node $i$ is a distant ancestor of a node $j$ in the PDG if there is a directed path of length two or more from $i$ to $j$. For example, node 6 is a distant ancestor of nodes 3 and 9 in Figure 3.38(b).

**Lemma 6** Let $G$ be a PDG of some rectangular floorplan and $i$, $j$ be two vertices in $V(G)$. If $i$ is a distant ancestor of $j$, then $i$ is not a parent of $j$ (Bhasker and Sahni, 1985c).

PROOF Since the floorplan is composed of rectangles only, then rectangle $i$ cannot possibly be directly on top of $j$ in one column and a distant ancestor of $j$ in another column.

Now let us examine the relationship between a given PTP graph and a corresponding PDG. A PTP graph usually has several PDGs. This is no cause of concern since anyway a PTP graph has several rectangular duals.

Suppose that the rectangular floorplan of Figure 3.38(a) is a rectangular dual of the PTP graph of Figure 3.38(c). Broken edges in this PTP graph represent edges not present in the PDG. An important relationship between the PTP graph and a PDG is captured by the following lemma.

**Lemma 7** If $(i, j)$ is an edge in a PTP graph, then neither $i$ nor $j$ is a distant ancestor of the other in the PDG (Bhasker and Sahni, 1985c).

PROOF Similar to that of previous lemma.

We now proceed to informally describe the basic steps of building a PDG from a given PTP graph (Bhasker and Sahni, 1985b). The main steps

**Algorithm** Construct_PDG;
**Begin**
    Identify the outer boundary of the PTP graph;
    Identify four vertices called NW, NE, SW, SE, (not necessarily distinct);
    Identify the top, left, right, and bottom segments of the outer boundary;
    $i \leftarrow 1$;
    **Repeat**
        Traverse the leftmost outer boundary of the PTP graph;
        **If** this boundary does not violate lemma 7
            **Then** this becomes the $i^{th}$ leftmost headnode to leaf path of the PDG
            **Else** make appropriate alternative choices;
        **EndIf**;
        Properly update the PDG and PTP graphs
    **Until** PDG contains all vertices
**End**.

**Figure 3.39** Algorithm to construct a PDG from a PTP graph.

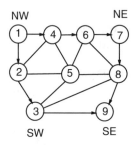

**Figure 3.40** Boundary orientation in the PTP graph.

of the algorithm are summarized in Figure 3.39. We illustrate the basic mechanics of this algorithm with the example of Figures 3.38 and 3.40. To construct a PDG, a northwest (NW), a northeast (NE), a southwest (SW) and a southeast (SE) vertex is identified. These four vertices are not necessarily distinct. In our example NW = 1, NE = 7, SW = 3 and SE = 9 (see Figure 3.40). In Figure 3.40, vertices 1, 4, 6, 7, 8, 9, 3, and 2 make-up the outer boundary. The NW to NE vertices define the top boundary (vertices 1, 4, 6, 7). And the SW to SE vertices define the bottom boundary (vertices 3,9). Similarly the NW to SW (1,2,3) and NE to SE (7,8,9) vertices define respectively the left and right boundary of the PTP graphs of Figure 3.40. The boundary orientations used by the algorithm are indicated by the arrowheads on the edges of the outer boundary of Figure 3.40.

    To obtain a PDG, we start with a headnode that has no descendants. Then the leftmost boundary $(1 \rightarrow 2 \rightarrow 3)$ of the PTP graph is traversed and becomes the leftmost 'headnode to leaf' path in the PDG. Then, the PTP graph is updated by removing those vertices and edges that have been

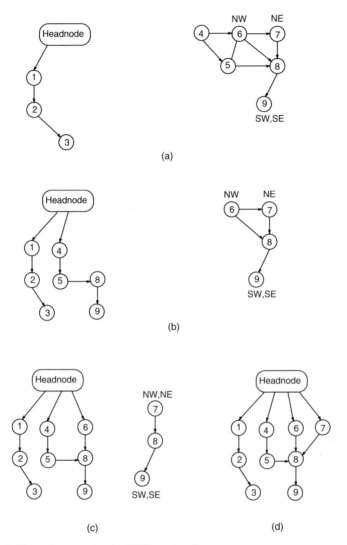

(a)

(b)

(c)                    (d)

**Figure 3.41** Execution steps for the PDG construction.

added to the PDG. The edges connecting the remaining vertices in the PTP graph to the vertices removed are also deleted (edges (2,4) and (2,5) of Figure 3.40). The new left boundary becomes $(4 \rightarrow 5 \rightarrow 8 \rightarrow 9)$ as shown in Figure 3.41(b). The stepwise execution of this greedy PDG construction algorithm on the PTP graph of Figure 3.40 is illustrated in Figures 3.41(a), (b), (c), and (d). Notice that the southwest (SW) vertex did not change when we went from Figure 3.41(a) to (b). This is because SW = SE and that all 'headnode to leaf' paths must always end at a bottom vertex.

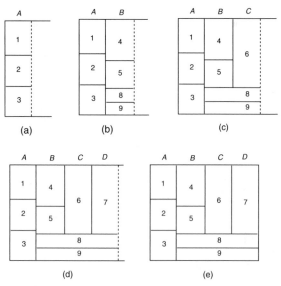

**Figure 3.42** Execution steps of the construction of a rectangular dual.

## Construction of a rectangular dual from PDG

Now that we have constructed a PDG from the PTP graph, obtaining a rectangular dual is fairly straightforward. The algorithm proceeds in a greedy fashion, processing the paths of the PDG one at a time from the leftmost to rightmost. We shall illustrate the working of the algorithm on the PDG of Figure 3.41(d).

The algorithm traverses the leftmost 'headnode to leaf' path and places rectangles of unit length in column A (Figure 3.42(a)). The next 'headnode to leaf' path is traversed ($4 \rightarrow 5 \rightarrow 8 \rightarrow 9$ of Figure 3.41(d)). Since nodes 1, 2, and 3 do not appear in the new leftmost path, rectangles 1, 2, and 3 are closed off. The PTP graph of Figure 3.40 indicates that rectangle 4 is adjacent to both 1 and 2, 5 is adjacent to 2 and 3, while 8 and 9 are adjacent to 3 only. Therefore, the resulting placement is that of Figure 3.42(b). This process continues, taking one path at a time, and obtaining the partial placements of Figures 3.42(a), (b), (c), (d), and (e). The algorithm stops when all paths are traversed. The rectangular dual obtained by the algorithm for the PTP graph of Figure 3.40 is that of Figure 3.42(e). The reader should be able to notice that the floorplan of Figure 3.42(e) is different from that of Figure 3.38(a). However, both are possible rectangular duals of the PTP graph of Figure 3.40. The reader can also easily check that all adjacency and on-top relationships are respected.

## 3.4 OTHER APPROACHES AND RECENT WORK

In this chapter, we examined traditional approaches to floorplan design. In these approaches, floorplanning is solved in two consecutive steps. In the first step, a topological placement is generated. In the second step, sizing is performed, where the actual sizes of cells, estimates of routing resources (from global routing), and overall area of the floorplan are determined. However, the following routing phase or performance adjustment phase may invalidate (at least in part) the outcome of the first step because of lack of routing space in the initial floorplan solution. This problem is even more severe for dense VLSI designs, which are demanding in terms of routability and performance. A number of papers have reported floorplan solutions which combine floorplanning with global routing. In one approach, during the floorplanning process, solutions which minimize both global routing area and total wirelength are preferred (Herrigel, 1990). The floorplan is assembled from bottom to top. During this process routing resources are estimated around each cell (or cluster of cells) and the cell/cluster shape and size are updated accordingly. In addition, pin positions are assigned so as to improve routability. In the TimberWolfMC package (Sechen, 1988), simulated annealing is used for floorplan design. Interconnect area around the cells are estimated and dynamically adjusted whenever cells are moved. A performance-driven approach to floorplan design has recently been reported (Brasen and Bushnell, 1990). The approach used is multi-start simulated annealing with force-directed cost functions. The forces exerted on a given cell depend on the positions assigned to other cells and on user supplied timing information. The timing data supplied consists of critical net weights, maximum wirelengths, and maximum path lengths. In the BEAR package (Dai and Kuh, 1987; Dai et al., 1989; Kuh and Ohtsuki, 1990), a connectivity clustering approach to floorplan design is adopted. The clustering algorithm proceeds bottom-up using a combined criterion of connectivity and geometry. Each node in the cluster tree represents a cluster of at most five cells. This limit is for the dual purpose of reducing the search space and allowing general non-slicing structures. Wiring is estimated in a top-down fashion. The cost of assigning clusters to rooms is a combination of connectivity and geometry criteria. The approach described by Luk performs floorplanning using a technique of multi-way partitioning, combined with global routing (Luk, 1989). Timing analysis is also incorporated during the floorplanning process so that nets that are timing critical are given high weights.

All approaches which integrated floorplanning with global routing reported superior results than traditional approaches.

There are several other approaches that have not been described in this chapter. Among these are force-directed, partitioning-based, and genetic algorithm approach. We shall examine these techniques in the context of the placement problem in the next chapter. For examples of applications of these techniques to floorplan design, the reader can consult Wipfler *et al.* (1982) for the force-directed approach, Luk (1989) for a partitioning based approach, and Cohoon *et al.* (1991) for the genetic technique.

Another iterative approach to floorplan design which makes use of the vertical and horizontal constraint graphs to determine the dimensions and positions of all blocks has recently been reported (Dong *et al.,* 1989). Based on experiments, the authors claim that their approach is much faster than simulated annealing. The algorithm reported by Vijayan and Tsay (1991) also relies on the usage of the vertical and horizontal constraints to construct a floorplan solution. Both these approaches require the existence of an initial floorplan from which the constraint graphs are derived.

The branch-and-bound technique (Horowitz and Sahni, 1984) has also been used to search an enumeration tree for an optimal floorplan (Wimer *et al.,* 1988). In this technique, each block is assigned to a level of the tree. For a problem with $n$ blocks, the tree will have $n$ levels. Therefore, if each block has $b$ implementations, then the tree will have $n^b$ leaf nodes. Each node in the enumeration tree corresponds to a partial floorplan and each root-to-leaf path represents a complete layout. The enumeration tree is used, together with the vertical and horizontal constraint graphs, to find a correct and optimal floorplan.

The difficulty and multi-objective nature of the floorplanning problem led some people to apply artificial intelligence techniques to the problem. We refer the reader to Chapter 9 of the book edited by Preas and Lorenzetti (1988) for a description of knowledge-based approaches to physical design automation in general.

## 3.5 CONCLUSION

Floorplanning is an important design step executed for the purpose of simplifying the physical design steps that follow (placement and routing). It is a preliminary step to placement, i.e., floorplanning is performed first, then followed by placement. Floorplanning helps designers make important decisions on some or all of the following:

*Block sizing:* decide the size and shape of each circuit component.
*Pin assignment:* decide the pin positions on the sides of each component.
*Chip sizing:* estimate the required size and best shape of the chip.
*Pad assignment:* decide the positions of the I/O pads.

*Interconnect resources:* estimate the routing resources needed to successfully connect the components as required by the circuit logic.

*Circuit performance:* floorplanning can also be used to provide early evaluation of whether the stated performance constraints can be met or not.

In this chapter, we described the four major techniques for floorplan design. These techniques are: cluster growth, simulated annealing, analytical approach, and the dual graph method.

Floorplanning is a generalization of the placement problem which will be covered in Chapter 4. Therefore, solution techniques that will be examined for the placement problem are also suitable for floorplan design.

All versions of the floorplanning problem are NP-hard. Hence, all solution techniques proposed for this problem are heuristics, whether constructive (such as cluster growth and dual graph) or iterative (such as simulated annealing). Furthermore, floorplanning is characterized by multiple objective functions such as, minimize area, minimize interconnection length, maximize routability, minimize delays, etc. Because of this multi-objective nature of the problem, there is no consensus as to what constitutes an optimal solution. In addition, the several possible formulations of the floorplanning problem make comparison between the reported solution techniques very difficult if not impossible. Only qualitative comparison is possible. One can say that most of the reported techniques lead to equally good solutions with respect to their target objective(s). However, they may vary widely in the amount of computation time expended to produce the desired solutions. Another important aspect of the solution approaches is how difficult they are to program and how easily they can include the various design constraints and objectives.

## 3.6 BIBLIOGRAPHIC NOTES

Among the most difficult and least understood techniques is the dual graph technique. Moreover, it is not clear how this technique can be made to include circuit performance or routability constraints.

Probably the most widely used and investigated technique for floorplan design is simulated annealing. The technique is very general and easy to program. Various objective functions and user constraints can easily be accommodated. However, besides its run time, the major problem with this technique is the choice of the appropriate cooling schedule, which may require several trials to tune the schedule parameters to the problem at hand. A good summary on the convergence of simulated annealing and the choice of suitable cooling schedules is given by Lengauer (1990).

The analytical approach is elegant and can accommodate various objectives and constraints. The major drawback of this approach is that it leads to very large optimization problems, thus requiring extensive computer run time. For large designs, hierarchical decomposition, such as the one reported by Sutanthavibul, Shragowitz, and Rosen (1990), is essential. However, this will affect the overall quality of the solution.

The easiest approach to floorplan design is the cluster growth approach. The price paid however, is usually a poor quality solution. This approach, though, can be used to generate an initial solution, that will be refined by an iterative improvement technique such as simulated annealing.

## EXERCISES

**Exercise 3.1** What are the main differences between floorplanning and placement?

**Exercise 3.2** List and briefly describe the various floorplanning solution approaches. Classify each approach as to whether it is constructive or iterative.

**Exercise 3.3** List and briefly explain the objective function(s) that are used to rate floorplan solutions.

**Exercise 3.4** For the rectangular floorplan of Figure 3.43 assume the cells have the sizes as given in Table 3.6.

(a) Draw the vertical and horizontal adjacency graphs corresponding to the above floorplan.

(b) Use the adjacency graphs to determine the minimum required width and height of the floorplan.

(c) Draw the skewed slicing tree corresponding to the above slicing floorplan.

(d) Determine the normalized Polish expression corresponding to the skewed slicing tree of (c).

(e) Assume that all cells are rigid and have fixed orientations. Use the slicing tree of (c) and the given cells sizes to find the area and dimensions of the smallest bounding rectangle of the given slicing floorplan.

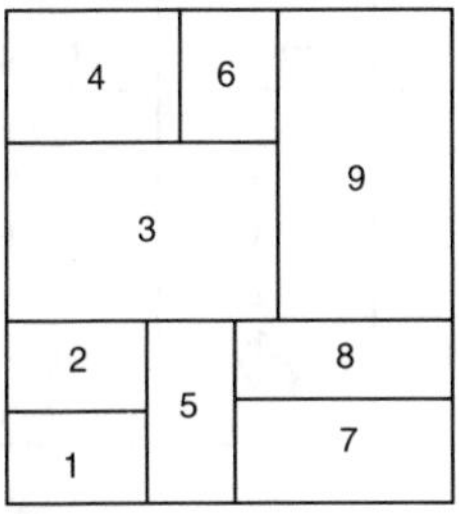

**Figure 3.43** Floorplan for Exercise 3.4.

| Modules No. | Width | Height |
|:-----------:|:-----:|:------:|
| 1 | 2 | 1 |
| 2 | 2 | 2 |
| 3 | 4 | 3 |
| 4 | 3 | 1 |
| 5 | 1 | 3 |
| 6 | 1 | 1 |
| 7 | 3 | 2 |
| 8 | 3 | 1 |
| 9 | 2 | 4 |

**Table 3.6 Table for Exercise 3.4.**

**Exercise 3.5** Determine the number of floorplan patterns (slicing or non-slicing) for $n = 3, 4$, and 5. Ignore rotation and mirroring.

**Exercise 3.6** How many cutlines are needed to generate a slicing floorplan with $n$ basic rectangles?

**Exercise 3.7** Given the following Polish expression, $E = 12H34V56VHV$

(a) Does the above expression have the balloting property? Justify your answer.
(b) Is the above expression a normalized Polish expression? Justify your answer.
(c) Why is it a desirable property to restrict ourselves to only normalized Polish expressions?
(d) Give the slicing tree corresponding to the Polish expression $E$.
(e) Assume that the modules 1, 2, 3, 4, 5, and 6 have the sizes and shapes indicated in Table 3.7. If all modules are rigid and have free orientations, what will be the size of the smallest bounding rectangle corresponding to the normalized Polish expression E? Show all the steps (with explanation) that led to your answer.

| Module No. | Width | Height |
|:---:|:---:|:---:|
| 1 | 2 | 1 |
| 2 | 2 | 2 |
| 3 | 3 | 1 |
| 4 | 2 | 3 |
| 5 | 1 | 2 |
| 6 | 2 | 2 |

**Table 3.7 Table for Exercise 3.7.**

**Exercise 3.8** Show that there is a one-to-one correspondence between the set of skewed slicing trees with $n$ leaves and the set of normalized Polish expressions of length $2n - 1$.

**Exercise 3.9** Show that there is a one-to-one correspondence between the set of normalized Polish expressions of length $2n - 1$ and the set of slicing floorplans with $n$ basic rectangles.

**Exercise 3.10** Let $N_k$ be the number of operators in the Polish expression $E = e_1 e_2 \ldots e_k, 1 \leq k \leq 2n - 1$. Assume that the $M_3$ move swaps the operand $e_i$ with the operator $e_{i+1}, 1 \leq i \leq k - 1$. Show that the swap will not violate the balloting property if and only if $2N_{i+1} < i$ and $e_{i-1} \neq e_{i+1}$.

**Exercise 3.11 Programming Exercise:** Implement a program which checks whether a given expression is a normalized Polish expression, and if so, builds and outputs a skewed slicing tree.

**Exercise 3.12** (*) **Programming Exercise:** Implement a program which checks whether a given expression is a normalized Polish expression, and if so, builds and outputs the corresponding slicing structure.

**Exercise 3.13** (*) **Programming Exercise:** Implement a program which finds a minimum area covering rectangle, together with sizes of its constituent basic rectangles. The input to the program should consist of the following:

(1) an expression with $n$ operands;
(2) the possible sizes of each of the $n$ cells.

**Exercise 3.14** Given the following netlist with 10 cells $[C_1, C_2, C_3, C_4, C_5, C_6, C_7, C_8, C_9, C_{10}]$ and 10 nets $N_1 = \{C_1, C_3, C_4, C_6\}$, $N_2 = \{C_1, C_3, C_5\}$, $N_3 = \{C_1, C_2, C_5\}$, $N_4 = \{C_1, C_2, C_6, C_7\}$, $N_5 = \{C_2, C_4, C_6\}$, $N_6 = \{C_2, C_5, C_8, C_{10}\}$, $N_7 = \{C_3, C_4, C_6, C_8\}$, $N_8 = \{C_4, C_8, C_9, C_{10}\}$, $N_9 = \{C_4, C_7, C_9, C_{10}\}$, $N_{10} = \{C_5, C_9, C_{10}\}$.

| Modules No. | Width | Height |
|:-----------:|:-----:|:------:|
| 1 | 2 | 3 |
| 2 | 2 | 2 |
| 3 | 3 | 1 |
| 4 | 2 | 3 |
| 5 | 2 | 4 |
| 6 | 3 | 6 |
| 7 | 2 | 2 |
| 8 | 2 | 5 |
| 9 | 1 | 2 |
| 10 | 3 | 5 |

**Table 3.8    Table for Exercise 3.14.**

Assume that all the cells are rigid but have free orientations.

(a) Apply the linear ordering heuristic given in Figure 3.10. Use cell $C_1$ as a seed.

(b) Assume that all modules are rigid and have fixed orientations. Use the cluster growth approach to grow a floorplan, starting from the lower left corner of the floorplan, and growing along the diagonal. The cell sizes are given in Table 3.8.

Use as a selection criterion for cell location the area of the rectangle covering the resulting partial floorplan. In case of a tie, select the location that will result in the minimum absolute difference $|W - H|$, where $W$ and $H$ are the width and height of the resulting partial floorplan. Break remaining ties as desired.

**Exercise 3.15** (*) **Programming Exercise:** Implement the cluster growth approach using a programming language of your choice. Test your implementation using the netlist of the previous exercise.

**Exercise 3.16** (*) **Programming Exercise:** Implement the simulated annealing approach described in this chapter. Test your implementation with the netlist of the previous exercise. Compare the simulated annealing solution with that of the cluster growth approach.

**Exercise 3.17** Formulate the corresponding mixed-integer linear programs, and determine the sizes of these programs for the following cases.

(a) All modules are rigid and have fixed orientations.

(b) All modules are rigid but have free orientations.

(c) All modules are flexible.

**Exercise 3.18** (\*) Modify the mixed integer formulation given in the text to include length bounds (upper bounds) on interconnections.

**Exercise 3.19** Given the graph $G = (V, E)$, where

$V(G) = \{1,2,3,4,5,6\}$, and
$E(G) = \{(1,2), (1,3), (1,4), (2,3), (2,5), (3,4), (3,5), (4,5), (4,6), (5,6)\}$.

(a) Determine the number of interior faces $f$ and number of triangles $t$ of $G$. What do you conclude?
(b) Seek a PTP graph drawing of $G$ using the algorithm of Bhasker and Sahni given in the text.
(c) Obtain a rectangular dual of the drawing obtained in (b).

**Exercise 3.20** (\*) **Programming Exercise:**

(a) Give a Pascal-like description of an heuristic algorithm, which takes as input a graph and seeks a drawing of the graph, while attempting to minimize the number of edge crossings. The algorithm should be of low polynomial time-complexity.
(b) Implement the heuristic described in (a).

**Exercise 3.21** (\*) **Programming Exercise:**

(a) Give a Pascal-like description of a planarization heuristic algorithm. The algorithm should be of low polynomial time-complexity.
(b) Implement the heuristic described in (a).

**Exercise 3.22** (\*) **Programming Exercise:**

(a) Give a Pascal-like description of an heuristic algorithm which checks whether a given planar graph is properly triangulated, and if not constructs a PTP drawing of the graph. The algorithm should be of low polynomial time-complexity.
(b) Implement the heuristic described in (a).

# REFERENCES

Baker, B. S. and J. S. Schwarz. Shelf algorithms for two-dimensional packing problems. *SIAM J. Compt*, 3, 1983
Baker, B. S. *et al*. Performance bounds for level-oriented two-dimensional packing algorithms. *SIAM J. Compt*, 9, 1980 (a).

Baker, B. S., D. J. Brown, and H.P. Katseff. A 5/4 algorithm for two-dimensional packing. *J. Algorithms*, 2, 1981.

Baker, B. S., E. G. Coffman Jr. and R. L. Rivest. Orthogonal packing in two dimensions *SIAM J. Compt*, 9:846–855, 1980 (b).

Bhasker, J. and S. Sahni. A linear algorithm to check for the existence of a rectangular dual of a planar triangulated graph. *Technical Report TR 85-21, Computer Science Department, University of Minnesota*, 1985 (a).

Bhasker, J. and S. Sahni. A linear algorithm to find a rectangular dual of a planar graph. *Technical Report TR 85-26, Computer Science Department, University of Minnesota*, 1985 (b).

Bhasker, J. and S. Sahni. A linear algorithm to find a rectangular dual of a planar triangulated graph. In *Proc. of the 23rd Design Automation Conference*, pages 108–114, 1985 (c).

Brasen, D. R. and M. L. Bushnell. MHERTZ: a new optimization algorithm for floorplanning and global routing. *Proc. of the 27th Design Automation Conference*, pages 107–110, 1990.

Burstein, M. and M. N. Youssef. Timing influenced layout design. *22nd Design Auotmation Conference*, pages 124–130, 1985.

Cohoon, J. P., S. U. Hegde, W. N. Martin and D. Richards. Distributed genetic algorithm for the floorplan design problem. *IEEE Transactions on CAD*, 10(4):483–492, April 1991.

Dai, W. and E. S. Kuh. Simultaneous floor planning and global routing for hierarchical building-block layout. *IEEE Transaction on CAD*, 6(5):828–837, September 1987.

Dai, W., B. Eschermann, E. S. Kuh and M. Pedram. Hierarchical placement and floor planning in BEAR. *IEEE Transaction on CAD*, 8(12):1335–1349, December 1989.

Dong, S., J. Cong and C. L. Liu. Constrained floorplan design for flexible blocks. *Proc. of ICCAD'89* pages 488–491, 1989.

Dunlop, A. E. *et al.* Chip layout optimization using critical path weighting. *21st Design Automation Conference*, pages 142–146, 1984.

Garey, M. R. and D. S. Johnson. *Computers and Intractability: a Guide to the Theory of NP-completeness*, W. H. Freeman, San Fransisco, 1979.

Goto, S., I. Cederbaum and B. S. Ting. Suboptimal solution of the backboard ordering with channel capacity constraint. *IEEE Transactions on Circuits and Systems*, pages 645–652, 1977.

Herrigel, A. Global routing driven floorplanning. *Proc. of the EUROASIC'90*, pages 214–219, 1990.

Horowitz, E. and S. Sahni. *Fundamentals of Computer Algorithms*, Computer Science Press, Rockville, 1984.

Jepsen, D. and C. Gelatt. Macro placement by Monte Carlo annealing. *Proc. of ICCAD*, 3:495–498, 1983.

Kang, S. Linear ordering and application to placement. *20th Design Automation Conference*, pages 457–464, 1983.

Kozminski, K. and E. Kinnen. An algorithm for finding a rectangular dual of a planar for use in area planning for VLSI integrated circuits. *Proc. of the 21st Design Auotmation Conference*, 655–656, 1984.

Kuh, E. S. and T. Ohtsuki. Recent advances in VLSI layout. *Proc. of the IEEE*, 78(2), February 1990.

Leiwand, S. M. and Y. Lai. An algorithm for building rectangular floorplans. In *Proc. of the 21st Design Automation Conference*, pages 663–664, 1984.

Lengauer, T. *Combinatorial Algorithms for Integrated Circuit Layout*, B.G. Teubner & John Wiley and Sons, 1990.

Lin, R. B. and E. Shragowitz. Fuzzy logic approach to placement problem. *29th Design Automation Conference*, pages 153–158, 1992.

Lokanathan, B. and E. Kinnen. Performance optimized floorplanning by graph planarization. *26th Design Automation Conference*, pages 116–121, 1989.

Luk, W. K. Multi-terrain partitioning and floorplanning for data-path chip (microprocessor) layout. *Proc. of ICCAD*, 3:492–495, 1989.

Markov, L. *et al.* Optimization techniques for two-dimensional placement. *Proc. of DAC*, pages 652–654, 1984.

Ohtsuki, T., N. Suzigama and H. Hawanishi. An optimization technique for integrated circuit layout design. *Proc. of ICCST, Kyoto*, pages 67–68, 1970.

Otten, R. and L. V. Ginneken. Floorplan design using simulated annealing. *Proc. of ICCAD*, 3, 1984.

Otten, R. H. J. M. Layout structures. *Proc. of IEEE Large Scale Systems Symposium*, pages 96–99, 1982.

Otten, R. H. J. M. Efficient floorplan optimization. *Proc. of ICCD*, 3, 1984.

Preas, B. and M. Lorenzetti, *Physical Design Automation of VLSI Systems*. Edited, The Benjamin-Cummings Publishing Company, Inc., 1988.

Rutenbar, R. A. Simulated annealing algorithms: An overview. *IEEE Circuits and Devices Magazine*, 3, 1989.

Schrage, L. *LINDO: Linear interactive Discrete optimizer. Copyright.* 1982.

Sechen, C. Chip planning, placement, and gobal routing of macro/custom cell intergrated circuits using simulated annealing. *Proc. of the 25th DAC*, pages 73–80, 1988.

Shahooker, K. and P. Mazumder. VLSI cell placement techniques. *ACM Computing Surveys*, 23(2):143–220, June 1991.

Sutanthavibul, S. and E. Shragowitz. JUNE: an adaptive timing-driven layout system. *27th Design Automation Conference*, pages 90–95, 1990.

Sutanthavibul, S. and E. Shragowitz. Dynamic prediction of critical path and nets for constructive timing-driven placement. *28th Design Automation Conference*, pages 632–635, 1991.

Sutanthavibul, S., E. Shragowitz and J. B. Rosen. An analytical approach to floorplan design and optimization. *Proc. of the 27th DAC*, 3, 1990.

Vijayan, G. and R. Tsay. A new method for floor planning using topological constraint reduction. *IEEE Transactions on CAD*, 10(12):1494–1501, December 1991.

Wang, T. and D. F. Wong. Optimal floorplan area optimization. *IEEE Transactions on CAD*, 11(8):992–1002, August 1992.

Wimer, S., I. Koren and I. Cederbaum. Optimal aspect ratios of building blocks in VLSI. In *Proc. of the 25th Design Automation Conference*, pages 66–72, 1988.

Wipfler, G. J., M. Wiesel and D. A. Mlynski. A combined force and cut algorithm for hierarchical VLSI layout. *Proc. of the 19th Design Automation Conference*, pages 671–677, 1982.

Wong, D. F. and C. L. Liu. A new algorithm for floorplan design. *Proc. of the 23rd DAC*, pages 101–107, 1986.

Wong, D. F. and K. The. An algorithm for hierarchical floorplan design. *Proc. of the ICCAD*, pages 484–487, 1989.

Ying. T. S. and J. S. Wong. An analytical approach to floorplanning for hierarchical building blocks layout, *IEEE Trans on CAD*, 8, 1989.

Youssef, H. and E. Shragowitz. Timing constraints for correct performance. *Proceedings of ICCAD 90*, pages 24–27, 1990.

Youssef, H., S. Sutanthavibul and E. Shragowitz. Pre-layout timing analysis of cell based VLSI design. *Computer Aided Design Journal*, 7:367–379, 1992.

# FOUR

# PLACEMENT

## 4.1 INTRODUCTION

*Placement* is the process of arranging the circuit components on a layout surface. As an example, consider the circuit of Figure 4.1(a); suppose that we need to place the gates on a two-dimensional surface. Figure 4.1(b) shows one such placement. (The same placement is shown in a symbolic format in Figure 4.1(c).) The symbolic placement shows gates as black boxes and nets as lines. Note that the actual details of routing are omitted from the symbolic placement. However, from a symbolic placement, it is possible to get an *estimate* of the routing requirements. Considering Figure 4.1(c) again, suppose that routing a net from one box to another takes up as much wire as the Manhattan distance between the boxes. For instance, net (1,5) must take up some horizontal and some vertical wiring, and the total sum of this must be at least 2 units of wire. The reader may verify that the placement of Figure 4.1(b) takes 10 units of wire. In Figure 4.1(d), we show another symbolic placement for which the total wirelength $\omega$ is 12. Finally, Figure 4.1(e) shows a one-dimensional placement of the circuit. The 1-D placement also requires 10 units of wiring. The total wirelength $\omega$ is a widely used measure of the quality of the placement. To illustrate the point, consider the symbolic placement of Figure 4.2(a). The same circuit may also be placed as shown in Figure 4.2(b). As you can see, the latter placement takes up more area. The area of a layout consists of two parts—the functional area, and the wiring area. The sum of the areas of the functional cells is known as the functional area. There are nine functional cells in the circuit of Figure 4.2. The functional area remains unchanged for

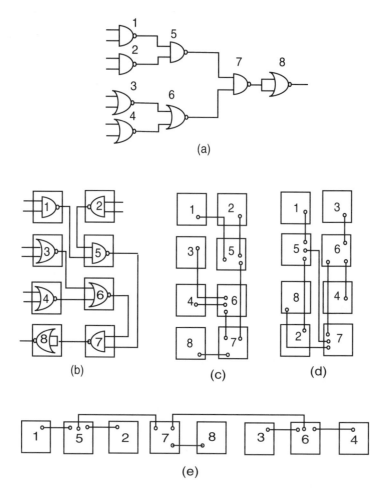

**Figure 4.1** (a) A tree circuit. (b) A 2-D placement of gates. (c) A 2-D symbolic placement. (d) A 2-D placement requiring 12 units of wiring. (e) A 1-D placement requiring 10 units of wiring.

all placements. It is the wiring area which changes with the placement. This is because of *minimum separation* that must be maintained between two wires and between a wire and a functional cell. For instance, consider the nets (2,5) and (8,9) in Figure 4.2(b). These nets must be placed on two separate vertical *tracks*. These tracks must be separated by a minimum distance to prevent any cross talk.

A placement which requires a large amount of wiring space must necessarily involve long wires and hence a large value of total wirelength. Thus total wirelength $\omega$ is a good measure of the area of the layout. The

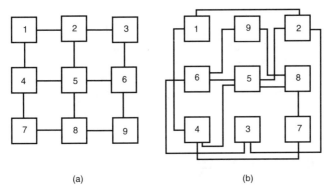

(a)                                    (b)

**Figure 4.2** (a) Optimal placement with $w = 12$. (b) Alternative solution with $w = 22$.

advantage of using $w$ as a measure is that it is easy to compute (see Exercise 4.1).

### 4.1.1 Complexity of placement

The placement of cells in order to minimize the total wirelength $w$ is an NP-complete† problem. Even the simplest case of the problem, namely one-dimensional placement, is hard to solve; there are as many as $\frac{n!}{2}$ linear arrangements of $n$ cells (see Exercise 4.2). In practice, the number of cells to be placed can be very large, in thousands. Therefore, it is impractical to take the brute force approach of enumerating all the placements and selecting the best one. Over the past three decades, a number of good heuristic techniques have been developed for solving the placement problem. Such heuristic algorithms give a *good* solution to the placement problem, not necessarily the best solution; however, the time requirements of the heuristic algorithms are modest, a polynomial function of $n$.

### 4.1.2 Organization of the chapter

In this chapter, we discuss several heuristic algorithms for the placement problem. Before we introduce the heuristics, it is necessary to define the placement problem more formally. This will be done in Section 4.2, where we will introduce methods to estimate various cost functions such as the total wirelength. We will also describe some of the constraints that may have to be handled by a placement algorithm. In Section 4.4, we look at some of the popularly used heuristic algorithms for circuit placement. We

†Decision version of the problem is NP-complete.

shall begin by describing the min-cut algorithm which has been used for gate-array placement as well as standard-cell placement. This will be followed by a discussion of a placement algorithm based on simulated annealing. Subsequently, we look at the force-directed placement technique, which attempts to solve the problem using numerical techniques. In the concluding section of the chapter, we present some current trends and recent algorithms for placement.

## 4.2 PROBLEM DEFINITION

Given a collection of cells or modules with ports (inputs, outputs, power and ground pins) on the boundaries, the dimensions of these cells (height, width, etc.), and a collection of nets (which are sets of ports that are to be wired together), the process of *placement* consists of finding suitable physical locations for each cell on the entire layout. By suitable we mean those locations that minimize given objective functions, subject to certain constraints imposed by the designer, the implementation process, or layout strategy and the design style. Examples of constraints include avoidance of overlap of layout cells and the requirement that the cells must fit in a certain rectangular surface. The cells may be standard-cells, macro blocks, etc.

Semi-formally the placement problem can be defined as follows. Given a set of modules $M = \{m_1, m_2, \cdots, m_n\}$ and a set of signals $S = \{s_1, s_2, \cdots, s_k\}$, we associate with each module $m_i \in M$ a set of signals $S_{m_i}$, where $S_{m_i} \subseteq S$. Similarly, with each signal $s_i \in S$ we associate a set of modules $M_{s_i}$, where $M_{s_i} = \{m_j \mid s_i \in S_{m_j}\}$. $M_{s_i}$ is said to be a *signal net*. We are also given a set of slots or locations $L = \{L_1, L_2, \cdots, L_p\}$, where $p \geq n$. The placement problem is to assign each $m_i \in M$ to a unique location $L_j$ such that some objective is optimized. Normally each module is considered to be a point and if $m_i$ is assigned to location $L_j$ then its position is defined by the coordinate values $(x_j, y_j)$. Sometimes a subset of the modules in $M$ are *fixed*, i.e., pre-assigned to locations, and only the remaining modules can be assigned to the remaining unassigned locations (Breuer, 1977a).

## 4.3 COST FUNCTIONS AND CONSTRAINTS

Layout design consists of *placement* followed by *routing*. Routing† is the process of assigning actual tracks to wires that connect ports. A placement

†Discussed in detail in Chapters 5-7

is acceptable if 100 per cent routing can be achieved within a given area. The objective function to be minimized can be written as a sum of $\gamma_1$ and $\gamma_2$. In most cases, $\gamma_1$ is the total estimated wirelength. $\gamma_2$ is generally penalties on non-feasible solutions and represents the cost for constraint violations such as overlap of cells to be placed. Performing actual routing to compare various placement solutions is impractical. Therefore, *estimates* are used. In the following paragraphs we present some commonly used techniques for estimation of wire-length required by a given placement.

### 4.3.1 Estimation of wirelength

The speed of estimation has a drastic effect on the performance of the placement algorithm. Thus a good estimation technique is central to any placement program. The estimation procedure must be as quick as possible. In addition the estimation error must be the same for all nets, that is, it must not be skewed.

One realistic assumption made in estimating the total wirelength is that routing uses Manhattan geometry, i.e., routing tracks are either horizontal or vertical (after Manhattan, NY, where the streets run either North–South or East–West).

For a two pin net connecting module $i$ to module $j$, the Manhattan length of this net is $r_{ij} + c_{ij}$, where $r_{ij}$ and $c_{ij}$ are the number of rows and columns separating the locations of the two modules. However, not all nets are two-pin nets. In this section, we explain how the assumption regarding two-point connectivity can be relaxed. What we need is a method to estimate the length of a multipoint net. There are various techniques available, and each one of these has its own advantages and disadvantages.

*Semi-perimeter method:* This is an efficient and the most widely used approximation to estimate the wirelength of a net. The method consists of finding the smallest bounding rectangle that encloses all the pins of the net to be connected. The estimated wirelength of the interconnects is half the perimeter of this bounding rectangle. Assuming no winding of paths in actual routing, for two and three pin nets this is an exact approximation.† This method provides a good estimate for the most efficient wiring scheme, which is the Steiner tree. For heavily congested chips this method always underestimates the wiring length.

*Complete graph:* For an $n$ pin net, the complete graph consists of $\frac{n(n-1)}{2}$ edges. Since a tree has $(n-1)$ edges which is $\frac{2}{n}$ times the number of

---

†Most practical circuits have either two or three terminal nets.

edges in the complete graph, the estimated tree length using this method is,

$$L = \frac{2}{n} \times \sum_{\forall pair \in net} (pair\ separation) \qquad (4.1)$$

*Minimum chain:* Here the nodes are assumed to be on a chain and each pin has at most two neighbours[†]. The method is to start from one vertex and connect to the closest one, and then to the next closest and so on, until all the vertices are included. This estimation technique is simpler than the minimum spanning tree but results in slightly longer interconnects.

*Source to sink connection:* Here the output of a cell is assumed to be connected to all other points of the net (inputs of other cells) by separate wires (star configuration). This method is the simplest to implement but results in excessive interconnect length. For heavily congested chips this might be a good approximation, but not for lightly congested ones. Hence this type of connection for estimation is seldom used.

*Steiner tree approximation:* A *Steiner tree* is the shortest route for connecting a set of pins. In this method, a wire can branch from any point along its length to connect to other pins of the net. The problem of finding the minimum Steiner tree is proven to be NP-complete. Lee algorithm (see Chapter 5) may be used to find an approximate Steiner tree by propagating a wave for the entire net (Lee, 1961).

*Minimum spanning tree:* Unlike the Steiner tree, in a minimum spanning tree branching is allowed only at the pin locations. For an $n$-pin net, the tree can be constructed by determining the distances between all possible pairs of pins, and connecting the smallest $(n - 1)$ edges that do not form cycles. A polynomial time complexity algorithm to find the minimum spanning tree is given by Kruskal (1956).

Figure 4.3 shows examples of the above wiring schemes and the respective wirelengths.

[†]The maximum degree of any vertex is 2.

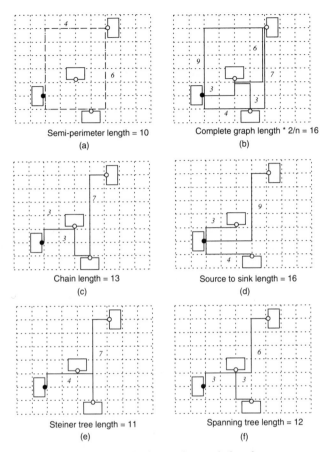

**Figure 4.3** Application of different methods to estimate wirelength.

## 4.3.2 Minimize total wirelength

The main objective of placement is to provide a solution that is completely routable. Also the area taken by the routing wires must be minimum. One way to accomplish this is to place strongly connected cells close to each other. A commonly used objective function that is minimized is $L(P)$, the total weighted wirelength over all signal nets, and is expressed as:

$$L(P) = \sum_{n \in N} w_n \cdot d_n \qquad (4.2)$$

where,

$d_n$ = estimated length of net $n$;
$w_n$ = weight of net $n$;
$N$ = set of nets.

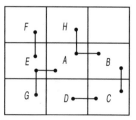

**Figure 4.4** Placement $P$ for Example 4.1.

In this estimate, the length of each net is computed independently of other nets. Therefore, the estimated area is only a rough approximation of the actual one.

**Example 4.1**   Consider a chip of size three rows by three columns. Given below are five signal nets and their corresponding weights ($A_i$ corresponds to pin $i$ of cell $A$). For the placement $P$ shown in Figure 4.4 where each cell occupies one grid unit, compute $L(P)$.

| Nets | Weights |
|---|---|
| $N_1 = (A_1, B_1, H)$ | $w_1 = 2$ |
| $N_2 = (B_2, C_1)$ | $w_2 = 4$ |
| $N_3 = (C_2, D)$ | $w_3 = 3$ |
| $N_4 = (E_1, F)$ | $w_4 = 1$ |
| $N_5 = (A_2, E_2, G)$ | $w_5 = 3$ |

SOLUTION   Ignoring the dimensions of cells (considering them as points), and defining the distance between adjacent slots as one unit length, $L(P)$ is computed using Equation 4.2 as follows:

$$L(P) = \sum_{n \in N} w_n \cdot d_n = 2 \cdot 2 + 4 \cdot 1 + 3 \cdot 1 + 1 \cdot 1 + 3 \cdot 2 = 18.$$

### 4.3.3 Minimize maximum cut

Consider the rectangular layout space shown in Figure 4.5. Assume that a circuit has been placed in this layout space. Consider the vertical line at $x = x_i$ which 'cuts' the layout area into a left region $L_i$ and a right region $R_i$. With respect to the cutline, we can classify nets as follows:

(a) Nets which lie entirely to the left of the cutline. All pins of such nets will reside in $L_i$.

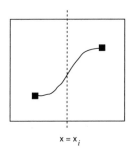

x = x_i

**Figure 4.5** A vertical cutline, and a horizontal net cut by the line.

(b) Nets which lie entirely to the right of the cutline. All pins of such nets will reside in $R_i$.

(c) Nets which are cut by the line. Each net in this class will necessarily have at least one pin in $L_i$ and at least one pin in $R_i$.

Let $\Phi_P(x_i)$ denote the number of nets of type (c) for placement $P$ cut by line $x_i$. It is clear that $\Phi_P(x_i)$ is a function of placement $P$. For a given placement $P$, let $X(P)$ indicate the maximum value of $\Phi_P(x_i)$ over all $i$, that is,

$$X(P) = \max_i [\Phi_P(x_i)] \qquad (4.3)$$

We can similarly define horizontal cutlines $y_j$ and the maximum vertical cut $Y(P)$ as

$$Y(P) = \max_j [\Phi_P(y_j)] \qquad (4.4)$$

Let us now reflect on the significance of $X(P)$ and $Y(P)$. Assume that the layout style under consideration is gate-array. Suppose that, for a given placement $P$, $X(P) = 10$ and $Y(P) = 15$. This means that at some vertical cutline $v$, 10 nets must cross the cutline $v$. Similarly, at some horizontal cutline $h$, 15 nets will cross the cutline $h$. The number of horizontal tracks along the cutline $v$ must therefore be at least 10, or else it will be impossible to route the circuit using the placement $P$. Similarly, there must be at least 15 vertical tracks along the cutline $h$ in order to be able to complete the routing. (Of course, making room for 10 horizontal and 15 vertical tracks does *not guarantee* that the routing will be completed. It is necessary, but not sufficient, to provide 10 horizontal and 15 vertical tracks.) From the previous discussion, it should be clear that $X(P)$ and $Y(P)$ are closely related to the routability of a gate-array. If we are given a gate array with $H_{\max}$ horizontal tracks and $V_{\max}$ vertical tracks per grid line, then it is necessary to find a placement which has $X(P) \leq H_{\max}$ and $Y(P) \leq V_{\max}$.

The cuts $\Phi_P(x_i)$ and $\Phi_P(y_j)$ are also closely related to the total wire length $L(P)$. In fact, assuming that the grid spacing is 1 unit, it may be shown that

$$L(P) = \sum_i \Phi_P(x_i) + \sum_j \Phi_P(y_j) \qquad (4.5)$$

where the summation is taken over all possible cutlines (see Exercise 4.10).

From the above discussion, it is clear that reducing the horizontal cut $X(P)$ and the vertical cut $Y(P)$ by selecting a good placement $P$ can increase the probability of routing a gate-array. In addition, minimizing $X(P)$ and $Y(P)$ can also have a beneficial influence on the total wirelength $L(P)$.

In macro-cell layout and standard-cell layout, minimization of $X(P)$ and $Y(P)$ is again important in order to reduce the total wirelength. In particular, a standard-cell placement which minimizes $Y(P)$ can be expected to have a smaller number of feed-throughs.

**Example 4.2** The placement $P$ shown in Figure 4.6 corresponds to the circuit whose signal nets are given below. Weights $w_i$ refer to the number of wires required for each net. Compute $X(P)$, $Y(P)$ and $L(P)$.

| Nets | Weights |
|------|---------|
| $N_1 = (A, B, C)$ | $w_1 = 1$ |
| $N_2 = (C, D, E)$ | $w_2 = 3$ |
| $N_3 = (D, F, G)$ | $w_3 = 4$ |

SOLUTION   Referring to Figure 4.6, we compute the number of nets crossing each horizontal and vertical line. The values are:

$$\Phi_P(x_1) = 4 + 1 = 5;$$
$$\Phi_P(x_2) = 4 + 3 + 1 = 8;$$
$$\Phi_P(y_1) = 4;$$
$$\Phi_P(y_2) = 3.$$

By definition,
$$X(P) = Max[\Phi_P(x_1), \Phi_P(x_2)] = 8;$$
$$Y(P) = Max[\Phi_P(y_1), \Phi_P(y_2)] = 4.$$

The total wirelength can thus be computed as

$$L(P) = \sum \Phi_P(x_i) + \sum \Phi_P(y_i)$$
$$L(P) = \Phi_P(x_1) + \Phi_P(x_2) + \Phi_P(y_1) + \Phi_P(y_2) = 5 + 8 + 4 + 3 = 20.$$

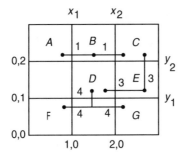

**Figure 4.6** Placement P for Example 4.2.

Note that moving cell G from location (2,0) to location (0,1) will reduce $X(P)$ from 8 to 5. This also causes reduction in wirelength estimate from 20 to 16 units.

### 4.3.4 Minimize maximum density

In the preceding section we introduced the idea of routability of a placement $P$. An alternative measure for routability is the *density* $D(P)$ defined as follows. Assume that the layout surface is divided into a grid. Figure 4.7(a) shows a gate-array with three rows by three columns. The wiring surface is shown in white, whereas the functional blocks are shaded. In Figure 4.7(b) a portion $A$ of the wiring surface (switchbox) is shown separately. A fixed number of horizontal wires can pass through this region. We call it the horizontal capacity of the region. Similarly, a vertical capacity is also defined for the switchbox. Figure 4.7(b) also shows a channel $B$ which permits vertical wiring. A vertical capacity is defined for this channel.

Given a placement $P$, it is possible to estimate the number of nets that must pass through each edge $e_i$ of a channel (or switchbox). If $\eta_P(e_i)$ indicates this estimate, and if $\psi_P(e_i)$ indicates the capacity of the edge $e_i$, then we define the density of edge $e_i$ as

$$d_P(e_i) = \frac{\eta_P(e_i)}{\psi_P(e_i)} \tag{4.6}$$

Clearly, $d_P(e_i)$ must be smaller than 1 (or at most equal to one) for routability. The routability measure of the placement is given by

$$D(P) = \max_i[d_P(e_i)] \tag{4.7}$$

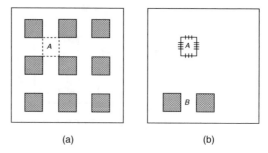

(a)                                    (b)

**Figure 4.7** Gate-array showing functional blocks and wiring surface. Region '*A*' is a switchbox.

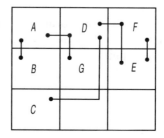

**Figure 4.8** Placement $P$ for Example 4.3.

where the maximum is taken over all edges $e_i$ of the routing regions (switchboxes).

**Example 4.3** Assuming that the capacity of each edge is three tracks, find $D(P)$ the maximum ratio of nets assigned to each edge to the channel capacity. The signal nets for the circuit whose placement $P$ is given in Figure 4.8 are listed below. Assume the weight of each net to be unity. Comment on routability of $P$.

Nets

$N_1(A, B)$
$N_2(E_1, F_1)$
$N_3(A, D_1, G)$
$N_4(C, D_2)$
$N_5(E_2, F_2, D_3)$

SOLUTION   Since $\eta_P(e_i)$, the maximum number of nets crossing any edge in Figure 4.8 is two, and given that $\psi_P(e_i)$ is three, then by definition, $D(P)$ is $\frac{2}{3}$. Since $D(P) < 1$, the placement *may be* routable.

## 4.3.5 Maximize performance

With the advances in integration technology, sizes of transistors have been decreasing and their switching speeds increasing. This trend has been so marked in recent years that wiring delays are becoming more noticeable when compared to switching delays. In technologies such as ECL (Emitter Coupled Logic), this effect is pronounced.

Since the switching time of gates has been lowered to the order of picoseconds, the clocking speed of VLSI chips has become more and more dependent on signal propagations through the interconnects. For many existing large computer circuits, the interconnect delays already account for more than a half of the clock cycle, and the portion of propagation time in the cycle continues to grow. This fact has made a great impact on the success of the design process. It has become impossible to verify the clock rate of a design in the early stages of the design process using only the logic characteristics and switching delays of circuits. Large propagation delays, which are due to electrical characteristics of interconnects can make it impossible to obtain the expected clock rate. To verify and improve the temporal properties of designs, designers use timing analysis tools. These tools commonly known as Delay Analysers, or Timing Verifiers help in checking for long and short path problems (Hitchcock, 1982). A path represents a sequence of circuit elements that the signal travels through to go from a start-point to an end-point. A start-point is an input pad or storage element's output pin. An end-point is an output pad or storage element's input pin. A design is free from long path timing problems if every path is shorter than the latest required arrival time (LRAT) of the signal at the path sink. Referring to Figure 4.9, there is no long-path problem on path $\pi$ if and only if:

$$T_\pi^{\max} \leq LRAT_\pi \qquad (4.8)$$

where $T_\pi^{\max}$ is the maximum delay observed on path $\pi$.

During the design process, the propagation delays on the interconnects are not known prior to layout. Long path timing problems registered after layout are very difficult to correct because they may require, not only new iterations of the physical design step, but possibly, many iterations of the logic design step.

Three general approaches have been suggested to correct long path timing problems. The first approach proceeds by making changes to the logic. For example, the delay of a path can be substantially decreased by reducing the loading on some of its circuits' elements. Also collapsing some

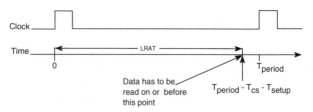

**Figure 4.9** Long path problems. $T_{cs}$ is the maximum clock skew. $T_{setup}$ is the setup time.


of the logic on the long paths can reduce some of the paths' delays (Darringer *et al.*, 1984; Micheli, 1986; Kick, 1987; Singh *et al.*, 1988).

The second approach relies on transistor sizing to speed up some of the circuit elements on the slow paths. By increasing the sizes of some of the driving transistors, the switching delays of the driving elements as well as the propagation delays along the nets driven by the resized transistors can be substantially lowered (Ruelhi *et al.*, 1977; Brayton *et al.*, 1981; Micheli, 1986; Jouppi, 1987).

One way to make a circuit faster without making any changes in its logic design is to reduce the propagation time to a minimum. This goal can be achieved by imposing timing constraints on the interconnects and paths of the design. The third approach adopts this strategy. Several attempts have been reported which tried to make the physical design sensitive to the timing requirements (Dunlop *et al.*, 1984; Burstein and Youssef, 1985; Nair *et al.*, 1989; Sutanthavibul and Shragowitz, 1989; Youssef *et al.*, 1992). Optimized for timing, a VLSI design can permit a 25–35 per cent increase in the clock rate without any changes in the logic or cell design (larger speed-ups have also been observed).

The speed performance of a circuit may be characterized by the longest combinational delay from an output pin to an input pin. If the path delays are to be kept below a maximum value then the wiring delays must be kept in check. Since placement affects the wiring requirements of a layout, the objective of the placement problem can be altered to satisfy the path timing requirements as stated in Equation 4.8.


### 4.3.6 Other constraints

Packaging considerations require that I/O pads must be placed on the periphery of a chip. Thus, if a placement program handles both I/O pads and functional cells, it must ensure that the pads are placed on the periphery and the functional cells are placed as internal cells. In a PCB placement, the power dissipation of individual chips is also a consideration. Chips which dissipate large amounts of power must not be placed close together so as to achieve thermal balance.

## 4.4 APPROACHES TO PLACEMENT

We have noted in the introduction that even the simplest version of the placement problem, namely arranging $n$ logic blocks in a row so as to minimize the total wire length, is NP-complete. The alternative cost functions and constraints discussed in the preceding section do not improve the situation. If the number of modules involved is small, say less than 10, it is possible to enumerate all the feasible solutions to the placement problem and pick an optimum one. For large problems, the solution space is too large to permit enumerative techniques. It is therefore natural to turn to *heuristic techniques* which require short execution times and find 'good' solutions, not necessarily the best solutions.

Heuristic algorithms for placement can be classified into two broad categories: constructive and iterative.

As the name suggests, a constructive placement algorithm *constructs* a solution by placing one cell at a time. Consider, for instance, the following algorithm to place $n$ cells in a row. The layout surface is imagined to be divided into $n$ slots. To begin with, all these slots are empty. During each iteration of the algorithm, one cell will be placed into one of the empty slots. At the end of each iteration, we have a *partial placement* of a subset of modules. There are two decisions to be made during each iteration; (a) which unplaced cell must be selected and added to the partial placement? and (b) where should the selected cell be placed?

Heuristics can be used to guide the above two decisions. For example, a possible selection heuristic would be to pick that cell which is 'most strongly connected' to the existing partial placement. Suppose the partial placement consists of cells $m_1, m_2, \cdots, m_i$. We examine each of the unplaced cells $m_j$ and compute the quantity

$$A_{mj} = \sum_{k=1}^{i} c_{m_j m_k} \tag{4.9}$$

$c_{m_j m_k}$ denotes the connectivity between the unplaced cell $m_j$ and a placed cell $m_k$. Therefore $A_{mj}$ denotes the number of connections from $m_j$ to the already placed cells $\{m_1, m_2, \cdots, m_i\}$. We select the cell for which $A_{mj}$ is maximum. This strategy is known as *maximum-connectivity* (maxcon) strategy.

**Algorithm** $Constructive\_Linear\_Placement(n, C, P)$;
**Begin**.
    (*$n$ is the number of cells.*)
    (*$C[1 \cdots n, 1 \cdots n]$ is the connectivity matrix.*)
    (*$P[1 \cdots n]$ is the placement vector.*)
    (*$P[i]$ indicates the slot in which module $i$ is placed.*)
    **For** $i = 1$ to $n$
        $P[i] = -\infty$; (*$P[i] = -\infty$ means slot $i$ is empty.*)
    **EndFor**
    $S \leftarrow$ Seed$(n, C)$; (*Determine the Seed cell.*)
    $P[S] \leftarrow \frac{n}{2}$; (*Place the Seed cell in the center.*)
    Mark $S$ as placed;
    **For** $i = 1$ to $n - 1$ **Do**
        $sc \leftarrow Select\_Cell(n, P, C)$;
        $ss \leftarrow Select\_Slot(n, sc, P, C)$;
        $P[sc] \leftarrow ss$;
        Mark $sc$ as placed;
    **End**;
**End**.

**Figure 4.10** Constructive one-dimensional placement procedure.

The selected cell may be placed in any of the $(n - i)$ empty slots. We can estimate the change in the cost function for each of the $(n - i)$ choices, and pick that choice which proves most beneficial. For example, if the cost function is total wirelength, the slot which results in minimum increase in wirelength is selected. We leave it to the reader to implement such an estimation procedure for the increase in wirelength (see Exercise 4.11).

To get the constructive placement algorithm rolling, we need an initial partial placement. For this, we can pick a single cell, called the 'seed' cell, and place it, say, in the centre slot. The seed cell may be selected randomly, or based on an heuristic criterion. It seems logical to pick the cell with largest connectivity as the seed cell. The complete algorithm *constructive-placement* is shown in Figure 4.10.

The constructive placement algorithm has modest requirements on execution time. The 'Select_Cell' procedure computes $A_{mj}$ for $(n - i)$ cells during the $i$th iteration, and it requires $O(i(n - i))$ time. The 'Select_Slot' procedure can be implemented in $O((n - i)i)$ time. Since each of these procedures is called $n - 1$ times inside the **For** loop, the algorithm would take $\sum_{i=1}^{n} i(n - i) = O(n^3)$ time.

The constructive placement procedure described above is *greedy*. At each step, it makes the best possible move. Does the procedure yield an optimum solution? Not necessarily. The reason is that, at any iteration the procedure makes a decision in the *absence of complete information*. For instance, when the cell is selected during $i$th iteration, the selection is made with respect to the modules in the partial placement; the unplaced modules are ignored. Once a slot has been chosen, the algorithm will not go back on

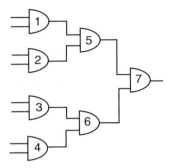

**Figure 4.11** Circuit for Example 4.4.

this decision. Indeed, this slot may not be the best choice for the module after the $(i + 1)^{st}$ iteration. Thus the final solution is not necessarily a global optimum.

**Example 4.4** Consider the circuit shown in Figure 4.11. The connectivity of module 1 is 1, since it is connected only to module 5. Similarly, we can compute the remaining connectivities as follows. $C_1 = 1$, $C_2 = 1$, $C_3 = 1$, $C_4 = 1$, $C_5 = 3$, $C_6 = 3$, $C_7 = 2$. Thus either module 5 or 6 may be selected as the seed cell. Let us break the tie by tossing a coin, and suppose that module 5 wins. We place the selected module in the centre slot, namely, slot 4.

We now decide which cell must be placed next. The modules 1, 2, and 7 are connected to the placed module 5. Further, $c_{15} = c_{25} = c_{75} = 1$. We must therefore pick any one of these randomly and place it in an appropriate slot. Suppose that module 1 wins the race. To minimize the increase in wirelength, module 1 must be placed as close to 5 as possible. Let us pick slot 3 for module 1. (We would have also picked slot 5.)

We continue in the manner described above and generate the placement shown in Figure 4.12. The wirelength requirement of the final placement is 16 units. (Connection $c_{15}$ takes 1 unit, $c_{36}$ takes 5 units, and so on.)

In the following subsections we discuss some of the approaches used to solve the placement problem. We begin with the partition-based method which is based on the min-cut heuristic. Next, we present the iterative technique that uses simulated annealing. As an example of the application of the annealing heuristic we discuss implementation details of the TimberWolf3.2 package. The section concludes with the discussion of the force-directed heuristic which is based on the analogy of attraction of masses connected by springs.

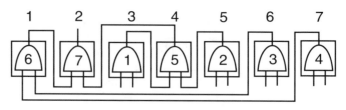

**Figure 4.12** Final placement for circuit of Example 4.4.

## 4.4.1 Partition-based methods

In Chapter 2 we discussed the circuit partitioning problem. A partitioning algorithm tries to group together closely connected modules. Such a grouping will also reduce the interconnection length and wiring congestion. We now discuss an algorithm which uses a partitioning procedure repeatedly to generate a placement.

### Min-cut placement

In Section 4.3 we introduced three objective functions, namely $X(P)$, $Y(P)$ and $L(P)$. It was pointed out that minimizing $X(P)$, the maximum horizontal cut, and minimizing $Y(P)$, the maximum vertical cut, would improve the routability of a gate-array placement. We also indicated that the number of feed-through cells in a standard-cell layout will be reduced by reducing $Y(P)$.

Reflecting on the function $X(P)$, it is apparent that minimizing $X(P)$ is closely related to the two-way partitioning problem which we studied in Chapter 2. Thus, we apply a partitioning algorithm to the given circuit to generate two blocks $A$ and $B$, place the modules in block $A$ on the left of an imaginary vertical cutline $c_1$, and the modules in block $B$ to the right of $c_1$. The cutset achieved by the partitioning algorithm is the number of horizontal nets cut by $c_1$ and is denoted by $\Phi_P(c_1)$ (see Section 4.3.3).

Suppose that we now repeat the process on blocks $A$ and $B$, i.e., we consider block $A$ as a circuit and partition it into two blocks $A_1$ and $A_2$ using a vertical cutline $c_2$. Similarly, block $B$ is partitioned into two blocks $B_1$ and $B_2$ using a vertical line $c_3$ (see Figure 4.13).

This process can be repeated by introducing more cutlines. Let us focus our attention on Figure 4.13. What can be said about $\Phi_P(c_2)$ or $\Phi_P(c_3)$ in relation to $\Phi_P(c_1)$? For the purpose of our discussion, let us assume that our partitioning procedure generates an optimum partition. With respect to the total circuit, $\Phi_P(c_1)$ is the minimum cut possible. Similarly, $\Phi_P(c_2)$ is the minimum possible cut with respect to the subcircuit $A$. It is not possible to

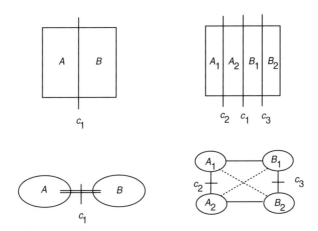

**Figure 4.13** Using partitioning to reduce $X(P)$ .

claim that $\Phi_P(c_{i+1}) \leq \Phi_P(c_i)$, $1 \leq i \leq r - 1$, although that would be desirable, leading to a minimization of $X(P)$ (see Exercise 4.7).

The procedure described above does not minimize $X(P)$, but minimizes $\Phi_P(c_2)$ subject to the constraint that $\Phi_P(c_1)$ is minimum. We write this function as $\Phi_P(c_2)|\Phi_P(c_1)$. The procedure also minimizes $\Phi_P(c_3)|\Phi_P(c_1)$.

Minimization of $X(P)$, $Y(P)$, or $L(P)$ are computationally very difficult. To simplify the problem, a *sequential objective function* denoted by $F(P)$ whose near minimal value is easier to achieve is used.

$$F(P) = \min[\Phi_P(c_r)]|\min[\Phi_P(c_{r-1})]|\cdots|\min[\Phi_P(c_1)] \qquad (4.10)$$

where $c_1, c_2, \cdots, c_r$, is an ordered sequence of vertical or horizontal cutlines (Breuer, 1977a).

## The algorithm

The min-cut placement algorithm assumes the availability of an ordered sequence of $r$ cutlines. These $r$ cutlines divide the layout into slots. Two key requirements of the algorithm are:

1. an efficient procedure to partition the circuit, and
2. the selection of cutlines.

$F(P)$ (Equation 4.10) is minimized by first partitioning the circuit into two, such that the number of nets that cross $c_1$ is minimized. If $c_1$ is a

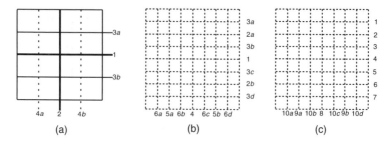

**Figure 4.14** Illustration of sequences of cutlines.

vertical cutline then the cells on the left of $c_1$ are fixed and cannot move to the right. The cells on the right of $c_1$ are also *constrained* and cannot move to the left. Then, the next cutline $c_2$ is used and the nets that cut $c_2$ are minimized subject to the constraint already imposed by $c_1$. The procedure is continued until all $c_r$ lines are used. Because of the *greedy* nature of the above procedure, the solution obtained is not guaranteed to be globally optimal.

Partitioning the circuit about the cutline $c_i$, so as to minimize $\Phi_P(c_i)$, can be accomplished by any one of the algorithms discussed in Chapter 2.

Three schemes for the selection of cutlines and the sequence in which they are processed are recommended by Breuer (1977a,b). In the first method, called the *Quadrature Placement Procedure*, the layout is divided into four units with two cutlines, one vertical and the other horizontal, both passing through the centre. The above division procedure is then recursively applied to each quarter of the layout cut until the entire layout is divided into slots. This sequence is illustrated in Figure 4.14(a) and is very suitable for circuits with high routing density in the centre.

In the second scheme called *Bisection Placement Procedure*, the layout is repeatedly divided into equal halves by horizontal cutlines yielding horizontal segments. This division procedure is continued until each horizontal segment is a row. The cells are assigned to rows. Next, each row is repeatedly *bisected* vertically until the resulting subregions contain one slot. This method is good for standard-cell placement and is illustrated in Figure 4.14(b).

In the third scheme, called *Slice/Bisection*, the $n$ cells of the circuit are divided using cut line $c_1$ into two sets of $k$ and $(n - k)$ cells, such that $\Phi_P(c_1)$ is minimized. The first $k$ cells obtained are assigned to the top-most (or bottom-most) row (one *slice* of the layout). The procedure is then applied on the remaining $(n - k)$ cells dividing them into $k$ and $(n - 2k)$ cells. The process is continued until all the cells are assigned to rows. The cells are then assigned to columns using vertical *bisection*. This sequence is

**Algorithm** $Min—cut(\aleph, n, C)$;
(*$\aleph$ is the layout surface.
$n$ is the number of cells to be placed.
$n_0$ is the number of cells in a slot.
$C$ is the connectivity matrix*).
    **Begin**
        **If** $(n \leq n_0)$ **Then** place-cells $(\aleph, n, C)$
        **Else Begin**
            $(\aleph_1, \aleph_2) \leftarrow$ $cut$-$surface(\aleph)$;
            $(n_1, c_1), (n_2, c_2 \leftarrow partition(n, C)$;
            Call Min-cut $(\aleph_1, n_1, c_1)$;
            Call Min-cut $(\aleph_2, n_2, c_2)$;
        **EndIf;**
    **End**.

**Figure 4.15**  Structure of a recursive min-cut algorithm.

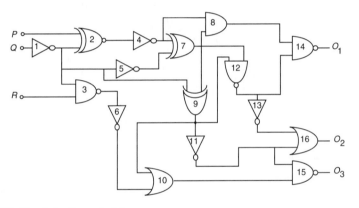

**Figure 4.16**  Circuit for Example 4.5.

illustrated in Figure 4.14(c), and is recommended for cells with high interconnection on the periphery.

The structure of a recursive min-cut algorithm is given in Figure 4.15. The min-cut placement procedure is further illustrated with the help of an example below.

**Example 4.5**  Given below is the netlist of the circuit shown in Figure 4.16. The gates of the circuit are to be assigned to slots on the layout with one gate per slot. Using the method of repeated partitioning, divide the circuit and the layout. Assign the subcircuits to vertical and horizontal partitions of the layout such that the number of nets crossing the cutlines is minimized. Use the method of Kernighan and Lin to partition and the scheme of *quadrature placement procedure* to generate and sequence the cutlines.

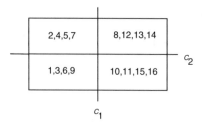

**Figure 4.17** Assignment of gates to four quadrants of Example 4.5.

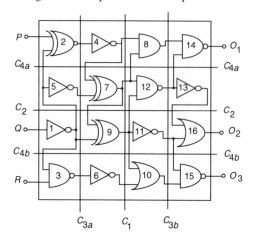

**Figure 4.18** Final placement $P$ for circuit of Example 4.5.

Nets

$$N_P = \{P, C_2\} \qquad\qquad N_Q = \{Q, C_1\} \qquad\qquad N_R = \{R, C_3\}$$
$$N_1 = \{C_1, C_2, C_3, C_5, C_9\} \quad N_2 = \{C_2, C_4\} \qquad\qquad N_3 = \{C_3, C_6\}$$
$$N_4 = \{C_4, C_7, C_8\} \qquad N_5 = \{C_5, C_7\} \qquad\qquad N_6 = \{C_6, C_{10}\}$$
$$N_7 = \{C_7, C_8, C_9, C_{12}\} \quad N_8 = \{C_8, C_{14}\} \qquad\qquad N_9 = \{C_9, C_{10}, C_{11}, C_{12}\}$$
$$N_{10} = \{C_{10}, C_{15}\} \qquad\quad N_{11} = \{C_{11}, C_{15}, C_{16}\} \quad N_{12} = \{C_{12}, C_{13}, C_{14}\}$$
$$N_{13} = \{C_{13}, C_{16}\} \qquad\quad N_{14} = \{C_{14}, O_1\} \qquad\qquad N_{15} = \{C_{15}, O_3\}$$
$$N_{16} = \{C_{16}, O_2\}$$

SOLUTION   Partitioning the circuit using the method of Kernighan and Lin to minimize the number of *nets* cut yields two sets of gates, namely $L$ and $R$, where $L = \{1,2,3,4,5,6,7,9\}$ and $R = \{8,10,11,12,13,14,15,16\}$. The cost of this cut is found to be 4. The cutline $c_1$ runs vertically through the centre of the layout. The gates of set $L$ are assigned to the left of $c_1$ and those of set $R$ are assigned to the right of $c_1$, and are constrained to those sides. Next the layout is divided with a horizontal

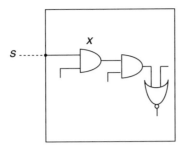

**Figure 4.19** Cell $x$ of a group connected to an external signal $s$.

cutline $c_2$ and the circuit is partitioned again but with the above constraint. This results in two partitions for the gates of set $L$ denoted by $LT$ and $LB$ and two for the gates of set $R$ denoted by $RT$ and $RB$. The number of nets of $L$ crossing the cutline $c_2$ is 2 and the number of nets of $R$ crossing the cutline $c_2$ is also 2. The elements of subsets are:

$LT = \{2, 4, 5, 7\};$      (* Top Left *)
$LB = \{1, 3, 6, 9\};$      (* Bottom Left *)
$RT = \{8, 12, 13, 14\};$      (* Top Right *)
$RB = \{10, 11, 15, 16\}.$      (* Bottom Right *)

The elements of the above subsets are assigned to each quadrant of the layout as shown in Figure 4.17. The procedure is repeated again with two cutlines running vertically ($c_{3a}$ and $c_{3b}$) and two cutlines running horizontally ($c_{4a}$ and $c_{4b}$) as shown in Figure 4.17. The final division of layout into slots and the assignment of gates to these slots of the layout is shown in Figure 4.18.

## 4.4.2 Limitation of the min-cut heuristic

Layouts obtained by merely partitioning cells and assigning them to regions are not nearly as good as they can be. A major component that was not taken into account is the location of external pin connections (which are generally fixed) and the probable locations of cells in the final placement.

To understand the above point refer to Example 4.5. The circuit is partitioned into two sets $L$ and $R$. In the next step we partitioned $L$ into two sub-sets $LT$ and $LB$. Cells $\{2,4,5,7\}$ were assigned to $LT$ and $\{1,3,6,9\}$ to $LB$. The other option is the reverse assignment, i.e., to assign $\{2,4,5,7\}$ to $LB$ and $\{1,3,6,9\}$ to $LT$. If this were done, then, since the positions of input pins $P$, $Q$ and $R$ are fixed as shown in Figure 4.18, longer wires would be required to connect input pins to gates.

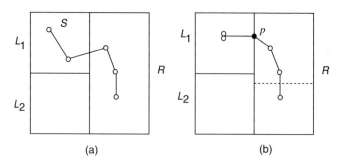

(a)                                           (b)

**Figure 4.20**   (a) Partitioning of $R$ following partitioning of $L$. (b) Propagating $s$ to the axis of partitioning.

Simply speaking the earlier procedure did not take into account the position of terminal pins and signals that enter a group of cells. These signals affect where the cells ought to be placed just as much as the internal connections within the group. The inclusion of these signals in partitioning-based placement is called *terminal propagation* (Dunlop and Kernighan, 1985).

## Terminal propagation

Consider a group of cells containing a cell $x$ connected to signal $s$ from outside the group (an I/O pad for instance) as shown in Figure 4.19. Clearly cell $x$ has to be nearest to the point where signal $s$ enters.

At the outermost level, signal positions are typically fixed by pad positions. Let us see what happens at an inner level of partitioning.

Referring again to Figure 4.18, line $c_{3a}$ divides $LB$ into $\{1,3\}$ and $\{6,9\}$. Line $c_{4b}$ divides $\{6,9\}$ into $\{6\}$ and $\{9\}$. The assignment shows gate 9 assigned to a slot which is above the slot of gate 6. Another valid solution would have been to assign gate 6 above gate 9. It is clear from the figure that the swap of gates 6 and 9 will result in longer wires. This is because gate 9 is receiving an input from above and the net is cut by the horizontal line $c_2$, therefore gate 9 must be placed close to $c_2$. But the min-cut heuristic does not favour one assignment over the other. To reiterate what was mentioned above consider the situation in Figure 4.20(a). The cells are partitioned into two groups $L$ and $R$, then $L$ into $L_1$ and $L_2$, and there is a signal net $s$ that connects two cells in $L_1$ with three cells in $R$. Now if we want to partition $R$ into $R_1$ and $R_2$, we would like to take into account the fact that signal net $s$ is in $L_1$ but not in $L_2$ and thus *bias* the partitioning process towards putting the cells into $R_1$ instead of $R_2$. This is done as follows. Assume all the cells of signal net $s$ in $L_1$ are at its geometrical

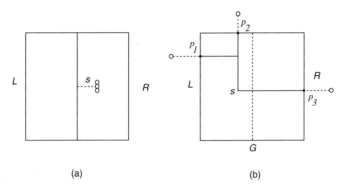

(a)                                              (b)

**Figure 4.21** First stage non-bias partition.

centre, and *propagate* that position to the closest point, say $p$, on $R$ as shown in 4.20(b). Imagine that a dummy cell with the same signal net $s$ is placed at $p$.

During partitioning, signal net $s$ is required to remain in set $R_1$. This biases the partitioning process as desired. The cost of partitioning $R$ will be one less if $s$ appears only in $R_1$ than if some or all cells containing $s$ are in $R_2$. Without this bias there would be nothing to *favour* $R_1$ over $R_2$.

The above example conveys one situation of terminal propagation. There are others. Assume that we are in the earlier stage of partitioning where neither $L$ nor $R$ has yet been partitioned as in Figure 4.21(a). When $L$ is partitioned the three elements of $s$ in $R$ are assumed to be concentrated at the centre of $R$ (geometrical centre), which propagates to the middle of $L$. External signals that propagate to a point near the axis about which partitioning is to be done *should not* be used to bias the solution in either direction. That is, elements of $s$ in $R$ should have no effect on how $L$ is partitioned. In the implementation presented by Dunlop and Kernighan (1985), the measure of 'near' is arbitrarily set at *'within the middle third of the side'*.

In the general case, suppose a group $G$ is to be partitioned into $L$ and $R$, and that some net has elements both inside and outside $G$. The elements may be individual cells or sets of cells at the centres of other groups. A low-cost Steiner tree is computed on the elements external to $G$, and the points of intersection $\{p_i\}$ with the border of $G$. This is illustrated in Figure 4.21(b). These points are treated as cells on $s$, but are fixed during partitioning, so they cannot move from whichever $L$ or $R$ they are in. As before, the middle third of the sides perpendicular to the axis of partitioning are excluded, so any $p_i$s that fall in these regions are simply ignored. Thus, in the situation illustrated in Figure 4.21(b), where a partition is created by a vertical cut, signal $p_2$ would be ignored, since it is within the middle third of the top side. But if partitioning were being

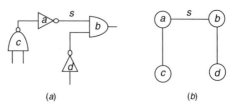

(a)                                    (b)

**Figure 4.22** (a) Circuit for Example 4.6. (b) Corresponding graph.

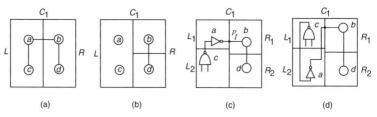

(a)                (b)                (c)                (d)

**Figure 4.23** Solution for Example 4.6. (a) Dividing the circuit into $L$ and $R$. (b) Unbiased partition of $R$. (c) Biased partition of $L$ producing $P$. (d) $L$ partitioned *without* terminal propagation.

carried about the other axis, $p_3$ would be ignored and signal net $s$ would be biased to the top half (Dunlop and Kernighan, 1985).

To do terminal propagation, the partitioning has to be done breadth first. There is no point in partitioning one group to finer and finer levels without partitioning the other groups, since in that case no information would be available about which group a module should preferably be assigned to.

The above algorithm has been tested on a chip with 412 cells and 453 nets. It yields areas within 10–20 per cent and track densities within 3 per cent of careful hand layouts.

**Example 4.6** The four gates of a circuit shown in Figure 4.22 are to be assigned to slots of a $2 \times 2$ array. Using the method of min-cut partitioning and terminal propagation find a suitable placement $P$.

SOLUTION   The graph corresponding to the circuit of Figure 4.22(a) is given in Figure 4.22(b). The result of the first step of partitioning is shown in Figure 4.23(a). Half the gates are assigned to $L$ and the other half to $R$. The cost of the cut is 1. In the second step, the cells in $R$ are partitioned. This step must be an unbiased partition. Cells in $R$ are assigned to $R_1$ and $R_2$ as shown in Figure 4.23(b). In the third step the cells in $L$ are partitioned. Before this is done, since the vertical line cuts a net (say $s$) connecting cells $a$ and $b$, it must be propagated to the cutline. A dummy cell with the same net $s$ is placed at point $p_1$. Now dividing the circuit in $L$ with net $s$ on the cutline will yield $a$ in $L_1$ and $c$ in $L_2$. The cost of this division is 1 (Figure 4.23(c)). If net $s$ is not considered, then the reverse assignment, that is, $c$ in $L_1$ and $a$ in $L_2$ would have also resulted in one net cut. But the actual cost of the latter assignment is 2 (Figure 4.23(d)).

### 4.4.3 Simulated annealing

Simulated annealing is the most well developed method available for cell placement. The general algorithm was discussed in detail in Chapter 2. In this section the simulated annealing algorithm is adapted for placement and explained with the help of an example. A place-and-route package called TimberWolf developed by Carl Sechen and Sangiovanni-Vincentelli (1986), was the earliest package to apply simulated annealing to the placement problem. We discuss the placement algorithm used in TimberWolf. This section makes use of the terminology and discussion of Section 2.4.4 of Chapter 2.

### The algorithm

The simulated annealing algorithm can be modified for cell placement by choosing a suitable *perturb* function to generate a new placement configuration (cell assignment to slots), and by defining a suitable *accept* function. For simplicity, let us use the *checker board model* of the layout, and assume that each cell of the circuit can be accommodated in one slot (a square of the checker board). A simple neighbour function is the pairwise interchange where two slots are chosen and their contents swapped. Other schemes to generate neighbouring states include displacing a randomly selected cell to a random location, the rotation and mirroring of cells if the layout strategy allows, or any other move that may cause a change in wirelength. Let $\Delta h = (Cost(NewS) - Cost(S))$ be the change in estimated wirelength due to a swap, where $Cost(S)$ is the old wirelength and $Cost(NewS)$ is the wirelength after perturbation. In simulated annealing, the swap is accepted if $\Delta h < 0$ (that is, $Cost(NewS) < Cost(S)$) or if the acceptance function ($random < e^{-\Delta h/T}$) is true, where $random$ is a uniformly generated random number between 0 and 1, and $T$ is the current value of the temperature. The procedure is further explained with the following example.

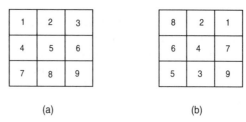

(a)                              (b)

**Figure 4.24** (a) Initial configuration for Example 4.7. (b) $P$ obtained by simulated annealing, wirelength using semi-perimeter estimate = 24.

**Example 4.7** Given the following netlist with 9 cells $C_1, \cdots, C_9$, and 13 nets $N_1, \cdots, N_{13}$. Assume that all cells are of the same size and that the layout surface is a checker board with three rows and three columns (9 slots). Write a placement program using the simulated annealing algorithm given in Chapter 2 in order to assign each cell to one of the 9 slots, while minimizing the total Manhattan routing length. Use the *semiperimeter* method to estimate the wirelength.†

Nets

$N_1 = \{C_4, C_5, C_6,\}$      $N_2 = \{C_4, C_3\}$      $N_3 = \{C_2, C_4\}$

$N_4 = \{C_3, C_7, C_8\}$      $N_5 = \{C_2, C_3, C_6\}$      $N_6 = \{C_4, C_7, C_9\}$

$N_7 = \{C_2, C_8\}$      $N_8 = \{C_1, C_7\}$      $N_9 = \{C_3, C_5, C_9\}$

$N_{10} = \{C_6, C_8\}$      $N_{11} = \{C_2, C_6, C_7\}$      $N_{12} = \{C_4, C_7, C_9\}$

$N_{13} = \{C_3, C_9\}$

Use sequential *pairwise exchange* as the *perturb* function and use the following annealing schedule:

Initial temperature: $T_0 = 10$;
Constants: $M = 20$; $\alpha = 0.9$; $\beta = 1.0$.

In sequential pairwise exchange, the cell in slot $i$ is trial-exchanged in sequence with the cells in slots $i+1, \cdots, n-1, n$, for $1 \le i \le n-1$. The termination condition is to halt the program if no cost improvement is observed at two consecutive temperatures.

SOLUTION    The initial placement is given in Figure 4.24(a). The output of the program is given in Table 4.1. The only entries shown are those where the new configuration was accepted. The pair $(a, b)$ under the 'swap' column indicates that cells $a$ and $b$ are selected for interchange; in this case $a$ and $b$ have been selected sequentially. The selection could also have been random. The new wirelength is computed as $Cost(NewS)$. If $Cost(NewS)$ is better than $Cost(S)$ then the swap is

†Since simulated annealing is a stochastic technique the solution is not unique.

| Count | $\alpha \times T$ | $(Swap)$ | Random | $Cost(S)$ | $Cost(NewS)$ | $e^{-\Delta h/T}$ |
|---|---|---|---|---|---|---|
| 1 | | (1,2) | 0.05538 | 34 | 36 | 0.81873 |
| 2 | | (1,3) | 0.37642 | 36 | 36 | 1.00000 |
| 3 | | (1,4) | | 36 | 35 | |
| 4 | | (1,5) | 0.11982 | 35 | 38 | 0.74082 |
| 5 | | (1,6) | | 38 | 36 | |
| 6 | | (1,7) | | 36 | 32 | |
| 7 | | (1,8) | 0.62853 | 32 | 32 | 1.00000 |
| 8 | | (1,9) | | 32 | 31 | |
| 10 | 10.000 | (2,3) | 0.75230 | 31 | 32 | 0.90484 |
| 11 | | (2,4) | 0.36827 | 32 | 32 | 1.00000 |
| 12 | | (2,5) | | 32 | 30 | |
| 13 | | (2,6) | 0.86363 | 30 | 30 | 1.00000 |
| 14 | | (2,7) | 0.76185 | 30 | 31 | 0.90484 |
| 15 | | (2,8) | 0.33013 | 31 | 32 | 0.90484 |
| 16 | | (2,9) | 0.65729 | 32 | 32 | 1.00000 |
| 17 | | (3,1) | 0.47104 | 32 | 33 | 0.90484 |
| 18 | | (3,2) | | 33 | 32 | |
| 19 | | (3,4) | 0.42597 | 32 | 32 | 1.00000 |
| 20 | | (3,5) | 0.86318 | 32 | 33 | 0.90484 |
| 21 | | (3,6) | | 33 | 27 | |
| 22 | | (3,7) | | 27 | 26 | |
| 24 | | (3,9) | 0.20559 | 26 | 28 | 0.80074 |
| 25 | | (4,1) | 0.58481 | 28 | 32 | 0.64118 |
| 26 | | (4,2) | 0.30558 | 32 | 36 | 0.64118 |
| 27 | | (4,3) | | 36 | 33 | |
| 28 | | (4,5) | 0.31229 | 33 | 33 | 1.00000 |
| 29 | | (4,6) | 0.00794 | 33 | 35 | 0.80074 |
| 30 | 9.000 | (4,7) | | 35 | 34 | |
| 31 | | (4,8) | | 34 | 33 | |
| 32 | | (4,9) | | 33 | 31 | |
| 33 | | (5,1) | | 31 | 30 | |
| 34 | | (5,2) | 0.28514 | 30 | 32 | 0.80074 |
| 35 | | (5,3) | 0.35865 | 32 | 34 | 0.80074 |
| 36 | | (5,4) | 0.87694 | 34 | 35 | 0.89484 |
| 37 | | (5,6) | | 35 | 34 | |
| 38 | | (5,7) | | 34 | 33 | |
| 39 | | (5,8) | 0.03769 | 33 | 35 | 0.80074 |
| 40 | | (5,9) | | 35 | 34 | |

**Table 4.1 Output of simulated annealing run for Example 4.7.**

made, and the *random* number is not generated. If $Cost(NewS)$ is greater than $Cost(S)$ then a random number is generated and compared with the value generated by the acceptance function.

Output of the program for only *two temperatures* is shown in Table 4.1. The final temperature $(\alpha \times T)$ was 0.581 when the program terminated, and this was at iteration *count* of 560. The corresponding

| Iterations | (Swap) | Cost(S) | Cost(NewS) |
|:---:|:---:|:---:|:---:|
| 7 | (1,8) | 34 | 33 |
| 15 | (2,8) | 33 | 32 |
| 20 | (3,5) | 32 | 30 |
| 21 | (3,6) | 30 | 28 |
| 49 | (7,1) | 28 | 27 |
| 60 | (8,4) | 27 | 26 |

**Table 4.2  Output generated by deterministic pairwise interchange algorithm.**

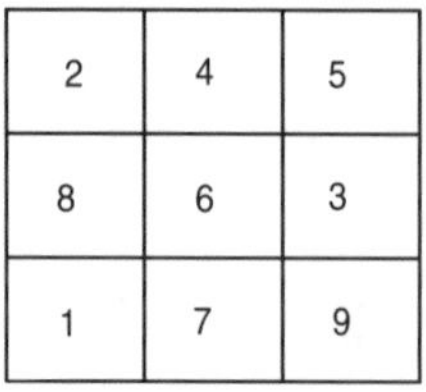

| 2 | 4 | 5 |
|:---:|:---:|:---:|
| 8 | 6 | 3 |
| 1 | 7 | 9 |

**Figure 4.25**  $P$ for Example 4.7 obtained by deterministic pairwise exchange.

wirelength obtained by using the semiperimeter estimate is 24 units. The placement $P$ given by the simulated annealing algorithm is shown in Figure 4.24(b).

The same program is executed again by *suppressing* the condition that probabilistically accepts bad moves. This transforms the simulated annealing algorithm to the deterministic pairwise exchange algorithm. The results of this execution are shown in Table 4.2 and the corresponding placement obtained is shown in Figure 4.25. The algorithm converges to a *local optimum* after 60 iterations. Note that the wirelength obtained by this solution using the same method for estimation is 26 units (greater by 2 units than that obtained by simulated annealing for this example).

An implementation that uses simulated annealing for placement and routing is the TimberWolf3.2 package (Sechen and Sangiovanni-Vincentelli, 1986). This package handles *standard-cell* circuit configurations. In the following section, we explain some of the features of the package and the algorithms employed by it.

## TimberWolf algorithm

Based on the input data and parameters supplied by the user, TimberWolf3.2 constructs a standard-cell circuit topology. These parameters, in conjunction with the total width of standard-cells to be placed,

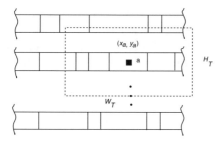

**Figure 4.26**   Limiter window centred around a cell.

enable TimberWolf3.2 to compute the initial position and the target lengths of the rows. Macro blocks (up to 11 are allowed) are placed next, followed by placement of pads. Pads and macro blocks retain their initial positions and only the placement of standard-cells is optimized. Following initial placement, the algorithm then performs placement and routing in three distinct stages. In the first stage, cells are placed so as to minimize the estimated wirelength. In the second stage, feed-through cells are inserted as required, wirelength is minimized again, and preliminary global routing is done. In the third stage, local changes are made in the placement to reduce the number of wiring tracks required. In the following discussion, we will be primarily concerned with the first stage that uses simulated annealing for placement.

The objective function which TimberWolf3.2 attempts to minimize during the placement is the estimated interconnect cost. The purpose of the first stage is to find a placement of the standard-cells such that the total estimated interconnect cost is minimized. A neighbour function called *generate* is used to produce new states by making a random selection from one of three possible perturb functions.

## Perturb functions:

1. Move a single cell to a new location, say to a different row.
2. Swap two cells.
3. Mirror a cell about the $x$-axis.

TimberWolf3.2 uses cell mirroring less frequently when compared to cell displacement and pairwise cell swapping. In particular, mirroring is attempted in 10 per cent of the cases only (where cell movement is rejected).

Perturbations are limited to a region within a window of height $H_T$ and width $W_T$. For example, if a cell must be displaced, the target location is found within a limiting window centred around the cell (see Figure 4.26).

Therefore, two cells $a$ and $b$, centred at $(x_a, y_a)$ and $(x_b, y_b)$ are selected for interchange only if $|x_a - x_b| \leq W_T$ and $|y_a - y_b| \leq H_T$. The dimensions of the window are decreasing functions of the temperature $T$. If current temperature is $T_1$ and next temperature is $T_2$, the window width and height are decreased as follows:

$$W(T_2) = W(T_1) \frac{\log(T_2)}{\log(T_1)} \tag{4.11}$$

$$H(T_2) = H(T_1) \frac{\log(T_2)}{\log(T_1)} \tag{4.12}$$

**Cost function**: The cost function used by the TimberWolf3.2 algorithm is the sum of three components

$$\gamma = \gamma_1 + \gamma_2 + \gamma_3 \tag{4.13}$$

$\gamma_1$ is a measure of the total estimated wirelength. For any net $i$, if the horizontal and vertical spans are given by $X_i$ and $Y_i$, then the estimated length of the net $i$ is $(X_i + Y_i)$. This must be multiplied by the weight $w_i$ of the net. Further sophistication may be achieved by associating two weights with a net—a horizontal component $w_i^H$ and a vertical component $w_i^V$. Thus,

$$\gamma_1 = \sum_{i \in Nets} [w_i^H \cdot X_i + w_i^V \cdot Y_i] \tag{4.14}$$

where the summation is taken over all nets $i$. The weight of a net is useful in indicating how *critical* the net is. If a net is part of a critical path, for instance, we want it to be as short as possible so that it introduces as little wiring delay as possible. We can increase the weights of critical nets to achieve this goal. Independent horizontal and vertical weights give the user the flexibility to favour connections in one direction over the other. Thus, in double metal technology, where it is possible to stack feed-throughs over the cell, vertical spans may be given preference over horizontal tracks. This can be accomplished by lowering the weight $w^V$. In chips where feed-throughs are costly in terms of area, horizontal wiring may be preferred by lowering $w^H$.

When a cell is displaced or when two cells are swapped, it is possible that there is an overlap between two or more cells. Let $O_{ij}$ indicate the area of overlap between two cells $i$ and $j$. Clearly, overlaps are undesirable and must be minimized. The second component of the cost function, $\gamma_2$, is interpreted as the penalty of overlaps, and is defined as follows:

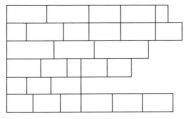

**Figure 4.27** Uneven row lengths in standard-cell design.

$$\gamma_2 = w_2 \sum_{i \neq j} [O_{ij}]^2 \qquad (4.15)$$

In the above equation $w_2$ is the weight for penalty. The reason for squaring the overlap is to provide much larger penalties for larger overlaps.

Due to cell displacements and pairwise exchanges of cells, the length of a row may become larger or smaller (see Figure 4.27).

The third component of the cost function represents a penalty for the length of a row $R$ exceeding (or falling short of) the expected length $\overline{L_R}$.

$$\gamma_3 = w_3 \sum_{rows} | L_R - \overline{L_R} | \qquad (4.16)$$

where $w_3$ is the weight of unevenness. Uneven distribution of row lengths results in wastage of chip area. There is also experimental evidence indicating a dependence of both the total wirelength and the routability of the chip on the evenness of distribution.

**Annealing schedule**: The cooling schedule is represented by

$$T_{i+1} = \alpha(T_i) \times T_i \qquad (4.17)$$

where $\alpha(T)$ is the cooling rate parameter which is determined experimentally. The annealing process is started at a very high initial temperature say $4 \times 10^6$. Initially, the temperature is reduced rapidly $[\alpha(T) \approx 0.8]$. In the medium range, the temperature is reduced slowly $[\alpha(T) \approx 0.95]$. Most processing is done in this range. In the low temperature range, the temperature is reduced rapidly again $[\alpha(T) \approx 0.8]$. The algorithm is terminated when $T < 1$.

**Inner loop criterion**: At each temperature, a fixed number of moves are attempted. The optimal number of moves depends on the size of the circuit. From experiments, for a 200-cell circuit, 100 moves per cell are recommended, which calls for the evaluation of $2.34 \times 10^6$ configurations in about 125 temperature steps. For a 3000-cell circuit, 700 moves per cell are recommended, which translates to a total of $247.5 \times 10^6$ attempts.

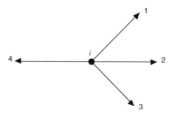

**Figure 4.28** Force on a cell $i$ connected to four other cells.

### 4.4.4 Numerical techniques

The placement problem can often be transformed into a numerical optimization problem. In this section, we describe a technique known as force-directed placement (Hanan and Kurtzberg, 1972). The placement problem is reduced to the problem of solving a set of simultaneous linear equations to determine equilibrium locations (ideal $x, y$ coordinates) for cells.

The basic idea behind the force-directed method is that cells connected by a net exert forces on one another. The magnitude of the force $F$ exerted by a cell $i$ on another cell $j$ is proportional to the distance separating them. This is analogous to Hooke's law in mechanics (force exerted on each other by two masses connected by a spring). If the spring constant is $k$ and the masses are separated by a distance $d$, the force with which the masses pull each other is $k \times d$. Suppose that a cell $a$ is connected to another cell $b$ by a net of weight $w_{ab}$. Let $d_{ab}$ represent the distance between $a$ and $b$. Then the force of attraction between the cells is proportional to the product $w_{ab} \times d_{ab}$. A cell $i$ connected to several cells $j$ at distances $d_{ij}$ by wires of weights $w_{ij}$, experiences a total force $F_i$ given by

$$F_i = \sum_j w_{ij} \cdot d_{ij} \tag{4.18}$$

Referring to Figure 4.28, the force $F_i$ on cell $i$ connected to four other cells is given by

$$F_i = w_{i1} \cdot d_{i1} + w_{i2} \cdot d_{i2} + w_{i3} \cdot d_{i3} + w_{i4} \cdot d_{i4} \tag{4.19}$$

If the cell $i$ in such a system is free to move, it would do so in the direction of force $F_i$ until the resultant force on it is zero. In the mechanics analogy, a free body which is connected by springs to other bodies will occupy a position such that the total tension from all springs is minimum. The location a cell would move to is called the *zero-force target location*. From Equation 4.18 above, we note that $F_i$ represents the total weighted length of wires that originate from module $i$. Thus, when all the cells move to their zero-force target locations, the sum of the square of distances is minimized. This is the principal idea behind the force-directed placement method.

The method consists of computing the forces on any given cell, and then moving it in the direction of the resulting force so as to put it in its zero-force target location. This location $(x_i{}^\circ, y_i{}^\circ)$ can be determined by equating the $x$- and $y$-components of the forces on the cell to zero, i.e.,

$$\sum_j w_{ij} \cdot (x_j^\circ - x_i^\circ) = 0 \qquad (4.20)$$

$$\sum_j w_{ij} \cdot (y_j^\circ - y_i^\circ) = 0 \qquad (4.21)$$

Solving for $x_i{}^\circ$ and $y_i{}^\circ$,

$$x_i{}^\circ = \frac{\sum_j w_{ij} \cdot x_j}{\sum_j w_{ij}} \qquad (4.22)$$

$$y_i{}^\circ = \frac{\sum_j w_{ij} \cdot y_j}{\sum_j w_{ij}} \qquad (4.23)$$

Care should be taken to avoid assigning more than one cell to the same location, or the trivial solution which assigns all the cells to the same $(x_i^\circ, y_i^\circ)$ location. Let us now illustrate the above idea on a small example.

**Example 4.8** A circuit with one gate and four I/O pads is given in Figure 4.29(a). The four pads are to be placed on the four corners of a $3 \times 3$ grid. If the weights of the wires connected to the gate of the circuit are $w_{vdd} = 8$; $w_{out} = 10$; $w_{in} = 3$; and $w_{gnd} = 3$; find the zero-force target location of the gate inside the grid.

SOLUTION  The zero-force location for the gate is given by

$$x_i{}^\circ = \frac{\sum_j w_{ij} \cdot x_j}{\sum_j w_{ij}} = \frac{w_{vdd} \cdot x_{vdd} + w_{out} \cdot x_{out} + w_{in} \cdot x_{in} + w_{gnd} \cdot x_{gnd}}{w_{vdd} + w_{out} + w_{in} + w_{gnd}}$$

$$= \frac{8 \times 0 + 10 \times 2 + 3 \times 0 + 3 \times 2}{8 + 10 + 3 + 3} = \frac{26}{24} = 1.083$$

$$y_i{}^\circ = \frac{\sum_j w_{ij} \cdot y_j}{\sum_j w_{ij}} = \frac{w_{vdd} \cdot y_{vdd} + w_{out} \cdot y_{out} + w_{in} \cdot y_{in} + w_{gnd} \cdot y_{gnd}}{w_{vdd} + w_{out} + w_{in} + w_{gnd}}$$

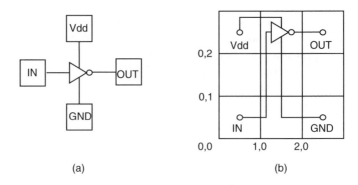

(a)  (b)

**Figure 4.29** (a) Circuit for Example 4.8. (b) Placement obtained.

$$= \frac{8 \times 2 + 10 \times 2 + 3 \times 0 + 3 \times 0}{8 + 10 + 3 + 3}$$
$$= \frac{36}{24} = 1.50$$

The zero-force location for the gate can be approximated to be at grid location (1,2). The final placement of pads and the gate is shown in Figure 4.29(b).

## Force-directed placement

An approach for determining the *ideal* locations of cells numerically was given above. This approach can be generalized into a constructive placement procedure as follows. Starting with some initial placement, a cell at a time is selected, and its zero-force location computed. The process can be iterated to improve on the solution obtained. The decisions to be made by such an algorithm include the order in which cells are selected and where the selected cell is to be put in case the zero-force target location is occupied.

The cell to be moved may be selected randomly or by using an heuristic technique. It seems logical to select the cell $i$ for which $F_i$ is maximum in the present configuration. If timing requirements are included, then those cells that lie on the critical path may be considered first.

If the zero-force location is occupied by another cell $q$, then several options to place the cell $p$ under consideration exist. They include the following:

1. Move $p$ to a free location close to $q$.
2. Evaluate the change in cost if $p$ is swapped with $q$. If the result is a decrease in cost, only then is the swap made. It is necessary to calculate the change in cost because it may so happen that the cell $q$ was already in its zero-force location.
3. 'Ripple move': The cell $p$ is placed in the occupied location, and a new zero-force location is computed for the displaced cell $q$. The procedure is continued until all the cells are placed.
4. 'Chain move': The cell $p$ is placed in the computed location and the cell $q$ is moved to an adjacent location. If the adjacent location is occupied, by another cell $r$, then $r$ is moved to its adjacent location, and so on, until a free location is finally found.
5. An alternative way to circumvent this problem is to compute the zero-force locations for all the cells. Search for cell pairs of the form $(p, q)$ such that the zero-force location of $p$ is the present position of $q$ and vice versa. Swap the cells $p$ and $q$; both the cells have found their zero-force locations.

There are several versions of the force-directed placement technique. Other variations for such algorithms are 'force-directed relaxation' and 'force-directed pairwise relaxation' (Preas and Lorenzetti, 1988). We now present one version of the algorithm that uses ripple moves.

## Force-directed algorithm

The force-directed algorithm shown in Figure 4.30 uses ripple moves. It is an iterative improvement algorithm. The formulation is based on the force-directed algorithm given in a recent survey paper by Shahookar and Mazumder (1991). The implementation uses two flags for each location of the placement surface. An OCCUPIED flag indicates if a cell is presently assigned to the location. A LOCKED flag indicates the status of the cell presently assigned to the location. (The LOCKED flag is *off* if the OCCUPIED flag is *off*.) If the cell which occupies the location has been displaced at least once, then the LOCKED flag for the location is *on* to prevent that cell from being displaced again.

In this algorithm a cell at a time is selected in order of connectivity and its zero-force target location is computed. Selection of cells is based on total connectivity and each selected cell is called seed cell. The implementation uses ripple moves where the selected cell is moved to the target location. If the target point is already *occupied* then the cell that previously occupied the computed location is selected to be moved next. To avoid infinite loops, any cell that is moved to its target location is *locked* for the current value of *iteration_count*. Infinite loops can occur if two cells $C_a$ and $C_b$ are competing

**Algorithm** *ForcedirectedPlacement*;
Compute total connectivity of each cell;
Order the cells in decreasing order of their connectivities and store them in a list L;
**While** (*iteration_count* < *iteration_limit*)
    Seed = next module from L;
    Declare the position of the seed *vacant*;
    **While** *end_ripple* = false
        Compute target point of the seed and round off to nearest integer;
        **Case** target point is
        *VACANT:*
            Move seed to target point and lock;
            *end_ripple* $\leftarrow$ *true*;
            *abort_count* $\leftarrow$ 0;
        *SAME AS PRESENT LOCATION:*
            *end_ripple* $\leftarrow$ *true*;
            *abort_count* $\leftarrow$ 0;
        *LOCKED* :
            Move selected cell to nearest vacant location;
            *end_ripple* $\leftarrow$ *true*;
            *abort_count* $\leftarrow$ *abort_count* + 1;
            **If** *abort_count* > *abort_limit* **Then**
                Unlock all cell locations;
                *iteration_count* $\leftarrow$ *iteration_count* + 1;
            **EndIf**;
        *OCCUPIED* : (*and *not* locked*)
            Select cell as target point for next move;
            Move seed cell to target point and lock the target point;
            *end_ripple* $\leftarrow$ *false*;
            *abort_count* $\leftarrow$ 0;
        **EndCase**;
        **EndWhile**;
    **EndWhile**;
**End**.

**Figure 4.30** Force-directed placement algorithm with Ripple Moves.

for the same target location. Once a seed cell is selected and its zero-force target location is computed, four cases are possible: the computed target location could be (1) the *same* as the initial location of the seed, (2) another *vacant* location, (3) a location that is *occupied* (but not locked), or (4) a location that is occupied and *locked*.

The inner **While** loop of the algorithm in Figure 4.30 is executed while *end_ripple* is *false*. If the computed location is the *same* as the present location or if it is another *vacant* location, then the flag *end_ripple* is set to *true*, *abort_count* is set to zero and the cell occupies the computed location. The next seed cell in the order of connectivity is selected and the inner loop continues.

If the computed location is *occupied* and not locked, then the cell is moved to the computed location, and the cell that occupied the location is

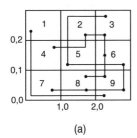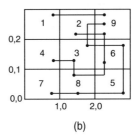

(a)                                      (b)

**Figure 4.31** Placement of Example 4.9. (a) Initial placement, wirelength estimate using chain connection = 16. (b) Final placement, wirelength estimate using chain connection = 14.

selected as the cell to be moved next; *end_ripple* is set to *false* and *abort_count* is set to zero.

In case the target point is occupied and *locked*, then the flag *end_ripple* is set to *true*, *abort_count* is incremented by one, and the cell is moved to the nearest vacant location. In this case, if *abort_count* is less than *abort_limit* then the next seed cell is selected and the locked locations remain as they are and the same iteration continues. However, if *abort_count* is greater than a certain preset value called *abort_limit* then all locked locations are unlocked, another seed cell is selected, the iteration count is incremented and a new iteration is started.

The procedure of selecting seeds based on connectivity and trying to place them in their ideal locations is continued until *iteration_limit* is reached (Shahookar and Mazumder, 1991).

**Example 4.9** Consider a gate-array of size three rows and three columns. A circuit with 9 cells and 3 signal nets is to be placed on the gate-array using force-directed relaxation. The initial placement is shown in Figure 4.31(a). The modules are numbered $C_1, \cdots, C_9$ and the nets $N_1, N_2, N_3$ are shown below. Show the final placement and calculate the improvement in total wirelength achieved by the algorithm.

Nets

$N_1 = (C_3, C_5, C_6, C_7, C_8, C_9)$
$N_2 = (C_2, C_3, C_4, C_6, C_8, C_9)$
$N_3 = (C_1, C_9)$

SOLUTION    The connectivity matrix for the given netlist and the total connectivity of cells is shown in Table 4.3. We will use the algorithm of Figure 4.30 to solve the problem. We select *abort_limit* = 3 and *iteration_count* = 2. Since $C_9$ has the largest connectivity, we select it as the seed cell in the first iteration. The target location for $C_9$ is

| Cells | 1 | 2 | 3 | 4 | 5 | 6 | 7 | 8 | 9 | $\sum$ |
|-------|---|---|---|---|---|---|---|---|---|--------|
| 1 | 0 | 0 | 0 | 0 | 0 | 0 | 0 | 0 | 1 | 1 |
| 2 | 0 | 0 | 1 | 1 | 0 | 1 | 0 | 1 | 1 | 5 |
| 3 | 0 | 1 | 0 | 1 | 1 | 2 | 1 | 2 | 2 | 10 |
| 4 | 0 | 1 | 1 | 0 | 0 | 1 | 0 | 1 | 1 | 5 |
| 5 | 0 | 0 | 1 | 0 | 0 | 1 | 1 | 1 | 1 | 5 |
| 6 | 0 | 1 | 2 | 1 | 1 | 0 | 1 | 2 | 2 | 10 |
| 7 | 0 | 0 | 1 | 0 | 1 | 1 | 0 | 1 | 1 | 5 |
| 8 | 0 | 1 | 2 | 1 | 1 | 2 | 1 | 0 | 2 | 10 |
| 9 | 1 | 1 | 2 | 1 | 1 | 2 | 1 | 2 | 0 | 11 |

**Table 4.3  Connectivity matrix for Example 4.9.**

calculated using the method illustrated in Example 4.8. The reader may verify that this target location is (1.1,1.1), which we round off to (1,1). Location (1,1) is locked.

Since (1,1) is occupied by cell $C_5$; $C_5$ is chosen as the next seed cell. Zero-force location for $C_5$ is (1.2,0.8), or (1,1) after rounding. However, this location is occupied by cell $C_9$ which is locked (refer to CASE:LOCKED in Figure 4.30). Therefore, $C_5$ must be moved to nearest vacant location. Since we have nine cells and nine locations, the nearest empty location is (2,0) vacated by moving cell $C_9$ to its zero-force location. Therefore, it is moved to (2,0) and *abort_count* is incremented. Since *abort_count* is less than *abort_limit* the same iteration is continued by selecting the next cell in order of connectivity.

$C_3$ is chosen as the next cell and its zero-force location is computed. The target point for $C_3$ is (1.1,0.7) which is rounded off to (1,1). Since location (1,1) is locked, $C_3$ is placed in the nearest vacant location that is (2,2) and *abort_count* is incremented to 2. The next cell selected in order of connectivity is $C_6$ and its zero-force location is computed and is found to be (1.1,0.9), rounded off to (1,1), which is a locked location. The cell is not moved and *abort_count* is incremented to 3, which is the *abort_limit*. This marks the end of the first iteration. All locations are unlocked and the next iteration is begun.

In the second iteration, the same procedure is repeated with this new placement as the initial placement. $C_9$ has the highest connectivity and is selected as the first seed cell. The computed target location for this cell is (1.2,1.0) which corresponds to location (1,1). Since this is the original location of the cell, it is not moved. The next cell chosen is cell $C_3$. Its computed target location is (1,1). Cell $C_3$ is moved to location (1,1) and the location is locked. The next cell is the one that occupied the location (1,1) before $C_3$ was assigned to it, and that is $C_9$. Therefore $C_9$ is chosen for computation of its target location and the computed location is (1,1), a locked location. The cell is moved to the nearest

| Iteration | Selected cell | Target point | Case | Placed at | Result |
|---|---|---|---|---|---|
| | 9 (Seed) | (1,1,1.1)=(1,1) | Occupied | (1,1) | 1 2 3<br>4 9 6<br>7 8 - |
| | 5 | (1.2,0.8)=(1,1) | Locked<br>abort_count = 1 | (2,0) | 1 2 3<br>4 9 6<br>7 8 5 |
| 1 | 3 (Seed) | (1.1,0.7)=(1,1) | Locked<br>abort_count = 2 | (2,2) | 1 2 3<br>4 9 6<br>7 8 5 |
| | 6 (Seed) | (1.1,0.9)=(1,1) | Locked<br>abort_count = 3 | (2,1) | 1 2 3<br>4 9 6<br>7 8 5 |
| | 9 (Seed) | (1.2,1.0)=(1,1) | Same | (1,1) | 1 2 3<br>4 9 6<br>7 8 5 |
| | 3 (Seed) | (1.1,0.7)=(1,1) | Occupied | (1,1) | 1 2 -<br>4 3 6<br>7 8 5 |
| 2 | 9 | (1.0,0.8)=(1,1) | Locked<br>abort_count = 1 | (2,2) | 1 2 9<br>4 3 6<br>7 8 5 |
| | 6 (Seed) | (1.1,0.9)=(1,1) | Locked<br>abort_count = 2 | (2,1) | 1 2 9<br>4 3 6<br>7 8 5 |
| | 8 (Seed) | (1.3,1.1)=(1,1) | Locked<br>abort_count = 3 | (1,0) | 1 2 9<br>4 3 6<br>7 8 5 |

**Table 4.4  First two iterations of force-directed placement of Example 4.9.**

vacant location (2,2), and abort_count is incremented to 1. The same situation results when cells $C_6$ and $C_8$, the next in the order of connectivity, are chosen. This increases the abort_count to 3 and marks the end of the second iteration.

The result of placement at the end of two iterations is shown in Figure 4.31(b). The wirelength computed before and after placement using the chain connection to estimate is 16 units and 14 units respectively. The results of the first two iterations are summarized in Table 4.4.

It should be mentioned that there is no systematic way of setting the parameters abort_count and abort_limit. These generally assume values that are increasing with the size of the problem.

# 4.5 OTHER APPROACHES AND RECENT WORK

## 4.5.1 Artificial neural networks

In the recent past, a paradigm known as *neural computing* has become popular for applications such as machine vision, robot control, and so on. Traditional computing methods have not been very successful in attacking these applications, despite the fact that today's computers have achieved speeds of hundreds of MIPS (Million Instructions Per Second). On the other hand, the human nervous system routinely solves problems such as pattern recognition, and natural language understanding. Artificial Intelligence (AI) techniques, which were predominantly the theme of fifth-generation computers, have been only partially successful in solving problems such as machine vision. Recently, there has been a revival of interest in neural computing and natural intelligence techniques. These techniques revolve around the concept of an *artificial neural network*, which is an ensemble of a large number of *artificial neurons*. One can think of an artificial neural network as the analogue of neural networks that are part of the human brain. It is believed by a large number of computer professionals that neural computing is the key to solving difficult problems like pattern recognition, computer vision, and hard optimization problems (Wasserman, 1989). Our interest in artificial neural networks in this chapter is in its application to the placement problem. We will focus our attention on a particular class of artificial neural networks introduced by Hopfield and Tank (1985).

The main component of an artificial neural network (ANN) is an artificial neuron. An artificial neuron receives several analogue inputs $X_1, X_2, \cdots, X_n$ and generates a single analogue output $OUT$. The output is computed as follows. Each input is weighted down by the neuron; let $W_i$ be the weight associated with input $X_i$. The net input, denoted $NET$, is given by

$$NET = \sum_{i=1}^{n} W_i \cdot X_i \qquad (4.24)$$

The output is a function $F$ of $NET$ (Figure 4.32). The function $F$ is also known as the activation function of the neuron. A popularly used activation function is the sigmoid function $F(x) = 1/(1 + e^{-x})$. If $x$ is a sufficiently large positive number, the sigmoid function approximates to unity. For sufficiently large negative values of $x$, the sigmoid function is close to 0. Another popular activation function is $F(x) = tanh(x)$. Several artificial neurons can be connected to form an artificial neural network. For

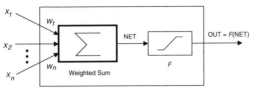

**Figure 4.32** An artificial neuron.

example, a single layer *feed-forward* network consists of $m$ neurons, each with $n$ inputs. The principal inputs to the network are denoted $X_1, X_2, \cdots, X_n$. The weights associated with neuron $i$ are denoted $W_{i1}, W_{i2}, \cdots, W_{in}$. The $m \cdot n$ weights of the network can be compactly represented by the $m \times n$ weight matrix $W = [W_{ij}]$. Figure 4.33(a) shows a feed-forward network with three neurons, each with three inputs. The output of neuron $i$ is denoted by $OUT_i$. A single layer *recurrent* network is similar to a feed-forward network, except that the outputs are fed back as inputs to the network. Figure 4.33(b) shows a recurrent network with three neurons, each with three feedback inputs and one external input. Hopfield and Tank (1985) used recurrent neural networks to solve optimization problems.

## Energy function and stability

Just as temperature plays an important role in simulated annealing energy plays an important role in Hopfield's neural networks. The set of all outputs $OUT_i$ is known as the state of the network. Suppose that the activation function of each neuron in the network is a *threshold* function, i.e.,

$$OUT_i = \begin{cases} 1 & \text{if } NET_i > T_i \\ 0 & \text{if } NET_i < T_i \\ unchanged & \text{if } NET_i = T_i \end{cases} \qquad (4.25)$$

where $T_i$ is the threshold level of neuron $i$. Since we are dealing with a recurrent network, $NET_i$ is given by

$$NET_i = (\sum_{j \neq i} W_{ij} \cdot OUT_i) + IN_i \qquad (4.26)$$

It is clear that the network can be in $2^n$ different states, since each of the $n$ neurons can output either 0 or 1. Each state is associated with an energy level. When the network changes state, there is a change in its energy level. It is known that the network will settle down to a state with minimal energy level if the weight matrix $W$ is a symmetric matrix and all the diagonal

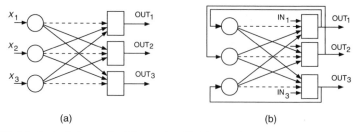

(a)                                    (b)

**Figure 4.33** (a) A single-layer, feed-forward artificial neural network with three neurons. (b) A single-layer, recurrent artificial neural network with three neurons.

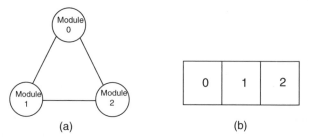

(a)                                    (b)

**Figure 4.34** (a) Circuit for Example 4.10. (b) Position definitions.

entries of the matrix are 0. The network is said to *converge* to the state of minimal energy. By constructing a neural network whose energy function is the objective function of a minimization problem, one can hope to solve the minimization problem.

**Example 4.10** Consider how an artificial neural network can be set up to solve the simplest case of the placement problem. Given $n$ circuit modules and a connectivity matrix $C = [C_{ij}]$, where $C_{ij}$ denotes the connectivity between module $i$ and module $j$; the objective is to put $n$ interconnected objects into $n$ slots of a 2-D array, such that the total Manhattan interconnection length is minimized. We shall use the circuit shown in Figure 4.34(a) to illustrate this approach. The slots to which these modules are to be assigned are shown in Figure 4.34(b).

SOLUTION    The solution to this problem presented below is due to Yu (1989), who used Hopfield's neural nets to solve the placement problem. To solve this problem a network with $n^2$ neurons is set up. This network consists of an $n \times n$ matrix of neurons as seen in Figure 4.35 (a 2-D array $\aleph$). Neurons are numbered from 0 to $n^2 - 1$, left to right, and top to bottom. The value of element $\aleph_{i,j}$ represents the 'chance' of module '$i$' being positioned at location '$j$'. Each row corresponds to a circuit module. The $n$ columns correspond to the $n$

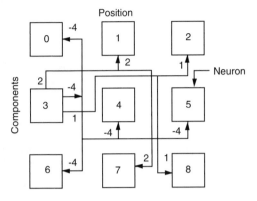

**Figure 4.35** An artificial neural network for placement.

| $T_{ij}$ | 0 | 1 | 2 | 3 | 4 | 5 | 6 | 7 | 8 |
|---|---|---|---|---|---|---|---|---|---|
| 0 | 0 | -4 | -4 | -4 |  | -4 |  |  |  |
| 1 | -4 | 0 | -4 | 2 | -4 |  | -4 |  |  |
| 2 | -4 | -4 | 0 | 1 |  | -4 |  |  | -4 |
| 3 | -4 |  |  | 0 | -4 | -4 | -4 |  |  |
| 4 |  | -4 |  | -4 | 0 | -4 |  | -4 |  |
| 5 |  |  | -4 | -4 | -4 | 0 |  |  | -4 |
| 6 | -4 |  |  | -4 |  |  | 0 | -4 | -4 |
| 7 |  | -4 | 0 | 2 | -4 |  | -4 | 0 | -4 |
| 8 |  |  | -4 | 1 |  | -4 | -4 | -4 | 0 |

**Table 4.5 Partial synapse strength matrix, offset = 3, inhibit = − 4.**

possible locations a circuit module can take. Therefore, in order to obtain a feasible solution, only one neuron in any row or any column can have its output 1. The output of the neuron is normalized and thus is always between 0 and 1.

The next step is to define the synapse (connection point) parity and strength. First the Manhattan distance between any two locations is computed. The value $T_{k_{i_1,j_1}, l_{i_2,j_2}}$, between neurons $k$ and $l$ defined as the connectivity between circuit modules $i_1$ and $i_2$ times $f(j_1,j_2)$, where $f$ is a function of the distance between locations $j_1$ and $j_2$, $k = i_1 \times \sqrt{n} + j_1$ and $l = i_2 \times \sqrt{n} + j_2$. After some experimentation $f$ was chosen to be ($offset$−Manhattan distance between $j_1$ and $j_2$), where the offset parameter is usually greater than $\sqrt{n}$.

As an example, the synapse strengths between neurons 2 and 3 ($T_{2,3}$) can be found as follows. Neuron 2 has $(i_1,j_1) = 0,2$; and neuron 3 has $(i_2,j_2) = 1,0$ (see Figure 4.35). Therefore $T_{2,3}$ by definition is equal to

$$C_{0,1} \times (offset - \text{Manhattan distance between 2 and 0})$$

$$= 1 \times (3 - 2) = 1$$

The partial synapse strength matrix is shown in Table 4.5 and the corresponding connections for neuron 3 are shown in Figure 4.35 (see Exercise 4.22).

In formalizing the above problem Yu (1989) modified the neural network solution to the TSP (Travelling Salesperson Problem) by Hopfield and Tank (1985). The energy function $E$ used by Hopfield has several minima, some of which are local minima; the network can converge to any one of them. As a result, there is no guarantee that the solution obtained will correspond to a global minimum. Moreover, how does one determine the parameters of the network (the weight matrix, thresholds, the constants involved in the energy function and the activation function)? How sensitive is the final solution to small variations in these parameters? How good is the final solution when compared to other known techniques for solving the same optimization problem? And finally, how fast does the network converge to the final solution? Since neural computing is still an active research area, the answers to these questions are still being investigated.

The impact of neural computing on VLSI CAD is still unclear. Yu's results on applying Hopfield neural networks to the placement problem were not promising. Some of the difficulties pointed out by him are long simulation times, poor solution quality and high sensitivity of the solution to network parameters. At this stage, it can only be concluded that more research is required in order to understand the applicability of neural networks to VLSI CAD problems.

### 4.5.2 Genetic algorithm

The genetic algorithm (GA) is a search technique which emulates the natural process of evolution as a means of progressing toward the optimum (Holland, 1975). It has been applied in solving various optimization problems including VLSI cell placement (Benten and Sait, 1994; Cohoon and Paris, 1987; Shahookar and Mazumder, 1990).

The algorithm starts with an initial set of random configurations called a *population*. Each individual in the population is a string of symbols. The symbols are known as *genes* and the string made up of genes is termed *chromosome*. The chromosome represents a solution to the optimization problem. A set of genes that make up a partial solution is called a *schema*. During each iteration (*generation*) the individuals in the current

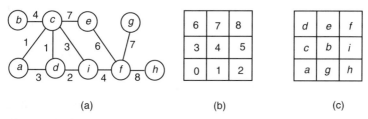

(a)                         (b)                         (c)

**Figure 4.36** (a) Graph of a circuit to be placed. (b) Position definition. (c) One possible placement.

population are evaluated using some measure of fitness. Based on the *fitness* value, two individuals at a time (called *parents*) are selected from the population. The fitter individuals have a higher probability of being selected. Then, a number of genetic operators are applied on the selected parents to generate new individual solutions called *offsprings*. These genetic operators combine the features of both parents. Common operators are *crossover*, *mutation*, and *inversion*. They are derived by analogy from the biological process of evolution. First we shall see how a solution can be mapped to a string. Then we will discuss the basic operators and the general structure of a GA.

**Example 4.11**    Consider the graph of Figure 4.36(a). The nine vertices represent modules and the numbers on the edges represent their weighted interconnection. Give a possible solution and express it as a string of symbols. Generate a population of four chromosomes and compute their fitness using the reciprocal of weighted Manhattan distance as a measure of fitness.

SOLUTION    The nine modules can be placed in the nine slots as shown in Figure 4.36(b). One possible solution is shown in Figure 4.36(c). Let us use a string to represent the solution as follows. Let the leftmost index of the string of the solution correspond to position '0' of Figure 4.36(b) and the rightmost position to location 8. Then the solution of Figure 4.36(c) can be represented by the string [aghcbidef] $(\frac{1}{85})$. The number in parenthesis represents the fitness value which is the reciprocal of the weighted wirelength based on the Manhattan measure.

If the lower left corner of the grid in Figure 4.36(b) is treated as the origin, then it is easy to compute the Cartesian locations of any module. For example the index of module $i$ is 5. Its Cartesian coordinates are given by $x = (5 \bmod 3) = 2$, and $y = \lceil \frac{5}{3} \rceil = 1$.

Any string (of length 9) containing characters $[a, b, c, d, e, f, g, h, i]$ represents a possible solution. There are 9! solutions equal to the number of permutations of length 9. Other possible solutions (chromosomes) are [bdefigcha] $(\frac{1}{110})$, [ihagbfced] $(\frac{1}{95})$, and [bidefaghc] $(\frac{1}{86})$.

For the above four solutions the fitness can be computed by calculating the Cartesian coordinates of the locations of the cells from their positions in the string. Using the interconnection weight matrix, the sum of the weighted Manhattan distances between pairs of connected components is found.

Next, we look at the genetic operators and their significance.

## Crossover

*Crossover* is the main genetic operator. It operates on two individuals and generates an offspring. It is an inheritance mechanism where the offspring inherits some of the characteristics of the parents. The operation consists of choosing a random cut point and generating the offspring by combining the segment of one parent to the left of the cut point with the segment of the other parent to the right of the cut.

**Example 4.12**   From our previous example (Example 4.11) consider the two parents [bidef|aghc] $(\frac{1}{86})$, and [bdefi|gcha] $(\frac{1}{110})$. If the cut point is randomly chosen after position 4, then the offspring produced is [bidefgcha]. Verify that the weighted wirelength of the offspring is reduced to 63 and therefore the fitness of the offspring is $\frac{1}{63}$.

In the example above the elements to the left of crossover point in one parent did not appear on the right of the second parent. If they had, then some of the symbols in the solution string would be repeated. In cell placement this does not represent a feasible solution. Modifications to the above crossover operations to avoid repetition of symbols are: (a) order crossover, (b) partially mapped crossover (PMX), and (c) cycle crossover (Shahookar and Mazumder, 1991). Here we will explain the operation of the PMX technique.

The PMX crossover is implemented as follows: select two parents (say 1 and 2) and choose a random cut point. As before, the entire right substring of parent 2 is copied to the offspring. Next, the left substring of parent 1 is scanned from the left, gene by gene, to the point of the cut. If a gene does not exist in the offspring then it is copied to the offspring. However, if it already exists in the offspring, then its position in parent 2 is

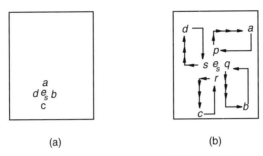

(a)                                  (b)

**Figure 4.37** (a) A random module and its neighbours. (b) The neighbours in (a) of parent 1 replace neighbouring modules in parent 2.

determined and the gene from parent 1 in the determined position is copied.

**Example 4.13**   As an example of PMX crossover consider the two parents [bidef|gcha] ($\frac{1}{86}$), and [aghcb|idef] ($\frac{1}{85}$). Let the crossover position be after 4. Then the offspring due to PMX is [bgcha|idef].

Observe that the right substring in parent 2, which is $idef$, is completely copied into the offspring. Then scanning the first parent from the left, since gene $b$ (position 0) is not in the offspring it is copied to position zero. The next gene, $i$ (position 1) exists in the offspring at position 5. The gene in position 5 in parent 1 is $g$, and this does not exist in the offspring, therefore gene $g$ is copied to offspring in position 1.

Finally we conclude with two crossover operations used in *Genie*, a genetic placement system for placing modules on a rectangular grid (Cohoon and Paris, 1987). The first crossover operator selects a random module $e_s$ and brings its four neighbours in parent 1 to the location of the corresponding neighbouring slots in parent 2. Then the modules that earlier occupied the neighbouring locations in parent 2 are shifted outwards one location at a time in a chain move in the direction of the old locations of the brought modules until a vacant location is found. This is shown in Figure 4.37. The result obviously is that a patch containing $e_s$ and its four neighbours is copied from parent 1 to parent 2 and that other modules are shifted by at most one position. The second crossover operator selects a square consisting of $k \times k$ modules from parent 1 and copies it to parent 2. The random number $k$ used has a mean 3 and variance 1. This method tends to duplicate some modules and leave out others. For example, referring to Figure 4.38, if modules in the square of parent 1 are copied to the square of parent 2, then those modules in the square of parent 1 and not in the square of parent 2 are duplicated. This problem is overcome as

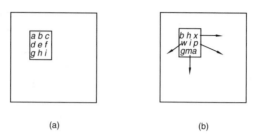

(a)                              (b)

**Figure 4.38** (a) A square is selected in parent 1. (b) Modules of square in parent 1 are copied to parent 2 and duplicate modules are moved out.

follows: let $SP_2 - SP_1$ be the set of modules in square of parent 2 but not in square of parent 1 ($SP_2 - SP_1 = \{x, w, p, m\}$). Similarly $SP_1 - SP_2$ is the set of modules in square of parent 1 but not in square of parent 2 ($SP_1 - SP_2 = \{c, d, e, f\}$). Each module in $SP_2 - SP_1$ is moved to a slot currently occupied by a module in $SP_1 - SP_2$. That is, in parent 2, modules $\{x, w, p, m\}$ are moved to slots currently occupied by modules $\{c, d, e, f\}$. Then all the modules in the square of parent 1 are copied into the square of parent 2 to yield a new offspring.

## Mutation and inversion

*Mutation* produces incremental random changes in the offspring generated by the crossover. In placement, the commonly used mutation mechanism is pairwise interchange. In terms of placement, a gene consisting of an ordered triple of a cell and its associated ideal coordinates may not be present in any of the individuals in the population. In that case crossover alone will not help because it is only an inheritance mechanism. The mutation operator generates new cell-coordinate triples. If the new triples perform well, then the configurations containing them are retained and these triples spread throughout the population. The mutation is controlled by a parameter referred to as the mutation rate $M_r$. A low ($M_r$) means that the infusion of new genes is very low. A high $M_r$ will cause the offsprings to lose resemblance to their parents, which mean the algorithm behaves like a memory-less process thus losing its ability to learn from the history of the search (Shahookar and Mazumder, 1991).

In the *inversion* operation, two points are randomly chosen along the length of the chromosome, the chromosome is cut, and the end points of the cut section switch places. For example, the string [bid|efgch|a] (cut after positions 2 and 7) will become [bid|hcgfe|a] after inversion. This operation is performed in such a way that it does not modify the solution represented by the chromosome, it modifies only the representation. Thus the symbols

**Algorithm** *Genetic_Algorithm*;
(*$N_p$ = Population Size*)
(*$N_g$ = Number of Generations*)
(*$N_o$ = Number of Offsprings*)
(*$P_i$ = Inversion Probability*)
(*$P_\mu$ = Mutation Probability*)
**Begin**
  (*Randomly generate the Initial Population*)
  Construct_Population($N_p$);
  **For** $j$ = 1 to $N_p$
  Evaluate *Fitness*(Population[$N_p$])
  **EndFor**;
  **For** $i$ = 1 to $N_g$
    **For** $j$ = 1 to $N_o$
    (*Choose parents with probability proportional to fitness value*)
    $(x, y)$ ← *Choose_parents*;
    (*Perform crosssover to generate offsprings*)
    offspring[$j$] ← *Generate_offspring(x,y)*;
    **For** $k$ = 1 to $N_p$
      With probability $P_\mu$ Apply *Mutation* (Population[$k$])
    **EndFor**;
    **For** $k$ = 1 to $N_p$
      With probability $P_i$ Apply *Inversion*(Population[$k$])
    **EndFor**;
    Evaluate *Fitness*(offspring[$j$])
    **EndFor**;
    Population← *Select*(Population, offspring, $N_p$)
  **EndFor**;
  Return highest scoring configuration in Population
**End.**

**Figure 4.39** General structure of a genetic algorithm.

of the string must have interpretation independent of their position (Goldberg, 1989; Shahookar and Mazumder, 1991).

The general structure of a genetic algorithm is given in Figure 4.39. The algorithm starts with an initial set of random configurations called a *population*. The size of the population is always fixed ($N_p$). Following this, a mating pool is established in which pairs of individuals from the population are chosen. The probability of choosing a particular individual for mating is proportional to its fitness value. Individuals with higher fitness values have a greater chance of being selected for mating. $N_o$ new offsprings are generated by applying crossover. Mutation and inversion are also applied with a low probability. Next the offsprings generated are evaluated on the basis of fitness, and a new generation is formed by selecting some of the parents and some of the offsprings. The above procedure is executed $N_g$ times, where $N_g$ is the number of generations.

After a fixed number of generations ($N_g$) the most fit individual, that is, the one with highest fitness value is returned as the desired solution.

Thus far we have seen operations on possible complete solutions. A schema is a set of genes that make up a partial solution. From our previous example (Example 4.1) the string [*i*efh*g*] indicates a subplacement, cells in positions 0, 2, 6 and 8 are 'don't cares'. Note that since cell $f$ (see Figure 4.36) is densely connected to cells $i, e, g$ and $h$, and that these cells are adjacent to cell $f$ in our solution, this schema represents a good subplacement. Similarly another good subplacement may consist of a cell at the input end of the network and a cell at the output end that are currently placed at opposite ends of the chip. Both these subplacements will improve the fitness of the individual that inherits them. Thus a schema is a topological placement of the corresponding cells, giving their relative positions rather than their detailed locations.

The genetic operators create a new generation of configurations by combining the schemata or subplacements of parents selected from the current generation. Due to the stochastic selection process, the fitter parents, which are expected to contain some good subplacements, are likely to produce more offsprings, and the bad parents, which contain some bad subplacements, are likely to produce less offsprings. Thus in the next generation, the number of good subplacements (or high fitness schemata) tends to increase, and the bad subplacements (or low fitness schemata) tend to decrease. Thus the fitness of the entire population improves. This is the basic mechanism of optimization by the genetic algorithm (Shahookar and Mazumder, 1991).

### 4.5.3 Other attempts

In GA, at any given instance a large number of possible solutions exist $(N_p)$. This is known as the size of the population. The choice of $N_p$ depends on the problem instance, size of the problem, and the available memory. Another technique that is similar to GA but operates on a *single* configuration at a time is known as ESP or *'evolution based placement'*. In this technique the modules in the placement are treated as the population. The evaluation function determines the goodness of placement of each module, that is, the individual contribution of each module to the cost of the solution. The technique was proposed by Kling and Banerjee (1987).

A multiprocessor-based simulated annealing algorithm for standard-cell placement was presented by Kravitz and Rutenbar (1987), and heuristics for parallel standard-cell placement with quality equivalent to simulated annealing were proposed by Rose *et al.*, (1988). A parallel implementation of the simulated annealing algorithm for the placement of macro-cells was reported by Casotto *et al.*, (1987).

The general force-directed placement technique was presented in this chapter. Another unique force-directed placement algorithm not discussed in this chapter but worth mentioning was proposed by Goto (1981) and Goto and Matsuda (1986). The algorithm consists of an initial placement phase and an iterative improvement phase. The initial placement phase selects modules for placement based on connectivity. The iterative improvement phase uses generalized force-directed relaxation technique in which interchange of *two or more* modules in a certain neighbourhood is explored.

The need for performance-driven placement arose as early as 1984 (Dunlop *et al.*, 1984). Since then a large number of approaches to timing-driven placement have been reported. The reported approaches fall into three general classes. The first class transforms the timing constraints on the critical paths (or sometimes all the paths) into *weights on nets*. These weights are used to categorize the nets and influence the placement procedure (Burstein and Youssef, 1985; Dunlop *et al.*, 1984). The second class transforms the path timing constraints into *timing bounds on the nets*. The net timing bounds are converted into length bounds and supplied to the placement program which tries to satisfy them (Nair *et al.*, 1989; Youssef *et al.*, 1992). The third approach consists of supplying to the placement procedure a set of the critical paths, together with their timing requirements. These paths are monitored during the placement process (Sutanthavibul *et al.*, 1993; Srinivasan *et al.*, 1992; Marek-Sadowska and Lin, 1989).

Placement is a multiobjective problem where many of the decisions that are made during the quest for a solution are qualitative and do not have clear yes/no answers. A suitable approach to such class of problems is to adopt a fuzzy logic formulation (Zimmermann, 1987). A description of the placement problem that relies on fuzzy logic has recently been reported (Rung-Bin Lin and Shragowitz, 1992).

A very good overview of placement techniques and a detailed bibliography is available in an excellent survey paper by Shahookar and Mazumder (1991), and in Chapter 4 of the book edited by Preas and Lorenzetti (1988).

## 4.6 CONCLUSION

In this chapter, we discussed a major VLSI design automation subproblem, namely placement. Wirelength is one of the most commonly optimized objective functions in placement. The actual wirelength is not known until the circuit modules are placed and routed. Different techniques that are commonly used to estimate the wirelength of a given placement were presented. Other commonly used cost functions, which are minimization of maximum cut, and minimization of maximum density were studied. A large number of good heuristics are available for solving the placement problem (Shahookar and Mazumder, 1991). A survey on commercially available automatic layout software in 1984 indicated that force-directed placement was the algorithm of choice (Robson, 1984; Hanan and Kurtzberg, 1972). A recent survey indicated an even mix of the use of min-cut based placement, force-directed algorithms, and simulated annealing (VLSI Systems Design Staff, 1987, 1988). The above three methods have been discussed in detail in this chapter.

When partitioning a circuit for placement, it is not sufficient to consider only the internal nets which intersect the cutline. Nets connecting external terminals or other modules in another partition must also be considered. To incorporate this, a method called terminal propagation was proposed by Dunlop and Kernighan. This extension considerably improves the quality of final solution. This technique was also presented with the help of an example (Dunlop and Kernighan, 1985).

Simulated annealing is currently the most popular technique in terms of placement quality, but it takes an excessive amount of computation time. The general algorithm was discussed in detail in Chapter 2. In this chapter the application of the algorithm to solve the placement was presented with the help of an example. We also discussed the TimberWolf3.2 package which uses simulated annealing for module placement (Sechen and Sangiovanni-Vincentelli, 1986).

Force-directed algorithms operate on the physical analogy of masses connected by springs, where the system tends to rest in its minimum energy state with minimum combined tension from all the springs. The basic algorithm was illustrated with an example.

Other recent attempts to solve the placement problem include use of neural networks and genetic algorithm. These approaches were presented in Section 4.5.

## 4.7 BIBLIOGRAPHIC NOTES

The three techniques popularly used for circuit placement are discussed in this chapter. They are, min-cut placement (Breuer, 1977a,b), simulated annealing approach (Kirkpatrick *et al.*, 1983; Sechen and Sangiovanni-Vincentelli, 1986), and force-directed techniques (Shahookar and Mazumder, 1990; Hanan and Kurtzberg, 1972). In this chapter we discussed approaches that ignore the dimensions of cells. In many popular methodologies, such as standard-cells and gate-array based design, these assumptions are valid and do not affect the quality of solution. In general-cell approach or macro-cell based design systems, the variation in sizes of different modules in a circuit is large, and the above assumptions will yield poor results. The problem is then approached differently. Lauther (1979) applied the min-cut placement procedure to the placement of general cells. In his method, blocks generated by the partitioning process are represented by a polar graph which retains information about relative positions of the blocks. His implementation also includes an improvement phase in which module rotation, mirroring and other compression techniques are used.

The general algorithm for simulated annealing was presented in Chapter 2. The application of this approach to the placement problem was presented in this chapter. A class of adaptive heuristics for combinatorial optimization and their application to solve combinatorial optimization problems was presented by Nahar et al. (1985, 1989). These authors also experimented with simulated annealing using different probablistic acceptance functions and cooling schedules.

## EXERCISES

**Exercise 4.1**  Consider the two-dimensional placement of $m \times n$ functional cells as an array of size $m \times n$. All the cells have identical shape and size. Let the cells be numbered $1, 2, \cdots, m \times n$. Cell $i$ is placed in row $r_i$ and column $c_i$ of the array.

1. Given two cells $i$ and $j$, derive an expression for the Manhattan distance $d_{ij}$ between the two cells in terms of $r_i$, $c_i$, $r_j$, and $c_j$.
2. Let $w_{ij}$ be the number of wires between cells $i$ and $j$. Derive an expression for the total wirelength of the placement.
3. Verify your formula for Figure 4.2(b).

**Exercise 4.2**  Given $n$ modules to be placed in a single row show that there are $\frac{n!}{2}$ unique placements of $n$ modules. When $n$ is large, show that the number of placements is exponential in $n$.

**Exercise 4.3**  A 5-point net is shown in Figure 4.40. The cost of connecting one point to another is measured by the Manhattan distance between the two points. Construct a minimum spanning tree to join all the five points of the net. What is the cost of the minimum spanning tree?

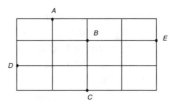

**Figure 4.40** Net of Exercise 4.3.

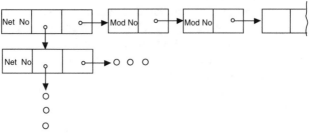

**Figure 4.41** Netlist data structure for Exercise 4.6.

**Exercise 4.4**  Repeat Exercise 4.3 using the minimum chain method to realize the 5-point net. Compare the resulting cost with that of the minimum spanning tree of Exercise 4.3.

**Exercise 4.5  Programming Exercise:** Write a procedure *CALC-LEN* to evaluate the total wirelength of a given placement. The inputs to the procedure are:

1. the number of modules $n$,
2. the connectivity matrix $C$; $C$ is an $n \times n$ matrix of non-negative integers, where $c_{ij}$ indicates the number of connections between modules $i$ and $j$,
3. the placement information. The placement surface is a uniform grid of dimensions $M \times N$. The array $P[1 \cdots M, 1 \cdots N]$ is used to represent the placement information. $P[i, j]$ contains the number of the module placed in row $i$ and column $j$.

You may assume that $M \cdot N = n$. What is the complexity of your procedure?

**Exercise 4.6  Programming Exercise:** Write a procedure *CALC-LEN2* to evaluate the wiring requirements of a placement. The inputs to the procedure are the same as in Exercise 4.5, except that the connectivity is represented by a *netlist* rather than a connectivity matrix. The number of nets is also an input to the procedure, and is

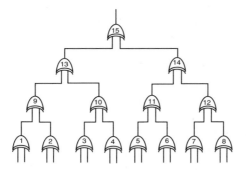

**Figure 4.42** A tree circuit with 15 EXOR gates.

indicated by $\eta$. For each net $k$, the netlist consists of a linked list of modules that are connected by the net $k$. The data structure for a netlist of three nets is shown in Figure 4.41. Net 1 connects modules 2, 1, 5. Net 2 connects modules 2, 3, 4, 5. Net 3 connects modules 1, 4, 5. What is the complexity of your procedure in terms of $n$ and $\eta$? What is the advantage of using a netlist representation rather than a matrix representation?

**Exercise 4.7** For min-cut partitioning, show that if $\Phi_P(c_{i+1}) \leq \Phi_P(c_i), 1 \leq i \leq r - 1$, then the application of the cut-lines in the following order, $c_1, c_2, \cdots, c_r$, leads to the minimization of $X(P)$ as defined by Equation 4.3.

**Exercise 4.8** Given $n$ modules to be placed using the min-cut placement technique. Alternate horizontal and vertical cutlines are to be used. The Kernighan–Lin algorithm of Chapter 2 is used at each stage to generate a two-way cut. What is the time requirement of the placement procedure? (Recall that the Kernighan–Lin algorithm takes $O(k^2 \log k)$ time to partition $k$ modules into two blocks.)

**Exercise 4.9** Apply the min-cut algorithm and place the circuit of Figure 4.1(a) on a $2 \times 4$ mesh. You may use the Kernighan–Lin algorithm for generating the partitions. Use alternate horizontal and vertical cutlines. What is the maximum wiring density of your solution?

**Exercise 4.10** Prove the result of Equation 4.5.

**Exercise 4.11 Programming Exercise:** Suppose that $n$ modules are placed in a row. The placement information is represented by array $p[1...n]$, where $p[i]$ indicates the module placed in the $i$th location. If the modules are numbered $1...n$, then $p$ is simply a permutation of $1...n$.

| 0000 | 0001 | 0100 | 0101 |
|------|------|------|------|
| 0010 | 0011 | 0110 | 0111 |
| 1000 | 1001 | 1100 | 1101 |
| 1010 | 1011 | 1110 | 1111 |

**Figure 4.43** Min-cut encoding.

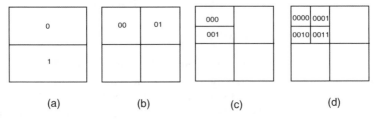

(a)   (b)   (c)   (d)

**Figure 4.44** Derivation of min-cut encoding. (a) $k = 1$. (b) $k = 2$. (c) $k = 3$. (d) $k = 4$.

Write a procedure $DELTA\text{-}LEN$ to compute the change in total wirelength when two modules in $p$ are swapped. Assume that the connectivity information is represented by a connectivity matrix $C$ as in Exercise 4.5.

**Exercise 4.12** (*) **Programming Exercise:** Implement the constructive placement algorithm of Figure 4.10. Use appropriate data structures for efficient implementation of 'select_cell' and 'select_slot' subroutines. Test your algorithm on the circuit of Figure 4.42.

**Exercise 4.13** The min-cut algorithm discussed in the chapter assumes that all the cells have identical shape and size. Extend the min-cut algorithm to handle unequal sized cells, i.e., outline a procedure which will place unequal sized cells by repeated two-way partitioning. Assume all cells are rectangular in shape. Apply your algorithm to the circuit of Figure 4.1(a). Assume that NAND gates are twice as large as NOR gates.

**Exercise 4.14** (*) **Programming Exercise:** Write a procedure $QMINCUT$ to implement the Quadrature min-cut algorithm. The inputs to the procedure are:

1. the number of modules $n$,
2. the connectivity matrix $C$ for the set of modules,

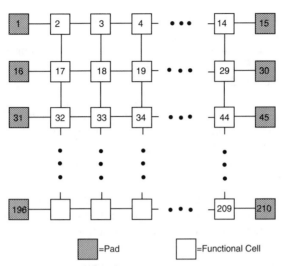

**Figure 4.45** 210 cell mesh for Exercise 4.16.

3. the type of cutline (1 = horizontal, 0 = vertical),
4. the level of bisection $k$.

The output of the program must be given as follows. For each module $i$, its placement must be encoded using a string of $\log_2 n$ bits. To illustrate, refer to Figure 4.43, where $n = 16$ and $\log_2 n = 4$. The position of the module placed in the top left corner is encoded as '0000'. Figure 4.44 shows how this encoding was obtained in four steps. (*Hint*: You may want to use recursion to code the *QMINCUT* procedure.)

**Exercise 4.15** Comment on the running time of the min-cut procedure for placement given in Figure 4.15.

**Exercise 4.16** (*) **Programming Exercise:** Implement a placement algorithm based on *simulated annealing*. Assume that there are 210 modules to be placed on a $15 \times 14$ mesh. There are two types of modules, functional blocks and I/O pads. The I/O pads must be placed only on the periphery of the mesh, whereas a functional block may be placed in any empty slot. Assume 28 I/O pads and 182 functional blocks. The connectivity of the modules is described in the format of a *netlist* (see Exercise 4.6). Note the following:

Generate a random initial placement which satisfies the pad position constraint. Use the following annealing schedule. $T_0 = 10.0$, $\alpha = 0.9$, $\beta = 1.0$, $T_f = 0.1$, $M = 200$. The *perturb* function must allow a circular shuffling of modules in $\lambda$ slots, where $\lambda$ is a user-specified constant. To test your program, you may set $\lambda = 2$ or $\lambda = 3$. The *perturb* function must respect the pad position constraint. Use the *DELTA-LEN* procedure of Exercise 4.11 to evaluate the change in cost function $\Delta h$.

1. Test your program for the sample circuit shown in Figure 4.45. In other words, synthesize the connectivity matrix for the circuit and give it as input to your program.
2. Run your program for several *random* initial placements. Does the initial solution influence the final solution?
3. Generate a 'good' initial solution for the circuit using either a constructive heuristic discussed in this chapter or one of your own heuristics. Does a good initial solution give better results than a random initial solution?
4. Study the influence of the $\lambda$ parameter on the quality of the final solution. Vary $\lambda$ in the range 2 to 5. Does the run time depend on $\lambda$?
5. In this book, we have been using the exponential function $e^{-\Delta h/T}$ in the acceptance criterion. Can you suggest an alternative function for this purpose? Experiment with your alternative and compare the results.

**Exercise 4.17** (*) **Programming Exercise:** Consider the annealing schedule used in the TimberWolf3.2 placement algorithm. The initial temperature is $T_1$, the mid-range temperature starts at $T_2$, the low-range temperature starts at $T_3$, and $T_4$ is the final temperature. The cooling rate in the high, middle and low temperature ranges are $\alpha_1$, $\alpha_2$, and $\alpha_3$ respectively. At each temperature, $M$ moves are made per cell. Calculate the number of moves attempted by the algorithm if there are $n$ cells in the circuit.

Compute the number of moves using your formula when $T_1 = 10^6$, $T_2 = 100$, $T_3 = 10$, $T_4 = 0.1$, $\alpha_1 = 0.8$, $\alpha_2 = 0.95$, $\alpha_3 = 0.8$, $M = 1000$, $n = 100$.

**Exercise 4.18** (*) **Gate-array placement:** Refer to the circuit shown in Figure 4.42. The I/O structure of a single EXOR gate is shown in Figure 4.46; here, $A$ and $B$ are inputs and $C$ is the output. Note that inputs and outputs are available on both sides of the module (e.g., the pin $A$ on the left-hand side is internally shorted to the pin $A$ on the right-hand side).

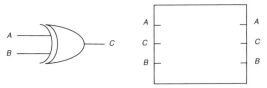

**Figure 4.46** The I/O interface of a two-input EXOR cell. The input $A$ may come from either side

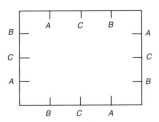

**Figure 4.47** A modified cell for two-input EXOR gate. Now the inputs $A$, $B$ and the output $C$ can come from all four sides.

The problem is to assign the EXOR gates to the logic blocks of a $4 \times 4$ gate-array. Show a gate assignment and sketch the wiring plan for each net. What is the required channel capacity (horizontal and vertical) of the gate-array? If the horizontal and vertical channel capacity of the gate-array is fixed to be 2, is it possible to implement the circuit?

**Exercise 4.19** In the previous problem, modify the I/O structure of the EXOR block as shown in Figure 4.47. Thus, inputs and output are available on all sides of the logic block. How does this I/O structure affect the requirement on channel capacities?

**Exercise 4.20** In Exercise 4.18, we ignored the pins VDD and GND. Modify the I/O structure of the EXOR gate to include the power and ground pins as well, and repeat Exercise 4.18.

**Exercise 4.21** (\*) **Standard-cell placement:** The description of a simple sequential circuit is shown in Figure 4.48. The circuit has four inputs, $G_0$, $G_1$, $G_2$, $G_3$, and a single output, $G_{17}$. It is relatively straightforward to interpret the circuit description. For example, the statement

$$G_{12} = \mathrm{NOR}(G_1, G_7)$$

means $G_{12}$ is the output of a NOR gate whose inputs are signals $G_1$ and $G_7$. The functional cells used in the circuit description are NOR, OR, NAND, AND, NOT, and DFF (D Flip-flop). The statement

INPUTS $(G_0, G_1, G_2, G_3)$
OUTPUT $(G_{17})$
$G_5 = DFF(G_{10})$
$G_6 = DFF(G_{11})$
$G_7 = DFF (G_{13})$
$G_{14} = NOT(G_0)$
$G_{17} = NOT (G_{11})$
$G_8 = AND(G_{14}, G_6)$
$G_{15} = OR(G_{12}, G_8)$
$G_{16} = OR(G_3, G_8)$
$G_9 = NAND(G_{16}, G_{15})$
$G_{10} = NOR (G_{14}, G_{11})$
$G_{11} = NOR(G_5, G_9)$
$G_{12} = NOR(G_1, G_7)$
$G_{13} = NOR(G_2, G_{12})$

**Figure 4.48** Netlist description of a sequential circuit for Exercise 4.21.

| Cell Name | Width | Functionality |
|-----------|-------|---------------|
| NOR | 3 units | 2-input NOR gate |
| OR | 4 units | 2-input OR gate |
| NAND | 3 units | 2-input NAND gate |
| AND | 4 units | 2-input AND gate |
| NOT | 2 units | Inverter |
| DFF | 9 units | D Flip-flop |

**Table 4.6 Table of cells and their dimensions for Exercise 4.21.**

$$G_5 = \text{DFF}(G_{10})$$

means $G_5$ is the $Q$ output of the D Flip-flop, whose $D$ input is $G_{10}$. Assume that a standard-cell library is available, and has all the required cells to implement the above circuit. All the cells are rectangular in shape, and their widths are summarized in Table 4.6. All the cells have the same height, namely, 7 units. The objective of this problem is to place the circuit using the standard-cells available in the library.

1. Which placement algorithms are suited for this problem? Justify your answer.

2. Represent the placement information using a one-dimensional array $p$. Apply the algorithm you selected above to generate a good one-dimensional placement. Use total wirelength as the objective function.

3. The one-dimensional placement will have an unacceptable aspect ratio, and must therefore be transformed into a two-dimensional placement which looks squarish. Propose a scheme for this transformation. Ensure that your transformation does not increase the wirelength achieved by the one-dimensional placement.

4. Give a rough estimate of the area of the circuit.

**Exercise 4.22** Complete Example 4.10 and obtain all the elements of the synapse matrix.

# REFERENCES

Aho, A. V., J. E. Hopcroft, and J. D. Ullman. *The Design and Analysis of Computer Algorithms.* Addison-Wesley, Reading, MA, 1986.

Benten, M .S. T. and Sadiq M. Sait. GAP: A genetic algorithm approach to optimize two-bit decoder PLAs. *International Journal of Electronics,* 76(1):99–106, January 1994.

Brayton, R., G. D. Hachtell and A. L. Sangiovanni-Vincentelli. A survey of optimization techniques of integrated circuit design. *Proceedings of IEEE,* 69(10):1334–1362, October 1981.

Breuer, M. A. A class of min-cut placement algorithms. *Proceedings of 14th Design Automation Conference,* pages 284–290, October 1977(a).

Breuer, M. A. Min-cut placement. *Journal of Design Automation and Fault Tolerant Computing,* 1(4):343–382, October 1977(b).

Burstein, M. and M. N. Youssef. Timing influenced layout design. *Proceedings of 22nd Design Automation Conference,* pages 124–130, 1985.

Casotto, A., F. Romeo and A. L. Sangiovanni-Vincentelli. A parallel simulated annealing algorithm for the placement of macro-cells. *IEEE Transactions on Computer Aided Design,* CAD-6(5):838–847, September 1987.

Cohoon, J. P. and W. D. Paris. Genetic placement. *IEEE Transactions on Computer Aided Design,* CAD-6:956–964, November 1987.

Darringer, J., D. Brand, J. V. Gerbi, W. H. Joyner Jr. and L. Trevillyan. LSS: A system for production logic synthesis. *IBM J. Res. Develop.,* 28(5):259–274, 1984.

Dunlop, A. E. and B. W. Kernighan. A procedure for placement of standard-cell VLSI circuits. *IEEE Transactions on Computer Aided Design,* CAD-4(1):92–98, January 1985.

Dunlop, A. E., V. D. Agraval and D. N. Deutsch. Chip layout optimization using critical path weighting. *Proceedings of 21st Design Automation Conference,* pages 133–136, 1984.

Goldberg, D. E. *Genetic Algorithms in Search, Optimization and Machine Learning.* Addison-Wesley Publishing Company, Inc., 1989.

Goto, S. An efficient algorithm for two-dimensional placement problem in electrical circuit layout. *IEEE Transactions on Circuits and Systems,* CAS-28:12–18, January 1981.

Goto, S. and T. Matsuda. Partitioning assignment and placement, in *Layout Design And Verification.* T. Ohtsuki, Ed. Elsevier North-Holland, New York, Chap.2, pp 55–97, 1986.

Hanan, M. and J. M. Kurtzberg. Placement techniques. In *Design Automation of Digital Systems.* M. A. Breuer, Ed. Prentice-Hall Inc, Englewood Cliffs, New Jersey, pages 213–282, 1972.

Hitchcock, R. B. Sr. Timing verification and the timing analysis program. *Proceedings of 19th Design Automation Conference,* pages 594–604, 1982.

Holland, J. H. *Adaptation in Natural and Artificial Systems.* University of Michigan Press, Ann Arbor, Mich., 1975.

Hopfield, J. J. Neural networks and physical systems with emergent collective computational abilities. *Proc. National Academy of Science,* 79: pages 2554–2558, 1982.

Hopfield, J. J. and D. W. Tank. Neural Computation of Decisions in Optimization Problems. *Biological Cybernetics,* 52:141–152, 1985.

Jouppi, N. P. Timing analysis and performance improvement of MOS VLSI design. *IEEE Transactions on Computer Aided Design,* CAD-6(4):650–665, July 1987.

Kick, B. Timing correction in logic synthesis. *Proceedings of ICCAD'87,* pages 299–302, 1987.

Kirkpatrick, S. Jr., C. Gelatt and M. Vecchi. Optimizaton by simulated annealing. *Science,* 220(4598):498–516, May 1983.

Kling, R. and P. Bannerjee. ESP: A new standard cell placement package using simulated evolution. *Proceedings of 24th Design Automation Conference,* pages 60–66, 1987.

Kravitz, S. A. and R. A. Rutenbar. Placement by simulated annealing of a multiprocessor. *IEEE Transactions on Computer Aided Design,* CAD-6(4):534–549, July 1987.

Kruskal, J. B. Jr. On the shortest spanning subtree of a graph and the travelling salesman problem. *Proc. AMS,* (7) pages 48–50, 1956.

Lauther, U. A min-cut placement algorithm of general cell assemblies based on a graph representation. *Proceedings of 16th Design Automation Conference,* pages 1–10, 1979.

Lee, C. Y. An algorithm for path connection and its application, *IRE Transactions on Electronic Computers,* EC-10, 1961.

Lin, Rung-Bin. and E. Shragowitz. Fuzzy logic approach to placement problem. *Proceedings of 29th Design Automation Conference,* pages 153–158, June 1992.

Marek-Sadowska, M. and S. Lin. Timing-driven placement. *Proc. of ICCAD'89,* pages 94–97, 1989.

Micheli, G. D. Performance-oriented synthesis in the yorktown silicon compiler. *Proc. of ICCAD' 86,* pages 138–141, 1986.

Nahar, S., S. Sahni and E. Shragowitz. Experiments with simulated annealing. *Proceedings of 22nd Design Automation Conference,* pages 748–752, 1985.

Nahar, S., S. Sahni and E. Shragowitz. Simulated annealing and combinatorial optimization. *International Journal on Computer Aided VLSI Design,* 1(1):1–24, 1989.

Nair, R. *et al.* Generation of performance constraints for layout. *IEEE Transactions on Computer Aided Design,* CAD-8(8):860–874, August 1989.

Preas, B. and M. Lorenzetti. *Physical Design Automation of VLSI Systems.* Edited, The Benjamin-Cummings Publishing Company, Inc., 1988.

Robson, G. Automatic placement and routing of gate arrays. *VLSI Design,* 5:35–43, 1984.

Rose, J. S., W. M. Snelgrove and Z. G. Vranesic. Parallel standard cell placement algorithms with quality equivalent to simulated annealing. *IEEE Transactions on Computer Aided Design,* 7(3):387–396, March 1988.

Ruelhi, A. E., P. K. Wolff, and G. Goertzel. Analytical power/timing optimization technique for digital system. *Proceedings of 14th Design Automation Conference,* pages 142–146, 1977.

Sechen, C. and A. L. Sangiovanni-Vincentelli. TimberWolf3.2: A new standard cell placement and global routing package. *Proceedings of 23rd Design Automation Conference,* pages 432–439, 1986.

Shahookar, K. and P. Mazumder. A genetic approach to standard cell placement using meta-genetic parameter optimization. *IEEE Transactions on Computer Aided Design,* 9(5):500–511, May 1990.

Shahookar, K. and P. Mazumder. VLSI cell placement techniques. *ACM Computing Surveys,* 23(2):143–220, June 1991.

Singh, K. J., *et al.* Timing optimization of combinational logic. *Proc. of ICCAD '88*, pages 282–285, 1988.

Srinivasan, A., K. Chaudhary and E. S. Kuh. Ritual: A performance-driven placement algorithm, *IEEE Transactions on Circuits and Systems - II*, 39(11):825–840, November 1992.

VLSI Systems Design Staff. Survey of automatic layout software. *VLSI System Design*, 8(4):78–89, 1987.

VSLI Systems Design Staff. Survey of automatic IC layout software. *VLSI System Design*, 9(4):40–49, 1988.

Sutanthavibul, S. and E. Shragowitz. June: An adaptive timing-driven layout system. *Technical Report TR-89-37*, 1989.

Sutanthavibul, S., E. Shragowitz and Rung-Bin Lin. An adaptive timing-driven placement for high performance VLSI's. *IEEE Transactions on Computer Aided Design*, 12(10):1488–1498, October 1993.

Wasserman, P. D. *Neural Computing – Theory and Practice*. Van Nostrand Reinhold, NY, 1989.

Youssef, H., Rung-Bin Lin and E. Shragowitz. Bounds on net delays for VLSI circuits. *IEEE Transactions on Circuits and Systems - II*, 39(11):815–824, November1992.

Yu, M. L. A study of the applicability of Hopfield Decision Neural nets to VLSI CAD. *Proceedings of 26th Design Automation Conference*, pages 412–417, 1989.

Zimmermann, H. J. *Fuzzy sets, decision making, and expert systems*. Kluwer Academic Publishers, Boston, MA, 1987.

# FIVE

# GRID ROUTING

## 5.1 INTRODUCTION

In the process of automatic design of VLSI layouts and printed circuit boards (PCBs), the phase following cell placement is *routing*. The number of cells on a VLSI chip, or IC chips on PCBs to be interconnected is generally very large. For this reason, routing is accomplished using computer programs called *routers*. The task of router consists of precisely defining paths on the layout surface, on which conductors that carry electrical signals are run. These conductors (also called wiring segments) interconnect all the pins that are electrically equivalent.

Routing takes up almost 30 per cent of design time and a large percentage of layout area. Automatic routing algorithms were first applied to design of PCBs. In the beginning PCBs were not very dense. None the less, basic ideas of automatic routing had been developed and some are still valid, and are being adapted to new emerging problems and larger PCBs. Recently, the main application of automatic routers has been in the automated design of VLSI circuits.

In this chapter we shall discuss algorithms for routing. The general definition of the problem is given in Section 5.2. In Section 5.3 cost functions and routing constraints are presented. One very common technique that is used to connect two points belonging to the same net considers the layout as a *maze*. Finding a path to connect any two points is similar to finding a path in the maze. Such routers that connect two points by finding a path in a maze are called *maze routers*. Maze routers assume the layout floor to be made up of a rectangular array of grid cells.

Functional cells to be interconnected fill up some slots in this grid and constitute the obstacles of the maze. Maze routing is discussed in Section 5.4. The most popular algorithm to find a path through the maze is Lee algorithm (Lee, 1961). This algorithm is discussed in detail in Section 5.4.1. Running time of this algorithm is large and memory requirement is high. Limitations of Lee algorithm and several improvements to the algorithm to decrease both the running time and memory required are presented in Section 5.4.2. Maze routers such as those based on the Lee algorithm connect a pair of points at a time. Multipin nets (number of pins greater than 2) can be connected in a spanning-tree like fashion. As will be seen, this procedure yields very poor solutions. An efficient technique that uses the standard Lee algorithm to connect multipin nets is presented in Section 5.4.3. A technique that finds a certain desirable path, not necessarily the shortest, is given in Section 5.4.4. Other algorithms based on variations of Lee algorithm to reduce its running time are presented in Section 5.4.5.

Another class of routers that do not use a physical but an imaginary grid to overcome the memory requirement of maze routers are based on *line search algorithms*. The algorithms used in these routers are introduced in Section 5.5. Other issues related to routing such as multilayer routing, effect of ordering of nets on the quality of solution, and routing of power and ground nets are discussed in Section 5.6. Current trends in routing and recent work are presented in Section 5.7.

## 5.2 PROBLEM DEFINITION

Routing can be defined as follows: given a set of *cells* with ports (inputs, outputs, clock, power and ground pins) on the boundaries, a set of *signal nets* (which are sets of points that are to be electrically connected together), and locations of cells on the layout floor (obtained from placement procedure, see Chapter 4), *routing* consists of finding suitable paths on the available layout space, on which wires are run to connect the desired sets of pins. By suitable is meant those paths that minimize the given objective functions, subject to constraints. Constraints may be imposed by the designer, the implementation process, or layout strategy and design style. Examples of constraints are minimum separation between adjacent wires of different nets, minimum width of routing wires, number of available routing layers, timing constraints, etc. Examples of objective functions include reduction in the overall required wirelength, and avoidance of timing problems due to delays of interconnects.

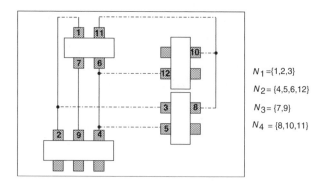

$N_1 = \{1,2,3\}$

$N_2 = \{4,5,6,12\}$

$N_3 = \{7,9\}$

$N_4 = \{8,10,11\}$

**Figure 5.1** Illustration of general routing.

The connections that are used in routing are made by first uniformly depositing metal on a carrier surface. Then unwanted metal is etched away to leave wires that carry signals. In PCBs this surface is usually fibreglass, while in VLSI it is silicon. In VLSI design, in addition to metal, polysilicon is also used to carry signals. Polysilicon wires carry signals on one layer and metal on the other (2-layer routing). These two layers are separated by an oxide insulating layer. Holes in this insulating layer called *contact-cuts* are used to connect conductors between two layers that belong to the same net. Certain VLSI technologies allow three layers for routing. Here, two layers are used to carry conductors in metal (commonly known as metal-1 and metal-2) and the third layer carries wires in polysilicon. Holes in the insulating layer that connect conductors between two metal layers are known as *vias*. (Holes in PCBs used to connect conductors on two sides of the board are also called vias.) An illustration of a general routing problem is given in Figure 5.1. The signal nets could be *two-point* or *multipoint* as shown.

## 5.3 COST FUNCTIONS AND CONSTRAINTS

The main objective of a routing algorithm is to achieve complete automatic routing with minimal need for any manual intervention. As mentioned earlier, the total area taken up by a circuit is the sum of functional area and the wiring area. In order to implement a circuit in minimal possible area, it is essential that the wiring area is reduced. Individual connections' lengths must also be kept small in order to satisfy performance criteria. Referring to Figure 5.2(a) the two points $a$ and $a'$ are connected using the shortest

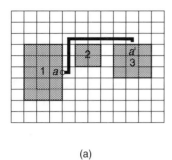 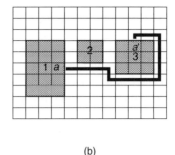

(a) (b)

**Figure 5.2** Two possible paths connecting a pair of points. (a) The shortest path. (b) A longer path with more bends.

path shown. The path is the shortest because its length is equal to the Manhattan distance between $a$ and $a'$. An alternative path that connects the same two points is shown in Figure 5.2(b). This path is longer, and it uses more routing cells (more area). It also has more bends. A signal travelling along this path will take longer time. Clearly, the path in Figure 5.2(a) is preferable to that of Figure 5.2(b).

The objective of automatic layout systems is to produce chips with smallest amount of area. Thus, a common goal of most routers is to accomplish complete automatic routing using as small a wirelength as possible.

Bends in a path result in holes for contact-cuts or vias. When two-layer routing is used, all horizontal segments are assigned to one layer, and vertical to the other. Connectivity between segments belonging to the same net is made using these holes. The *yield* of the manufacturing process decreases with the increase in the number of cuts or vias. Delays due to impedance of cuts and vias also degrade performance (reduce speed). Therefore, reducing the number of cuts/vias is crucial.

In addition, it is required that performance must not be affected due to signal delays. Currently, the switching delay of gates has been so much reduced that delays due to interconnects or wiring can no longer be ignored. In fact, sometimes the delays due to wiring dominate. The objective of the router then is not only to reduce the overall wirelength but to do so keeping the maximum delay of any net within a certain minimum.

Finally the routing programs must be capable of handling very large circuits. Thus it is imperative that the algorithms used are time efficient and use minimal memory.

Routing programs have to perform their task within a set of given constraints. The basic type of constraints are: (a) placement constraints, (b) the number of routing layers, and (c) geometrical constraints.

### 5.3.1 Placement constraints

Many routing programs are based on fixed placement. This means that all cells are in predefined locations and are not movable. This concept leads to imperfect solutions and may leave some nets unconnected. These constraints apply to PCBs, gate-arrays and to some extent to standard-cells too.

Full-custom and general-cell layout styles allow *both* the size and placement of cells to be specified. In *gate-arrays*, all cells in the layout are of the same size. Cells are placed in fixed locations in rows with horizontal and vertical separation between them being available for routing. Thus the available routing space in gate-arrays is fixed. As in the gate-array approach, *standard-cell* design methodology is characterized by fixed height rows of cells. The size of the layout surface in *not* predetermined. The problem is to make all connections between nets using the smallest width of horizontal space between rows called *channels* (see Chapter 7).

### 5.3.2 Number of routing layers

Another constraint to a routing program is the number of layers which are available for routing. Single-layer routing is most economical and is often useful in PCB design. Planar routing algorithms are necessary in single-layer layouts. But they are of high computational complexity and generally do not accomplish complete routing.

In most applications routing is constrained to two layers. The usual approach in this case is called $H-V$ routing; one layer carries wires in the horizontal direction only, the other carries wires in the vertical direction. Connectivity between layers is achieved by vias or contact holes. Using this concept, it is guaranteed that all connections of a circuit can be wired provided that the routing area is large enough and that the vias are not restricted. The disadvantage of $H-V$ routing is that the number of vias or cuts is high. Since the yield of a product decreases with increasing number of vias, the elimination of redundant holes is achieved in most applications by a post processor. This post processor assigns vertical wire segments not crossed by horizontal segments of other nets into the horizontal layers, and assigns horizontal segments not crossed by vertical segments of other nets to the vertical layer.

Due to speed requirements on signal paths and power distribution, two-layer techniques are not sufficient for all applications. Advanced technologies use more than two layers. But high production costs and yield considerations still give preference to two-layer routing. This holds for standard MOS technologies which use one metal layer and one polysilicon layer.

### 5.3.3 Geometrical constraints

During routing, minimal geometries must be maintained. These include minimum width and spacing of routing paths, which are dictated by the technological process. In manual designs, design-rule-checking (DRC) programs are used for this task. Automatic routing tools must be able to consider all geometrical constraints, thus abolishing the need for design rule checking. For routing purposes, only those design rules must be considered which define geometries of wires and contact holes. Commonly, this is achieved by using a proper equidistant grid. Wires are represented by lines and restricted to grid line positions. Hence wire widths and separation between wires is constant for all nets and design rules are avoided. However, in this approach, a realization of variable wire width, which may be required for power nets or other special nets, is not easy.

## 5.4 MAZE ROUTING ALGORITHMS

A class of general purpose routing algorithms which use a grid model are termed *maze routers*. The entire routing surface is represented as a rectangular array of grid cells. All ports, wires, and edges of bounding boxes that enclose the cells are aligned on the grid. The segments on which wires run are also aligned with the grid lines. The size of grid cells is defined such that wires belonging to other nets can be routed through adjacent cells without violating the width and spacing rules of wires. Two points are connected by finding a sequence of adjacent cells from one point to the other. Maze routers connect a single pair of points at a time.

In any routing problem, a certain minimum width of conductors, and a minimum separation between them must be maintained. If $w$ represents the minimum width, and $s$ the minimum separation between two adjacent conductors, then the two requirements can be combined into a single constraint called $\Delta$, where $\Delta = w + s$. Now by defining a uniform grid, with the edge length of each grid cell being $\Delta$, and allowing wires to run on grid lines (or parallel to them shifted by a fixed amount), ensures that the above constraints are satisfied. This is illustrated in Figure 5.3.

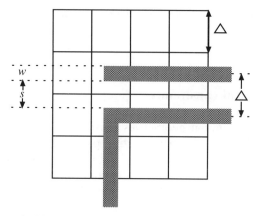

**Figure 5.3** Illustration of grid cell size.

## 5.4.1 Lee algorithm

The most widely known maze routing method for finding a path between two points on a grid that has obstacles is Lee algorithm (Lee, 1961). An excellent characteristic of Lee algorithm is that if a path exists between two points, then it is surely found. In addition it is guaranteed to be the shortest available one. The algorithm can be divided into three phases.

The first phase of Lee algorithm consists of labelling the grid, and is called the *filling* or *wave propagation* phase. It is analogous to dropping a pebble in a still pond and causing waves to ripple outwards. The pair of grid cells to be connected are labelled $S$ and $T$. In this phase, non-blocking grid cells at Manhattan-distance $i$ from grid cell $S$ are all labelled with $i$ (during step $i$). Labelling continues in an expanding diamond fashion until the target grid cell $T$ is marked in step $L$, where $L$ is the length of the shortest path, and each grid cell on that path contributes one unit length. The process of filling continues until on the $i$th step:

1. the grid cell $T$ is reached; or
2. $T$ is not reached and at step $i$ there are no empty grid cells adjacent to cells labelled $i - 1$; or
3. $T$ is not reached and $i$ equals $M$, where $M$ is the upper bound on a path length.

If grid cell $T$ is reached in step $i$ then there is a path of length $i$ between points $S$ and $T$. On the other hand, if at step $i$ there are no empty cells adjacent to the cells labelled with $i - 1$, then the required path is not found. An upper bound on path length may be placed. If $i = M$ where $M$ is the

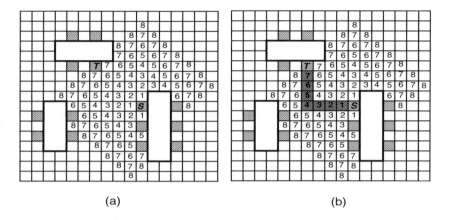

(a)                                                    (b)

**Figure 5.4** Lee algorithm. (a) Filling. (b) Retrace.

upper bound on the path length, and if $T$ is not reached, then the path with the requirement $L \leq M$ is not found.

The filling phase begins by entering a '1' in all empty cells adjacent to the source cell $S$. Next, 2s are entered in all empty cells adjacent to those containing 1s. Then, 3s are entered adjacent to 2s and so on. This process continues and is terminated when one of the above three conditions occurs. The process of filling is illustrated in Figure 5.4(a).

The second phase of the algorithm is called the *retrace* phase. This procedure is the reverse of the procedure for filling. The actual shortest path is found in this phase as follows. If grid cell $T$ was reached in step $i$, then surely there will exist at least one grid cell adjacent to it which contains label $i - 1$. Likewise, a grid cell containing $i - 2$ will be adjacent to one containing label $i - 1$ and so on. In the example of Figure 5.4(a), since the target $T$ was reached on the 8th step, surely there must be a cell with label 7 adjacent to it. Likewise, there is at least one cell with label 6 adjacent to 7. By tracing the numbered cells in descending order from $T$ to $S$, the desired shortest path is found. The cells of the retraced path for the filled grid of Figure 5.4(a) are shaded in Figure 5.4(b). In the retrace phase, there is generally a choice of cells. That is, there may be two or more cells with label $i - 1$ immediately adjacent to the cell with label $i$. Theoretically, any of the cells may be chosen and the shortest path is still found. In practice, the recommended rule is not to change the direction of retrace unless one has to do so. This is because bends in paths result in vias which is undesirable. Once the desired path is found, the cells used for the route connecting $S$ and $T$ are regarded as occupied for subsequent interconnections.

The final phase is called *label clearance*. In this phase all labelled cells except those used for the path just found, are cleared for subsequent interconnections. In the label clearance phase, searching for all the labelled cells is as involved as the wave propagation itself (see Exercise 5.3).

## 5.4.2 Limitations of Lee algorithm for large circuits

In Lee algorithm, if $L$ is the length of the path to be found, then the processing time for filling is proportional to $L^2$, while the processing time for retrace is proportional to $L$ (Why?). Therefore, the algorithm has a time complexity of $O(L^2)$ for each path. In addition, for an $N \times N$ grid plane, the algorithm requires $O(N^2)$ memory. Also, some amount of temporary storage is required to store the positions of cells that are on the wavefront. The worst case running time is also of $O(N^2)$.

Several extensions to the basic Lee algorithm exist that aim at reducing the running time of the algorithm. To reduce the problem of storage, coding schemes for labelling of grid cells have been proposed. We now present some schemes proposed for reducing the running time and memory requirement of Lee algorithm.

### Coding schemes to reduce memory requirement

A non-trivial storage problem in the Lee algorithm is that a unit of memory space is needed for every grid cell. Suppose a $250 \times 250$ grid plane is to be filled. Then, if maximum path lengths of up to 250 ($M$) grid units are allowed, during the filling phase, a cell may be labelled by a number as large as 250. This implies that at least 8 bits must be allocated for each grid cell.

Now consider the filled grid of Figure 5.4. Observe that for each cell labelled $k$, all adjacent cells are labelled either $k - 1$ or $k + 1$. Therefore, during retrace, it is sufficient if we can distinguish the *predecessor* cells from the *successor* cells. Labelling schemes based on this idea are widely used. Two are listed below.

1. In the first scheme labelling begins as in Lee algorithm with cells adjacent to $S$ labelled with 1s, and those adjacent to 1s with 2s, and those adjacent to 2s with 3s. Then cells adjacent to 3s are again labelled with 1s, and the process continues. Therefore the sequence 1,2,3, 1,2,3 ... is used for labelling. Only three bits per memory cell are required since a grid cell may be in one of the five states which are labelled with 1, 2, or 3, *empty*, or *blocked*.

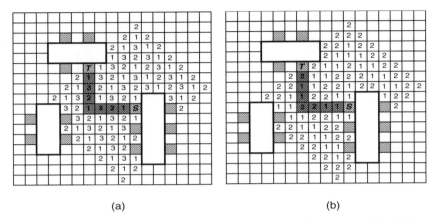

(a)                                                    (b)

**Figure 5.5** Filling sequences that reduce memory requirement. (a) Sequence 1,2,3, 1,2,3 ....
(b) Sequence 1,1, 2,2, 1,1, 2,2 ....

2. The second scheme proposed by Akers is to use the sequence 1,1, 2,2, 1,1, 2,2 .... This scheme is most economical, since each cell will be in one of the four states: *empty*, *blocked*, labelled with 1, or labelled with 2 (Akers, 1967). Thus, independent of the grid size, two bits per memory cell are sufficient for the implementation of the labelling phase of Lee algorithm.

**Example 5.1**    Fill the grid of Figure 5.4 with sequences 1,2,3, 1,2,3 ...,
and 1,1, 2,2, 1,1, 2,2 .... Comment on the running time of the algorithm that uses these schemes of filling.

SOLUTION    The grids filled with the above sequences are shown in Figure 5.5(a) and (b) respectively. Since the number of grid cells labelled using these sequences are the same as in the case of the standard Lee algorithm, the running time is unchanged.

## Reducing running time

Running time is proportional to the number of cells searched in the filling phase of Lee algorithm. Thus, to reduce the number of cells filled, the following speed-up techniques may be used.

1. *Starting point selection:* In the standard Lee algorithm either of the two points to be connected can be chosen as the source or starting point. The number of cells filled is less if the point chosen as starting point is the one that is furthest from the centre of the grid. Since the source is closer

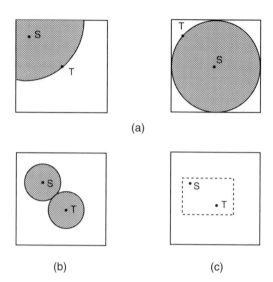

(a)

(b)                                    (c)

**Figure 5.6** Schemes to reduce running time of Lee algorithm. (a) Starting point selection. (b) Double fan-out. (c) Framing.

to the frame of the grid, the area of wave propagation is bounded by it. This is illustrated in Figure 5.6(a). The shaded portion in the figure indicates the number of cells filled if either point is chosen as the source.

2. *Double fan-out:* In this technique, during the filling phase, waves are propagated from both the source and target cells. Labelling continues until a point of contact between the two wavefronts is reached. This technique approximately halves the number of cells filled (Figure 5.6(b)).

3. *Framing:* In this technique, an artificial boundary is made around the terminal pairs to be connected, as shown in Figure 5.6(c). The size of this boundary may be 10 per cent–20 per cent larger than the size of the smallest bounding box containing the pair of points to be connected. This approach speeds up the search considerably. If no path is found then the bounding frame is enlarged and the search is continued.

### 5.4.3 Connecting multipoint nets

Lee algorithm as seen above connects two terminal pins using the shortest path on the grid. A multipin net consists of three or more terminal pins to be connected. The optimal connection of these terminal pins belonging to the same net which gives the least wirelength is termed as the *Steiner tree*

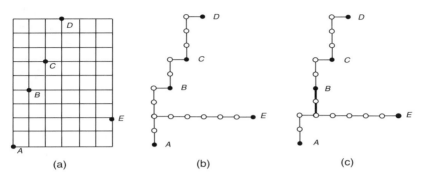

**Figure 5.7** Routing a multipoint net. (a) Five points of a net. (b) Interconnection tree found by repeated application of modified Lee algorithm. (c) A shorter interconnection found by deleting an edge and rerouting.

*problem.†* This problem has been proven to be NP-hard. Our objective is *not* to find the minimum solution but one that satisfies constraints. In this case it is sufficient for the wirelength to be within a certain given maximum. A sub-optimal solution to connect a multipoint net can be obtained using Lee algorithm. The method is explained below followed by an example.

In the classical Lee algorithm the two points to be interconnected were labelled as *S* (source) and *T* (target). To connect a multipoint net, one of the terminals of the net is treated as source, and the rest as targets. A wave is propagated from the source until any one of the targets is reached. Then the path connecting the source with this target is found in the retrace phase. Next, *all* the cells in the determined path are labelled as source cells and the remaining unconnected terminals are labelled as targets and the process is repeated until one of the remaining targets is reached. Again, the shortest path from the determined target to one of the sources is obtained. This process is continued until all the terminals are connected.

**Example 5.2** Consider the five points belonging to the same net shown in Figure 5.7(a). Explain how this multipin net is routed using Lee algorithm. Suggest a technique to improve the wirelength of the generated sub-optimal tree.

SOLUTION Initially one terminal, say *A*, is chosen as the source. Then the wave is propagated starting from *A* with the other four terminals *B*, *C*, *D*, and *E* as targets. The filling phase continues until one of the targets (the closest), in this case *B*, is reached. The path obtained from

†See Chapter 1 for definition.

$A$ to $B$ is laid out. Now all the cells in this laid out path are labelled as source cells and the remaining three $(C, D, E)$ as target cells. Again the wave is propagated starting from the cells on path $A$–$B$ as the sources until any one of the target points is reached. In this case it is point $C$. A path from one of the cells of path $A$–$B$ to terminal $C$ is laid out. Next, all the cells on the path $A$–$B$–$C$ become source cells, and this process continues. Figure 5.7(b) shows the final pattern of the interconnection.

The interconnection obtained by this process is not guaranteed to be of minimum length. A shorter interconnection between two points can be found using the following simple technique. Removing any segment (edge) from a tree will result in two subtrees. The shortest path between subtrees can be found by applying the Lee algorithm again with all cells in one subtree serving as source cells and all cells in the other as target cells. If the shortest path obtained is smaller than the length of the deleted segment, then by inserting this new path, a shorter length interconnection is obtained. Applying this technique to the segment between subtree $A$–$E$ and subtree $B$–$C$–$D$, a shorter interconnection, as shown in Figure 5.7(c), is found.

### 5.4.4 Finding more desirable paths

Often practical situations require a more desirable path, not necessarily the shortest, to be found. An example of such a situation is of finding a path that will cause least amount of difficulty for finding subsequent path connections. The filling phase of the Lee algorithm can be easily modified to accommodate such and other constraints. The basic requirement of any modified filling phase is that the *desired* path be unambiguously traced back. Akers (1967) observed that a path running along obstructions would leave more room for subsequent interconnections. Suppose, for example, that a net $x$ has already been routed as shown in Figure 5.8, and that pin $S$ is to be connected to $T$. The standard Lee algorithm will result in the shortest path $z$. However, the longer path $y$ could be preferable. Before proceeding to find the desired path let us understand the motivation for finding it. Four grid segments make a grid cell. A grid segment is common to at most two grid cells (see Figure 5.9). Once a grid cell is occupied by a path or blocked, its segments are not usable. A segment is said to be eliminated or *unusable* if a path cannot cross it. The number of grid segments eliminated by the path $y$ are less than that by path $z$. Also, paths such as $y$ tend to reduce dead space on the layout grid.

In any case, if the objective is to accomplish the desired path such as $y$, then the required path selection can be accomplished by preparing a

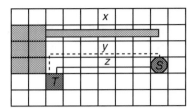

**Figure 5.8** Illustration of desirable paths.

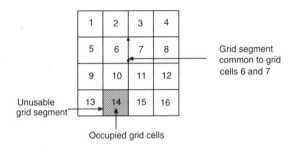

**Figure 5.9** Usable and unusable grid segments.

weighted array as shown in Figure 5.10. The weight assigned to a grid cell in this case is the number of usable grid cell segments (or edges) minus one.

The desired path may be generated by routing a net so as to minimize the total weight of used cells. For path $y$ this weight is 13, and for path $z$ it is 15. The wave propagation phase in Lee algorithm is modified to minimize the total weighted sum of grid points. The modified procedure is explained below followed by an example.

Refer to Figure 5.10. The process begins by selecting source pin $S$ (with weight zero) and assigning each cell adjacent to $S$ with a value equal to its weight. Each such cell is marked as the *latest cell* (circled in the figure). At each step from here on, each latest cell $C$ (with value $V$) is selected and every empty adjacent cell (with weight $W$) receives a value $V + W$. In addition, the newly filled cells serve as the list of latest cells. In the above

| 2 | 2 | 2 | 2 | 2 | 2 | 2 | 2 | 2 | 3 |
|---|---|---|---|---|---|---|---|---|---|
|   |   |   |   |   |   |   |   |   | 2 |
|   |   | 1 | 2 | 2 | 2 | 2 | 2 | ① | 3 |
|   |   | 1 | 3 | 3 | 3 | 3 | ② | S | ② |
| 2 | 1 | *T* | 2 | 3 | 3 | 3 | 3 | 2 | 3 |
| 3 | 3 | 2 | 3 | 3 | 3 | 3 | 3 | 3 | 3 |

**Figure 5.10** A weighted cellular array.

procedure, if an adjacent cell has already a value greater than $V + W$, then the old value is replaced and the cell with the replaced value is added to the list of latest cells. Therefore, some old latest cells may become new latest cells again if their value is decreased.

**Example 5.3**   Using the technique described in Section 5.4.4 find a path of minimum weight between $S$ and $T$ in the weighted grid shown in Figure 5.10.

SOLUTION   Latest cells in each step are indicated by circles in the grid. Initially, cells adjacent to $S$ become latest cells. This is shown in Figure 5.10. The procedure just explained is then applied and the next eight steps are shown in Figure 5.11(a)–(h).

In step 2 (Figure 5.11(a)), the wavefront expands and latest cells are again marked with circles. Think of these circles as moving towards the target. In each step the circles cause adjacent cells to become latest cells (encircled), and also pick up their weights. In step 7 (Figure 5.11(f)) one circle has reached the target $T$ and its weight (which is equal to the weight of the path it followed) is 15. Although the target cell is reached, it does not mean that the desired path has been found. Why? Observe that there are other latest cells (circles) on the wavefront which have weights less than that of the one that has reached the target. The first circle to reach the target does not constitute the desired path because our objective is *not only* to find the path, but to find the one with the least weight. And as long as there are latest cells with weights less than the weight of the target, there is hope that when they reach the target, the total weight of the path they generate may be less than the weight of the path already found. Filling must therefore continue. In step 7 (Figure 5.11(f)), cell with weight 12 is one such contender. In step 8 (Figure 5.11(g)), the value of cell above $T$, which was earlier 15 is updated to 13, and since this is still less than the current value of cell $T$ the procedure is continued.

Step 9 (Figure 5.11(h)) is the last and final step of the procedure. The value of target cell is renewed to 13. The minimum value of the latest cells (other than the target) is 19 and this value is not smaller than the value of any old latest cell, hence further filling is abandoned.

The desired path is retraced as follows. Starting at the target cell, the adjacent cell with the minimum weight that is in the direction of the retrace is selected. This process continues until the source cell is reached. For this example the desired path is indicated in Figure 5.11(h) (see Exercises 5.14 and 5.15).

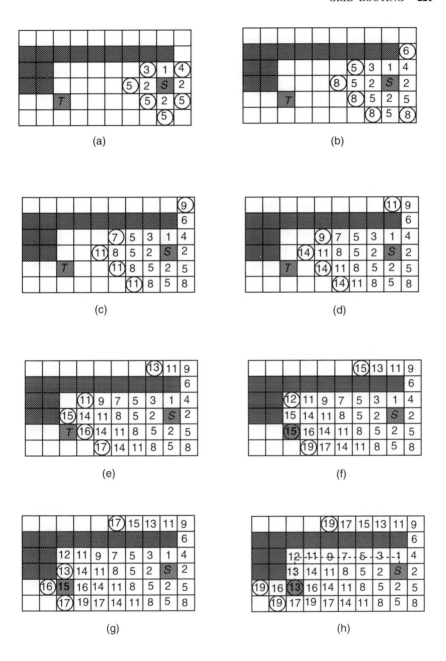

(a)

(b)

(c)

(d)

(e)

(f)

(g)

(h)

**Figure 5.11** Filling in a weighted array.

## 5.4.5  Further speed improvements

The *filling* phase of Lee algorithm is similar to the *breadth first search* technique in graph theory. It can also be thought of as construction of a tree with each node having at most four children. Each node corresponds to a cell and the four children correspond to its four neighbours. As the filling proceeds, each node of the tree further produces (at most) three more children, and this process continues for all nodes in each level of the tree, until the leaf node corresponding to the target cell is reached. Figure 5.12(a) shows a grid being filled where adjacent cells are searched in the following order: top($T$), left($L$), right($R$), and bottom($B$). Figure 5.12(b) shows the corresponding tree. As described earlier, the running time for a particular instance of source–target pairs is proportional to the number of cells being searched until the target is reached. In Section 5.4.2 few speed-up techniques were discussed. The common idea behind speed-up techniques is to advance the wavefront with a higher priority towards the target direction. In this section we present two more speed-up techniques based on this idea.

### Hadlock's algorithm

This is a shortest path algorithm with a new method for cell labelling called *detour numbers*. The algorithm was suggested by Hadlock (1977). It is a goal directed search method. The detour number $d(P)$ of a path $P$ connecting two cells $S$ and $T$ is defined as the number of grid cells directed away from its target $T$. If $MD(S,T)$ is the Manhattan distance between $S$ and $T$, then it can be proved that the length $l(P)$ of a path $P$ is given by

$$l(P) = MD(S,T) + 2 \times d(P). \tag{5.1}$$

Figure 5.13 illustrates how path length is represented by the detour number. Note that $P$ is the shortest path if and only if $d(P)$ is minimized among all paths connecting $S$ and $T$. Note also that in Equation 5.1 $MD(S,T)$ is fixed, independent of the path connecting $S$ and $T$. Based on this idea, the filling phase of the Lee algorithm is modified as follows:

1. Instead of filling a cell with a number equal to the distance from the source, the detour numbers with respect to a specified target are entered.
2. Cells with smaller detour numbers are expanded with higher priority.

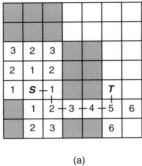

(a)

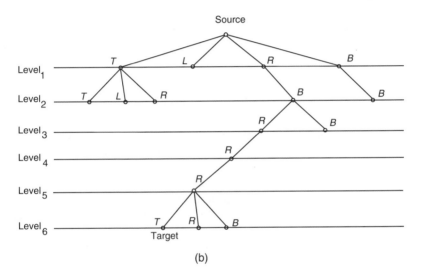

(b)

**Figure 5.12** (a) Filled grid. (b) Tree corresponding to filled grid of (a).

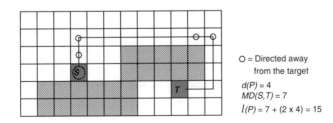

O = Directed away
from the target

$d(P) = 4$
$MD(S,T) = 7$
$l(P) = 7 + (2 \times 4) = 15$

**Figure 5.13** Path length and detour number.

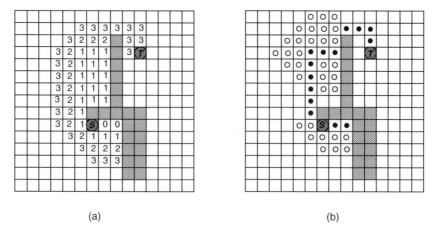

(a)                                          (b)

**Figure 5.14** (a) Hadlock's detour algorithm. (b) Soukup's fast algorithm.

Figure 5.14(a) shows the filling of a grid using the detour algorithm. Observe that for any cell filled with $i$, if the adjacent cell is towards the target, then it is filled with the same number, and if it is away from the target then it is filled with $i + 1$. Thus the cell to the right of $S$ is filled with a 0 because target $T$ is to the right of source $S$. And the cell to the left of $S$ is filled with 1 because it is away from target $T$.

Path retracing is slightly different from the standard Lee algorithm (see Exercise 5.20). The number of grid units filled in Hadlock's algorithm is considerably smaller than in Lee algorithm. Therefore, the speed improvement due to this algorithm is obviously remarkable. Running time for an $N \times N$ grid ranges from $O(N)$ to $O(N^2)$ and depends on the position of the source–target pairs and the locations of obstructions.

## Soukup's algorithm

The previous algorithms performed filling in a breadth first manner. Soukup (1978) suggested adding *depth* to the search. In Soukup's algorithm a line segment starting from the source is initially extended towards the target. Then the cells on this line segment are searched first. The line segment is extended without changing direction unless it is necessary. When the line hits an obstacle, Lee algorithm is applied to search around the obstacle. During the search, once a cell in the direction of the target is found, another line segment starting from there is extended towards the target. Figure 5.14(b) shows the set of searched cells. The darkened circles in the figure indicate the cells directed towards the target.

Although this algorithm guarantees finding the path if one exists, it does not guarantee that it is the shortest. Its disadvantage is that it generates sub-optimal paths (both in terms of length and number of bends). However, it is extremely fast, especially when the routing space is not congested. Soukup claims that it is 10–50 times faster than the Lee algorithm on typical two-layer routing problems.

## 5.5 LINE SEARCH ALGORITHMS

As seen above, one of the major drawbacks of Lee algorithm and its variations is the amount of memory required for the grid representation of the layout. Line search algorithms overcome this drawback.

The idea behind these algorithms is as follows. Suppose two points $S$ and $T$ on the layout are to be connected. For the moment assume that there are no obstacles. If a vertical line is drawn passing through $S$ and a horizontal line passing through $T$, the two lines will naturally intersect giving a Manhattan path between $S$ and $T$. Will this always work? In the absence of obstacles, yes. But in general, more has to be done to find a path between $S$ and $T$.

As opposed to Lee's maze running algorithm and its variations, which proceed in a breadth-first manner, line search algorithms perform a depth-first search. Therefore, maze running guarantees finding the shortest path even if it is most expensive in terms of vias. However, because of their depth-first nature, line search algorithms do not guarantee finding the shortest path, and may need several backtrackings.

In practice, line search algorithms produce completion rates similar to Lee algorithm, with the difference that both memory requirements and execution times are considerably reduced. This is because the entire routing space is *not* stored as a matrix as in the case of Lee algorithm but the routing space and paths are represented by a set of line segments. Line search algorithms were first proposed by Mikami and Tabuchi (1968) and Hightower (1969). A brief description of these two algorithms is presented below.

### 5.5.1 Mikami–Tabuchi's algorithm

Let $S$ and $T$ be a pair of terminals of a net located on some intersection of an *imaginary* grid. The first step is to generate four lines (two horizontal and two vertical) passing through $S$ and $T$. These lines are extended until they hit obstructions (a placed cell for example) or the boundary of the

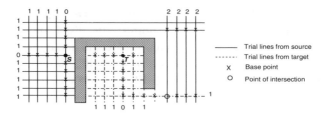

**Figure 5.15** Routing using Mikami–Tabuchi line search algorithm.

layout. If a line generated from $S$ intersects a line generated from $T$ then a connecting path without any bend or with one bend has been found. If the four generated lines do not intersect, then they are identified as *trial lines* of level *zero* and stored in temporary storage. Then at each step $i$ of the iteration the following operations are done.

1. Trial lines of level $i$ are picked one at a time. Along each trial line all its grid points (*base-points*) are traced. Starting from these *base-points* new *trial lines* are generated perpendicular to trial line $i$. Let the generated line segments be identified as *trial lines* of level $i + 1$.
2. If trial line of level $(i + 1)$ intersects a trial line (of any level) originated from the other terminal point, then the required path is found by backtracing from the point of intersection to both points $S$ and $T$. Otherwise all trial lines of level $(i + 1)$ are added to the temporary storage and the procedure is repeated from Step 1.

The above algorithm guarantees to find a path if one exists, provided that all trial lines up to their deepest possible level of nesting are examined.

Figure 5.15 illustrates an example of finding a path by the application of the procedure. In this example the trial line of level 1 originating from $T$ intersects a trial line of level 2 generated from $S$.

## 5.5.2 Hightower's algorithm

This algorithm is similar to the Mikami–Tabuchi's algorithm. The difference is that instead of generating all line segments perpendicular to a trial line, Hightower algorithm considers only those lines that are extendable beyond the obstacle which blocked the preceding trial lines.

The procedure is best explained with the illustration in Figure 5.16. As before, let $S$ and $T$ be two points to be connected. The shaded regions $p$, $q$, and $r$ constitute obstacles around which the path is to be found. We need to

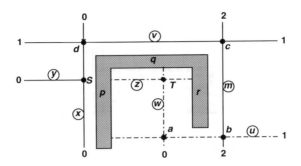

**Figure 5.16** Routing using Hightower's line search algorithm.

define a few terms here. The procedure begins by constructing horizontal and vertical *escape lines* from the source and target.

*Blockage:* A *blockage* is a cover of a point if a horizontal or vertical line from it passes through the point. For example, in Figure 5.16 blockage *p* is a cover of point *S* because a horizontal line from *p* passes through point *S*.

*Escape line:* *Escape line* passing through a point is a line that is perpendicular to the two nearest blockages on both sides (either left and right, or top and bottom). For example, lines *x* and *y* are escape lines through *S*.

*Escape point:* Given a point *k* and its escape line *L*, an *escape point* on *L* is a point that is not covered by at least one cover of *k*. Thus point *d* is an escape point for escape line *x* passing through *S*. Similarly point *a* is an escape point for escape line *w* passing through *T*.

Now we shall see how a connection between two points *S* and *T* is found. The procedure begins by constructing horizontal and vertical *escape lines* from the source and target. First, for each escape point, the longest escape line is found. In case of a tie the escape line nearest to the starting terminal is taken. Thus escape lines *u* and *v* are drawn through escape points *a* and *d*. Proceeding in this way, escape line *m* through escape point *c* on line *v* intersects line *u* at *b* (labels of lines are encircled). Escape points are then retraced to find the required path which in this case is *TabcdS*.

When the routing area is not congested, the above algorithms are expected to run fast. Particularly, Hightower algorithm is expected to run in time proportional to the number of bends. A conservative estimate of running time in a complicated maze is $O(N^4)$ (see Exercise 5.23). Thus the memory saving in line search algorithm is dramatic, but the running time does not improve very much. We might also need to backtrack from dead ends (resulting from bad sequences of trial lines).

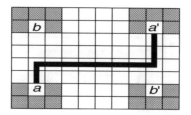

**Figure 5.17** Multi-net routing.

# 5.6 OTHER ISSUES

Maze routers connect two points at a time. The entire set of nets can be divided into a set of pairs of points to be connected. Multiple points belonging to the same net can be connected using the technique suggested in Section 5.4.3. Once a net is placed, it becomes an obstruction to other nets to be routed. The *order* in which pairs of points are chosen for routing may affect the quality of final solution. Due to the large amount of current carried by power/ground nets, and problems of delay and skew on clock nets, they are to be treated differently. In the following discussion we will look at the requirements for successful multi-net routing, the effect on ordering of nets, and routing of power/ground nets.

### 5.6.1 Multilayer routing

In the previous discussion (Section 5.4), we have seen how points belonging to a net can be connected either with shortest paths or paths that meet our requirements. If a single net is to be routed, then, given enough separation between cells, one layer is sufficient. However, if several nets are to be connected, which is generally the case, then the problem of finding complete connectivity becomes very difficult. This is because the path found will occupy grid cells which may become obstacles to future paths. Consider two nets $N_1 = \{a, a'\}$ and $N_2 = \{b, b'\}$ of Figure 5.17. Net $N_1$ is connected using one of the algorithms discussed earlier. If we have a single layer for wiring, then net $N_2$ can never be connected. However, if another layer was available (for example the second side of the PCB), then it is possible to route $N_2$. Clearly, the more the number of layers, the greater is the chance of accomplishing complete routing of multi-nets. In this section we will see how we can use Lee algorithm to interconnect pins belonging to different nets on the same layout using two layers.

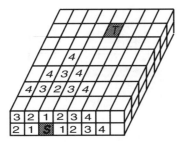

**Figure 5.18** Three-dimensional cellular array for two-layer routing.

## Three-dimensional grid

In general, multilayer interconnection can be accomplished by extending the standard Lee algorithm to a three-dimensional grid. Here, a cellular array consisting of unit cubes is used. To find a path in a three-dimensional cellular array the same technique of filling, as proposed in the original Lee algorithm is followed, except that all adjacent empty cells including those accessible from above or below are labelled at each step. The process leads to a path which occupies a minimum number of cells. Note that inter-layer connection through vias or contact-cuts is assumed to have the same weight as unit length of one layer. For a specific case of two-layer routing, the cellular array is as shown in Figure 5.18.

## Two planar arrays

Another implementation of two-layer routing is to use two planar cellular arrays, one for each layer. The filling process is the same as in Lee algorithm except (a) filling both layers is done simultaneously and (b) whenever an empty cell in one layer is labelled, the same label is entered in the corresponding cell on the other layer, unless that cell is already occupied. This procedure is illustrated in the example below.

**Example 5.4** Two routing layers, layer-1 and layer-2, are shown in Figure 5.19(a) and (b) respectively. The shaded cells in these layers indicate that the cells are occupied by other paths on that layer and cannot be used for routing. Apply the method discussed above which uses two planar arrays (one for each layer) and determine the path between points $S$ and $T$.

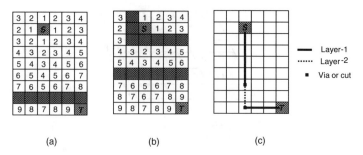

(a)                    (b)                    (c)

**Figure 5.19** Two-layer routing using two arrays. (a) Layer-1. (b) Layer-2. (c) Retrace path.

SOLUTION    The grids corresponding to the two layers are filled using the required scheme. Labels in cells are shown in Figure 5.19(a) and (b). In the 10th step the target $T$ is reached. Retrace can begin in any layer. Let's choose layer-1. During the retrace in layer-1, when the cell labelled 7 is reached, we find no adjacent cells with label 6. A via is assigned to this location and a switch to layer-2 is made. The retrace now continues in the array corresponding to layer-2 (Figure 5.19(b)). In layer-2, when the cell with label 5 is reached we find that there is no cell with label 4 adjacent to it. A via is placed again in this location and a switch back to layer-1 is made. Retrace continues in this manner and the obtained path is shown in Figure 5.19(c).

An efficient approach generally adopted to two-layer routing is to assign all horizontal wiring segments to one layer and all vertical segments to the second layer. Two-layer routing which strictly follows this horizontal –vertical rule can be achieved by blocking vertical runs on one layer and horizontal runs on the other during the filling phase of Lee algorithm. Of course this approach is going to create a large number of unnecessary vias. A post processing step may be used to assign all horizontal segments of the horizontal layer not crossed by vertical wires to the vertical layer. Similarly, some vertical segments are assigned to the horizontal layer. Each such movement of segments between layers causes a reduction of two vias or contact-cuts.

### 5.6.2 Ordering of nets

Both the maze router and the line-search router (Section 5.5) presented in this chapter are intended for routing a *single* net. Of course, in practice an entire netlist needs to be routed. Let there be $n$ nets in the netlist, and let us order them in some fashion so that we can use the maze or line search router to route the individual nets sequentially, thus:

> **For** $i = 1$ to $n$ **Do**
> Use maze router to route net $p_i$;

The order $p_1, p_2, \cdots, p_n$ is a permutation of $1, 2, \cdots, n$. After any net $p_i$ has been routed, the area occupied by its wiring must be marked as *obstacles* to the unrouted nets. The permutation $p$ influences the routability of the nets (see Exercise 5.24). Thus, even though all the nets are indeed routable, the routing order may prevent successful completion of routing.

Maze routers which connect one net at a time suffer from the shortcoming that they do not provide any feedback or anticipation in the routing process to avoid conflict between wires to assure that some early wire routing will not prevent successful routing of some later connections. Because of this, it is felt that the order in which a set of nets is routed is of crucial importance to the successful completion of routing.

Consider the situation in Figure 5.20(a). It is required to connect two two-point nets $a-a'$ and $b-b'$. Which net should be connected first? Since there are only two nets to consider, a brute force approach may be used to determine the effect of ordering. Say net $a-a'$ is the first net chosen for interconnection. The two possible paths which may be found by our router are as shown in Figure 5.20(b) and (c). From the figure it is clear that if the path chosen by the router is as in Figure 5.20(b) the length of path $b-b'$ is equal to its Manhattan distance. But if the path $a-a'$ is determined as in Figure 5.20(c) then the length of path $b-b'$ is far greater than its Manhattan distance.

Now let us see what would happen if $b-b'$ is chosen as the first pair of points to be connected. The two possible paths between $b$ and $b'$ are as shown in Figure 5.20(d) and (e) respectively, and both paths do not cause any obstacle to the shortest path between $a$ and $a'$ to be determined.

Now referring to Figure 5.20(f) which consists of the bounding boxes of pins $a-a'$ and $b-b'$, it appears that this two wire case will yield the following result: *If a pin p belonging to one net lies within the bounding box formed by pins belonging to another net q then net p must be laid out first.*

Let us extend this observation to three two-point nets $a-a'$, $b-b'$, and, $c-c'$, of Figure 5.21. The bounding boxes around these pins are shown. Applying the same logic as before for this example results in the following conditions to be satisfied:

*a must be routed before  b*

*b must be routed before  c*

*c must be routed before  a*

Clearly, due to lack of transitivity, the rule cannot be universally applied. But it is also clear that if a certain point $p$ is in rectangle formed by

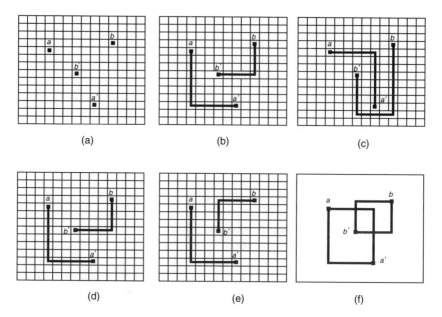

(a)                              (b)                              (c)

(d)                              (e)                              (f)

**Figure 5.20** Figure illustrating ordering of two 2-point nets.

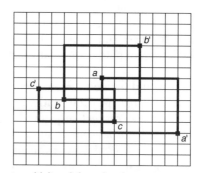

**Figure 5.21** Illustrating non-transitivity of three 2-point nets.

net $q-q'$ it is desirable to lay out $p$ before $q$. This rule can be generalized as follows: *The greater the number of pins in a certain rectangle formed by pins $q-q'$, the greater the number of wires which should be laid out before $q-q'$ is laid out.* The rule can be used to assign priorities to nets, where the priority number of net $p$ is given by the number of pins in the rectangle formed by $p-p'$. The larger the number, the lower the priority of the corresponding net.

**Example 5.5** Consider the four pin pairs $a-a'$, $b-b'$, $c-c'$, and $d-d'$ as shown in Figure 5.22. Order the nets using the priority scheme discussed above and show the layout.

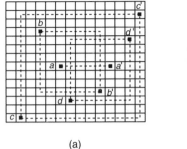 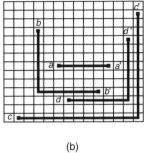

(a)                                    (b)

**Figure 5.22** An example of four 2-point nets to be ordered (Example 5.5).

SOLUTION   When the four bounding boxes with nets $a-a'$, $b-b'$, $c-c'$, and $d-d'$ are drawn, the number of pins within the bounding box formed by pins of these nets (that is $a$, $b$, $c$, and $d$) are found to be 0, 1, 6, and 2 respectively. Thus the order for this set of nets is $a$, $b$, $d$, and $c$. The layout is shown in Figure 5.22(b).

In the above paragraphs we presented a method to order two point nets on a single layer. Other most common methods for interconnection ordering which can be used for two-layer and multi-layer routing are:

1. order the nets in the ascending order of their lengths;
2. order the nets in the descending order of their lengths; and
3. order the nets based on their timing criticality.

The first two criteria are conflicting. Proponents of the first criterion argue that it is easier to route a long wire around a short wire than vice versa; whereas supporters of the second criterion believe that longer wires are more difficult to lay out, and hence should be attempted first.

The topic of net ordering is controversial. There are several arguments against the importance of even investigating an optimal ordering. First, there are $n!$ possible orderings for connecting $n$ nets. Heuristics aimed at obtaining sub-optimal ordering are disappointing and no effective means is known to compare the goodness of orderings. Secondly, an experiment has been conducted to show that the performance of a router is almost independent of the order in which connections are attempted (Abel, 1972). In spite of this, it is possible to find examples where one particular ordering leads to a higher completion rate than the other.

Whatever scheme is used to order the nets, since the actual length of the net is not known until routing is complete, a technique to estimate the net length must be adopted. One possibility is to define the net length as the length of the minimum rectilinear Steiner tree (MRST). A sub-optimal

**Algorithm** Generate_MRST;
**Begin**
  **ForEach** $n$-terminal net **Do**
  **Begin.**
    Find the pair of terminals with shortest rectilinear distance $L_0$;
    Construct a shortest rectilinear path (MRST);
    $L \leftarrow L_0$;
    $k \leftarrow 1$;
    **Repeat**
      Find terminal $k$ with shortest distance $L_k$ to MRST;
      Connect it to MRST by a shortest path;
      $L \leftarrow L + L_k$;
      $k \leftarrow k + 1$;
    **Until** all terminals are connected;
  **EndFor;**
**End.**

**Figure 5.23** Constructive algorithm for MRST generation.

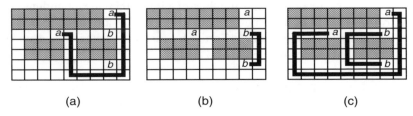

(a)    (b)    (c)

**Figure 5.24** (a) Optimal routing of $a$ prevents routing of $b$. (b) Optimal routing of $b$ prevents routing of $a$. (c) Non-optimal routing of nets $a$ and $b$.

but quick constructive algorithm that can be used for MRST generation and net length calculation is given in Figure 5.23.

The third ordering criterion targets the performance of the circuit. In high speed circuits it may be required that the lengths of all nets must be less than a certain critical value. If this is the case then it may be preferable to check during the ordering step for those wires whose pin-pairs are too far apart. These wires may have to be laid out early in the process to ensure that their actual lengths will meet the required constraints.

Next, an example which illustrates a condition where net ordering is of no help is presented. Refer to Figure 5.24. It is assumed that the router generates a shortest path with least amount of bends. If net $a$ is connected first (Figure 5.24(a)) it blocks net $b$ and if net $b$ is connected first (Figure 5.24(b)) it blocks net $a$. A solution in which both nets $a$ and $b$ are routed is given in Figure 5.24(c). Note that the connectivity here is made by non-optimal paths.

### 5.6.3 Rip-up and rerouting

For most complex layouts, it is usually the case that the detailed router fails at connecting all pins required by the netlist. There are two general approaches to this problem: (1) the manual approach, where an expert human designer attempts to complete the connection left by the router, and (2) the automatic procedure, which consists of identifying the congested routing area that caused the unconnects, ripping-up a selected number of connections, then rerouting them.

The manual strategy is very time consuming and impractical for designs of reasonable complexity. The automatic procedure can be interactive or activated in batch-mode. However, the interactive procedure is more attractive and is the one most widely used. It consists of two steps: (1) identification of bottleneck regions—completed connections going through these regions are responsible for the unconnects, therefore a selected number of these completed connections are ripped off using some built-in criteria or user input; (2) the blocked connections are routed, and finally the ripped-up connections are rerouted.

These steps are repeated until all connections are completed, or a time limit is exceeded. In this latter case a human designer intervenes to complete the few remaining unconnects. Numerous papers are available which report similar as well as other more elaborate strategies (Dees and Smith, 1981; Dees and Krager, 1982; Ohtsuki et al., 1985).

### 5.6.4 Power and ground routing

Routing of power and ground nets requires special consideration. First, power and ground nets must be routed preferably on a single layer. This is because parasitic capacitance and resistance of contact-cuts and vias are high. Second, power and ground nets are much wider than other nets because they carry larger amounts of current. As a consequence of their varying widths routing of these nets cannot be done with an equidistant grid. If an equidistant grid is used then the programme must be modified to accommodate variable width routing.

In determining the widths of power net segments, the power index for each module indicating its power consumption is used. The process of routing can be achieved in three steps: (1) planar topological representation of power nets, (2) computation of widths of net segments, and (3) embedding of segments into the routing layer.

1. *Planar topological representation of nets:* In this step, each net is represented by a tree. Branches of the tree correspond to net segments. For power and ground net routing the minimization of total length of a net is not the objective but rather the planar routing of both the power supply and ground in a single layer. There are two conditions that guarantee a planar routing of two nets in one layer (Rothermel and Mlynski, 1981):

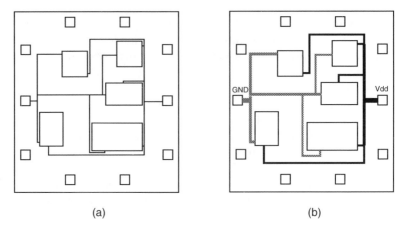

(a)                                                    (b)

**Figure 5.25** (a) Topological trees for power and ground nets. (b) Actual widths of routing layers.

(a) Each macro cell must have only one connection point for each of the power and ground nets.

(b) The cells must be spaced such that both nets may be routed next to any macro cell edge.

Two trees corresponding to two nets (power and ground) are generated, one from the left edge of the chip and the other from the right edge. The two trees are inter-digitated into each other. The routing order of terminals influences the tree topology. By selecting a routing order based on the horizontal distance of each connecting point from the edge of the chip, both nets are grown simultaneously. This is illustrated in Figure 5.25(a).

2. *Computation of widths of net segments:* Once a planar solution has been found, the wire widths need to be determined based on the electrical demands of the circuit. Given the trees of both nets as determined above, and the power consumption of each module as indicated by its power index, the widths of each segment can be calculated depending on local current values as follows. Each segment corresponding to a branch in the tree is weighted by the total supply currents of those modules which are connected to the root via this segment. This calculation can be based on a simple tree analysis starting at the power input pad of the chip as the root of the tree. Then the necessary width of the segments is scaled proportionally to the calculated individual current values.

3. *Embedding of segments into the routing layer:* Given the width of all segments of both nets as calculated before, the embedding into the routing layer is done by widening each line segment to a rectangle of

proper width. The translation of the topology into actual widths and the embedding of the segments into the routing layer is illustrated in Figure 5.25(b).

## 5.7 OTHER APPROACHES AND RECENT WORK

The maze running technique explained in this chapter is very general and can be modified to route complex circuits with obstacles. It has been successfully used in two routers, **Mighty** and **Beaver**. The Mighty router is based on an incremental routing strategy (Shen and Sangiovanni-Vincentelli, 1986). It employs maze running but has the special feature of modifying the already-routed nets. The cost function penalizes long paths and those which require vias. Rip-up and rerouting may be necessary to complete routing. The other recently developed heuristic that uses maze routing is Beaver (Cohoon and Patrick, 1988). The algorithm consists of three successive parts: corner routing, line-sweep routing and maze routing. A priority queue is used to determine net ordering. In addition, track control is employed to prevent routing conflicts. The maze routing technique has also been used in a routing phase of Magic's layout system (Ousterhout, *et al.*, 1984).

In the past decade new heuristics for routing standard-cells and gate-array methodologies have been developed. They are known as channel routing and switchbox routing heuristics. We shall discuss these in Chapter 7. The other type of routers that have been used in PCB design are commonly known as *pattern routers*. The idea of pattern router is to enumerate all possible patterns according to cost functions and then to determine the best path for a two-terminal net (Soukup and Fournier, 1979; Asano, 1982). This strategy has been combined with others to develop *template based routers* (Srinivasan and Shenoy, 1988).

Recently there has been an increased activity in the development of special-purpose hardware architectures for solving a number of design automation problems. These special purpose hardware processors are commonly known as *hardware accelerators, backend processors*, and *special purpose engines*.

A rectangular grid has a natural mapping into a two-dimensional array of processors. Specialized array processor machines dedicated to the implementation of shortest path grid routing algorithms have been reported in the literature (Breuer and Shamsa, 1981; Iosupovici, 1981; Carrol, 1981). A class of two-dimensional SIMD (single instruction multiple data) array processors for the implementation of grid routing algorithms was proposed by Iosupovici (1986). More general array processors which can be programmed to implement routing along with other DA tasks have also been reported (Blank *et al.*, 1981; Hong and Nair, 1983; Adshead, 1982; Rutenbar *et al.*, 1984).

## 5.8 CONCLUSION

In this chapter we examined two types of grid routers, the maze router and line-search router. The maze router uses a regular physical grid, and line search routers use an imaginary grid. The basic grid router that uses Lee algorithm has a large memory requirement and also may require a large amount of running time (Lee, 1961). Techniques to reduce the running time and memory requirement were discussed in detail. Other algorithms that modify the filling phase of Lee algorithm to reduce the running time are Hadlock's algorithm and Soukup's algorithm. Their techniques were illustrated with examples. Line search algorithms overcome the high memory requirement of Lee algorithm. Two line search heuristics, one due to Mikami and Tabuchi (1968), and the other due to Hightower (1969) were presented.

Maze running algorithms guarantee finding a shortest path if one exists, even if it is the most expensive in terms of the number of vias. Line search algorithms guarantee finding a path if one exists (not necessarily the shortest). But they may require several backtracks for all dead ends that are reached. In practice, however, line search algorithms can be significantly faster than maze running algorithms.

The major advantage for which maze running algorithms are preferred over line search algorithms is that the former are *grid-cell* oriented. This gives more flexibility to the weighting of routing area of the chip. This is of extreme importance since proper weighting of cells enables finding superior routes.

Both the maze router and line-search router connect a single net at a time. Modifications to the basic routing technique to accommodate multipoint nets, and use of multi-layers were also presented. Other issues such as multi-layer routing, power and ground routing, and issues related to ordering for successful completion of routing were also examined.

## 5.9 BIBLIOGRAPHIC NOTES

The most popular and widely referenced algorithm for maze routing is Lee algorithm (Lee, 1961). Several modifications to Lee algorithm have been

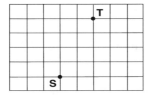

**Figure 5.26**  Routing grid for Exercise 5.1.

presented. Modifications to labelling during the filling phase of the Lee algorithm were proposed by Akers (1967).

Modifications, which add breadth and depth to reach the target while searching in a maze are due to Hadlock (1977) and Soukup (1978), respectively. Line search algorithms have been successfully applied for automating PCB design and also in routers for VLSI designs. The two line search algorithms discussed in this chapter are due to Hightower (1969) and Mikami and Tabuchi (1968).

## EXERCISES

**Exercise 5.1**  $S$ and $T$ are two points on a routing grid, as shown in Figure 5.26. The horizontal separation between $S$ and $T$ is $m$ units, and the vertical separation is $n$ units. Show that there are $R(m, n) = \binom{m+n}{m}$ ways of routing a two-pin net from $S$ to $T$. Assume that only vertical and horizontal routing are permitted, and assume that there is a single routing layer. What happens to $R(m, n)$ when $m = n$? Plot $R(n, n)$ for $n = 2, 4, 8, 16, \cdots$ and write your conclusion. Derive a close form expression for $R(n, n)$ by using Stirling's approximation for $n!$.

**Exercise 5.2**  In the above problem, if we permit $45°$ and $135°$ routing also, how many ways are there to route a net from $S$ and $T$?

**Exercise 5.3**  The label clearing phase of Lee algorithm is as complex as the filling phase. Suggest a technique to speed up this process.

**Exercise 5.4**  (*) **Programming Exercise:** Assume a maze routing procedure $MROUTE2$ that can route a given two-pin net. The inputs to $MROUTE2$ are the source and destination points in the maze. The procedure returns $SUCCESS$ or $FAILURE$ depending on whether or not it can find a route. Upon success, the procedure updates a global data structure, $MAZE$, by writing the net number in each cell used by the route found by *MROUTE2*.

Construct a procedure which uses *MROUTE2* to route a $k$-pin net, $k > 2$.

**Exercise 5.5**  (*) Explain how you will use the maze router explained in this chapter to handle multiple routing layers.

**Exercise 5.6**  (*) How will you extend the maze router explained in the chapter to handle 45° and 135° paths?

**Exercise 5.7**  Suppose that $n$ nets, numbered $1, 2, \cdots, n$ are to be routed one after another using a maze router. Which of the following strategies do you think will be most successful in completely routing all the nets?

1.  Order the nets randomly.
2.  Sort the nets in the descending order of their length, i.e., *longest net first*.
3.  Sort the nets in the ascending order of their length, i.e., *shortest net first*.

**Exercise 5.8**  Continuing the above exercise, give an example where the *longest net first* strategy succeeds but the *shortest net first* strategy fails.

**Exercise 5.9**  In a certain technology that allows two-layer routing, the width of the wires that are used to connect pins is 3 units. The minimum separation to be maintained between wires is 4 units. The size of a contact-cut is $3 \times 3$ units and it is required that metal must extend beyond the cut on all sides by at least 1 unit. What must be the size of a grid unit in order to implement a maze router in this technology?

**Exercise 5.10**  The number of grid cells marked during the filling process of Lee algorithm depends on the distance between the points to be connected.

1.  Suppose that the distance between the two points to be interconnected is **L**. Develop an expression giving the number of marked cells as a function of **L**. You may make any necessary assumptions.

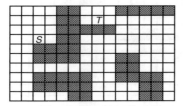

**Figure 5.27** Grid of Exercise 5.14.

2. With the help of the expression you developed in the first part of this question prove that the double fan-out method will be faster than the regular method where fan-out is done only from the source node.

**Exercise 5.11**  What is the percentage saving in memory when the conventional filling sequence as proposed by Lee is modified to the sequence 1-1-2-2-1-1-2-2- ...? Assume that the layout contains 100 cells each of size $8 \times 8$ grid units and the average routing area is twice the area occupied by the cells. Clearly state any assumptions made.

**Exercise 5.12**  Given two points $(x_1, y_1)$ and $(x_2, y_2)$ to be connected using Lee algorithm, write the pseudo-code of a procedure that will assign one of the points as the source and the other as the target. Remember that proper selection of starting point affects the number of grid cells filled and thus the speed of the algorithm. What other information is required by your procedure?

**Exercise 5.13**  (*) **Programming Exercise:** Design and implement a two-layer maze router. Use the router to experiment with different ordering schemes.

**Exercise 5.14**  Fill the grid given in Figure 5.27 with weights equal to the number of usable grid segments as explained in Section 5.4.4, and find the least weighted path between $S$ and $T$.

**Exercise 5.15** (*)
1. The labelling procedure given in Section 5.4.4 and illustrated in Example 5.3 does not run as fast as Lee algorithm. Why?
2. Determine the time complexity of the algorithm to fill a weighted grid as discussed in Section 5.4.4.
3. Show that using priority queues for storing the set of latest cells will result in $O(N^2 \log N)$ running time.

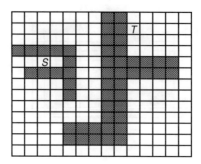

**Figure 5.28** Grid of Exercise 5.17.

4. Another implementation to find desirable paths is *not* to obtain the minimum value of latest cell at each step but to continue labelling until no cell on the wavefront receives a smaller value. Show that the worst case running time in this case is $O(N^4)$. This implementation seems to be more efficient than the one of the previous question with priority queues unless cell weights are distributed over a wide range. Is this true?

**Exercise 5.16**   Derive Equation 5.1.

**Exercise 5.17**   Fill the grid of Figure 5.28 using the technique suggested by 'Hadlock' for the minimum detour router.

**Exercise 5.18**   Write an algorithm that uses two stacks to manage the efficient filling of the grid for Lee algorithm. One stack always contains the locations of those cells on the wavefront to be examined and the second stack contains the cells that will constitute the wavefront in the next filling phase.

**Exercise 5.19**   Modify the algorithm of the previous problem to perform filling using detour numbers to implement the Hadlock router.

**Exercise 5.20**   (*) Path retracing in the Hadlock algorithm is not as straightforward as in the case of Lee algorithm. Why? Suggest a suitable data structure to store the filled values and write the pseudo-code of the procedure that will generate the retraced path efficiently.

**Exercise 5.21**   In the Mikami–Tabuchi procedure, if a level $i$ trial line

**Figure 5.29** Maze of Exercise 5.24.

originated from the source intersects a level $j$ trial originated from the target, what is the number of bends (vias) of the generated path?

**Exercise 5.22**  Discuss the merits and drawbacks of routers based on the maze running algorithm as compared to those that use the line-search algorithm.

**Exercise 5.23**  For the implementation of line-search algorithm, assume a linked list data structure to manage lines. Let each line segment be defined by three integers to specify the $(x,y)$ coordinates of its two end points. The lines are sorted according to their $(x,y)$ coordinates, and the co-horizontal (co-vertical) lines are grouped together. For each $y$ ($x$) coordinate there is a pointer to the first horizontal (vertical) line on that coordinate and for every line there is a pointer to the line that comes next. (The lines generated while searching a connection can also be stored in a temporary storage with a similar data structure.)

For an $N \times N$ grid plane show that the memory requirement is of order $O(N^2)$ and running time is of order $O(N^4)$.

**Exercise 5.24**  Consider the maze shown in Figure 5.29. There are 5 two-point nets to be routed. The nets are numbered 1, 2, $\cdots$, 5. Assume that a single layer is available for routing. Suppose that the maze router is used to route the nets in the order 1, 2, 3, 4, 5. Is it possible to complete the routing? Assume that the router visits adjacent cells in the order $W$, $E$, $N$, $S$.

In what order should the nets be routed to complete the routing using the maze router?

**Exercise 5.25**  What restrictions are usually placed when routing power/ground nets on general cells?

# REFERENCES

Abel, L. C. On the ordering of connections for automatic routing. *IEEE Transactions on Computers*, C-21:1227–1233, November 1972.

Adshead, H. G. Employing a distributed array processor in a dedicated gate array layout system. *IEEE International Conf. Circuits and Computers*, pages 411–414, September 1982.

Akers, S. B. A modification of Lee's path connection algorithm. *IEEE Transactions on Electronic Computers*, pages 97–98, February 1967.

Asano, T. Parametric pattern router. *Proceedings of 19th Design Automation Conference*, pages 411–417, 1982.

Blank, T. *et al.* A parallel bit map processor architecture for DA algorithms. *Proceedings of 18th Design Automation Conference*, pages 837–845, April 1981.

Breuer, M. A. and K. Shamsa. A hardware router. *J. Digital Systems*, 4(4):393–408, 1981.

Carrol, C. R. A smart memory array processor for two layer path finding. In *Proceedings of Second Caltech Conference on VLSI*, pages 165–195, January 1981.

Cohoon, J. P. and L. H. Patrick. Beaver: A computational-geometry-based tool for switchbox routing. *IEEE Transactions on Computer-Aided Design*, CAD-7(6):684–696, June 1988.

Dees, W. A. and P. G. Krager. Automated rip-up and reroute techniques. *Proceedings of 19th Design Automation Conference*, pages 432–439, 1982.

Dees, W. A. and R. J. Smith. Performance of interconnection rip-up and reroute strategies. *Proceedings of 18th Design Automation Conference*, pages 382–390, 1981.

Hadlock, F. O. A shortest path algorithm for grid graphs. *Networks*, 7:323–334, 1977.

Hightower, D. W. A solution to line-routing problem on the continuous plane. *Proceedings of 6th Design Automation Workshop*, pages 1–24, 1969.

Hong, S. J. and R. Nair. Wire Routing Machines – New Tools for VLSI Design, *Proceedings of IEEE*, 71:57–65, January 1983.

Iosupovici, A. Design of an iterative maze router. In *Proceedings of International Conference on Circuits and Computers*, pages 908–911, 1981.

Iosupovici, A. A class of array architectures for hardware grid routers. *IEEE Transactions on Computer-Aided Design*, CAD-5(2):245–255, April 1986.

Lee, C. Y. An algorithm for path connection and its application. *IRE Transactions on Electronic Computers*, EC-10, 1961.

Mikami, K. and K. Tabuchi. A computer program for optimal routing of printed circuit connectors. *Proceedings of IFIP*, H47:1475–1478, 1968.

Ohtsuki, T., M. Tachibana and K. Suzuki. A hardware maze router with rip-up and reroute support. *Proceedings of ICCAD*, 1985.

Ousterhout, J. K., G. T. Hamachi, R. N. Mayo, W. S. Scott and G. S. Taylor. Magic: A VLSI layout system. *Proceedings of 21st Design Automation Conference*, pages 152–159, 1984.

Rothermel, H. and D. A. Mlynski. Computation of power supply nets in VLSI layout. *Proceedings of 18th Design Automation Conference*, pages 37–42, 1981.

Rutenbar, R. A. *et al.* A class of cellular architecture to support physical design automation. *IEEE Transactions on Computer-Aided Design*, CAD(3):264–278, October 1984.

Shen, H. and A. L. Sangiovanni-Vincentelli. Mighty: A rip-up and reroute detailed router. *Digest of Technical Papers, ICCAD*, pages 2–5, November 1986.

Soukup, J. Fast maze router. *Proceedings of 15th Design Automation Conference*, pages 100–102, 1978.

Soukup, J. and J. Fournier. Pattern router. *In proceedings of IEEE ISCAS*, pages 486–489, 1979.

Srinivasan, A. and N. Shenoy. Template based pattern router. *EECS244 Class Report, University of California, Berkeley*, 1988.

# SIX

# GLOBAL ROUTING

## 6.1 INTRODUCTION

The two principal steps of physical design are placement and routing. Placement consists of assigning the cells of the circuit to fixed locations of the chip, while routing consists of interconnecting the cells consistently with the circuit netlist. All of the mathematical models of the placement problem gave rise to NP-hard problems. It is for this reason that heuristic approaches as well as hierarchical decomposition are used to find a solution (usually sub-optimal) to the placement problem.

Similarly, all of the mathematical formulations of the routing problem led to NP-hard problems. Here also routing is solved in a stepwise fashion as a hierarchy of easier problems, which are sometimes polynomially solvable, or are small enough that a full enumeration approach is practical. The main objective is to quickly find a solution which satisfies the constraints. This decomposition is of course at the expense of global optimality.

The accepted practice to routing consists of adopting a two-step approach: global routing is performed first, then followed by detailed routing. The objective of the global routing step is to elaborate a routing plan so that each net is assigned to particular routing regions, while attempting to optimize a given objective function (usually an estimate of the overall wiring length). Then, detailed routing takes each routing region and, for each net, particular tracks within that region are assigned to that net. In the previous chapter, we described one class of routing algorithms, that is maze running. This routing technique was described in the context of detailed routing. However, as we will see in this chapter, it is also used for global routing.

Global routing approaches belong to four general categories: (1) sequential approach, (2) mathematical programming approach, (3) stochastic iterative approach, and (4) hierarchical approach.

For the sequential approach, nets are selected one at a time in a specific order and routed individually. The routing space may or may not be updated after the routing of each net. When the available routing space is updated after each net, the approach is order dependent, otherwise it is order independent. For the mathematical programming approach, global routing is formulated as a 0-1-integer optimization program, where a 0-1 integer variable is assigned to each net and each possible routing tree of that net. The stochastic iterative approach, such as simulated annealing, iteratively updates current solution by ripping up and rerouting selected nets, until an acceptable assignment of the nets is found. Hierarchical approaches can be bottom-up or top-down. For the bottom-up approach, grid cells are clustered into bigger cells until the entire chip is seen as a super cell. At each level of the hierarchy, global routing is performed between the individual cells considered for grouping. For the top-down approach, the hierarchy proceeds from super cells to cells, until each cell is an individual grid cell or a small group of individual grid cells. The top-down approach is usually guided by the structure of the design floorplan.

## 6.2 COST FUNCTIONS AND CONSTRAINTS

Global routing is slightly different for different design styles. For the gate-array design style the routing regions consist of horizontal and vertical channels. Channels are rectangular regions with pins on the opposite sides of the region. The available routing capacities within the channels are fixed. A feasible global routing solution should not exceed the channel capacities. Among possible feasible solutions, the one that optimizes the given cost function is selected. The cost function is usually a function of the global routes of all nets, and/or function of overall performance (interconnect delays on the critical paths). Since the array has a fixed size and fixed routing space, the objective of global routing in this case is to check the feasibility of detailed routing.

For the standard-cell design style the routing regions are horizontal channels with pins at their top and bottom boundaries. Global routing consists of assigning nets to these channels so as to minimize channel congestion and overall connection length. Interchannel routing is provided

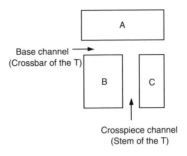

**Figure 6.1** T-intersection.

by feedthrough cells inserted within the cell rows. Here, the channels do not have pre-fixed capacities. Channels can be made wider to achieve routability.

In building-block design style the cells are of various shapes and sizes. This leads to irregular routing regions. These routing regions may be decomposed into horizontal and vertical channels, and sometimes switch-boxes (rectangular regions with pins on all four sides). The identification of these routing regions is a crucial first step to global routing. Here again, the routing regions do not have pre-fixed capacities. For both the standard-cell and building-block layout styles the objective of global routing is to minimize the required routing space and overall interconnection length while ensuring the success of the following detailed routing step. Therefore the cost function is a measure of the overall routing and chip area. Constraints could be a limit on the maximum number of tracks per channel and/or constraints on performance.

An important problem we are faced with in all design styles is the identification of the shortest set of routes to connect the pins of individual nets. For two-pin nets, the problem is trivial and amounts to finding the shortest path connecting the two pins. For multipin nets, the problem consists of finding the shortest Steiner tree spanning all pins of the net. Steiner tree problems are generally NP-hard. However, there are several heuristic algorithms which find reasonable short Steiner trees.

## 6.3 ROUTING REGIONS

Routing regions definition consists of partitioning the routing area into a set of non-intersecting rectangular regions called channels. Two types of channels may arise: horizontal and vertical. A channel is horizontal (vertical) if and only if it is parallel to the $x$- ($y$-) axis. In most cases, horizontal and vertical channels can touch at T-intersections (Figure 6.1). The channel representing the stem of the T is called the *crosspiece* and the other is called the *base*.

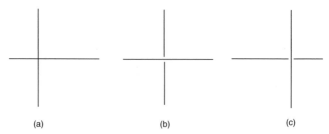

(a)            (b)            (c)

**Figure 6.2** Conversion of cross junctions. (a) Cross junction. (b) Horizontal conversion. (c) Vertical conversion.

Channel definition and ordering is an essential part of layout design. It is the knot that ties placement, global routing, and detailed routing together.

### 6.3.1 Routing regions definition

In a building-block layout style, three types of channel junctions may occur: an L-type, a T-type, and a +-type. L-type junctions occur at the corners of the layout surface. For such junctions, the ordering does not have an impact on the final detailed routing. For T-type junctions, the stem channel (crosspiece) must be routed before the base channel (the crossbar). The +-type junctions are more complex and require the use of switchbox routers. On the other hand, L-type and T-type channels can be completely routed using channel routers. This is of extreme importance since channel routers are the best and most widely investigated routing approaches. Therefore it is advantageous to transform all +-type junctions into T-type junctions so that a channel router can be used for the following detailed routing step. However, this conversion should be carried out carefully so as not to create cycles in the corresponding order constraint graph.†

In the following paragraphs we describe a conversion approach due to Cai and Otten (1989). This approach assumes that the layout has a slicing structure. Since a general building-block layout is not necessarily a slicing structure, the authors also give a polynomial time algorithm to convert a general layout into a slicing structure (Cai and Otten, 1989). Slicing structures are preferred topologies because they can be internally represented using a simple and flexible data structure (the slicing tree). Moreover, such structures lead to computationally efficient manipulation algorithms.

---

†The order constraint graph (OCG) indicates the order of performing the detailed routing of the individual channels. It is discussed later in this section.

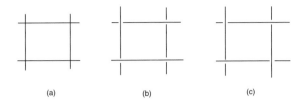

(a)    (b)    (c)

**Figure 6.3** Conversion of cross-intersections. (a) A channel configuration. (b) A cycle-free conversion. (c) A conversion introducing cycles.

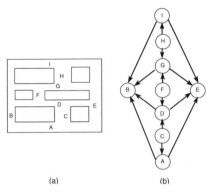

(a)                              (b)

**Figure 6.4** Order constraint graph. (a) Channel structure. (b) Its corresponding order constraint graph.

## Criteria for channel crossing conversion

There are two ways of converting a +-junction into a T-junction: (1) vertical conversion where the horizontal channel is split, and (2) horizontal conversion where the vertical channel is split (see Figure 6.2). This conversion must be carefully performed so as to avoid creating cycles in the order constraint graph. This is illustrated in Figure 6.3.

After converting all cross-junctions into T-channels, the channel ordering constraints are captured by a directed graph called the order constraint graph (see Figure 6.4). Another positive property of slicing structures is that a slicing structure is guaranteed to have at least one conflict-free channel structure, i.e., a cycle-free order constraint graph.

To minimize the negative side-effects that channel conversions might have on the wireability of the layout as a whole, the following two criteria are used (Cai and Otten, 1989):

1. *Critical path isolation criterion*: The objective of this criterion is to protect the critical paths of the channel position graphs (see Chapter 3) from neighbouring channels. Recall that a vertical (horizontal) channel

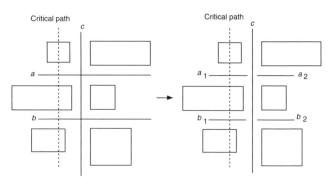

**Figure 6.5** Illustration of the critical path criterion.

position graph is a bipartite graph representing the vertical (horizontal) adjacencies between the blocks and the routing channels. The vertical (horizontal) channel position graph has a vertex for each block and each horizontal (vertical) channel. There is an edge from vertex $b$ to vertex $c$ if and only if the bottom (right) side of block $b$ is bordering channel $c$. Each block-vertex in the horizontal (vertical) channel position graph is assigned a positive weight indicating the width (height) of the corresponding block. Also, each channel vertex is assigned a positive weight indicating the width of the corresponding channel. The length of the critical path in the vertical (horizontal) position graph is equal to the height (width) of the design.

The critical path criterion attempts to perform the conversion in the direction of the critical path. This is in order to make neighbouring channels perpendicular to the direction of the critical path as short as possible, thus splitting them. This is illustrated in Figure 6.5. Such criterion will also lead to a reduction in the widths of the channels along the critical paths, thus, reducing the overall layout size.

2. *Major flow criterion:* Channel conversion is carried out after global routing. Therefore, the number of wires flowing across all channels is known before the conversion process starts. In order to minimize the number of wire bends across channels, among the two channels of the cross-junction, we split the thinner of the two, i.e., the channel with the lesser number of flowing nets (see Figure 6.6).

For each cross-junction, the above two criteria are used to compute a positive gain function. This function is a bonus rewarding conversions of cross-junctions that favour the critical path isolation and major flow criteria. Therefore, for each cross-junction adjacent to a channel segment that is on some critical path, a bonus is added to the conversion in the

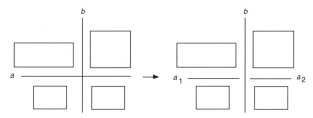

**Figure 6.6** Illustration of the major flow criterion. Channel *b* is assumed to be having more nets than channel *a*.

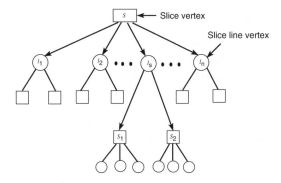

**Figure 6.7** A generic slice graph.

direction of the path. Furthermore, for each cross-junction, a bonus is added to the direction of the channel with the largest wire flow. For all other cases, a zero bonus is assigned to the crossing conversion in either direction.

The optimal channel structure is the one with the largest sum of crossing conversion bonuses.

## Conversion algorithm

The algorithm assumes that the layout has already been converted to a slicing structure. A bipartite directed graph called the slicing graph is constructed. The slicing graph has a vertex for each candidate slice-line and a vertex for each slice. There is a directed edge from a slice-vertex to each of its candidate slice-line vertices. There is a directed edge from a slice-line vertex to the two resulting slice vertices (see Figure 6.7).

Cai and Otten (1989) made the clever observation that we need not have a vertex for each possible slice-line. This is due to what they referred to as the *locality property*.

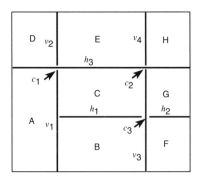

**Figure 6.8** Channel conversion example.

The preferred conversion direction of a cross-junction is the direction with the maximum bonus. When both directions have the same bonus, then both are equally good choices, and both should be considered.

Suppose that for a particular slice there is a slice-line which converts all crossings on that line in their preferred directions (for example slice-line $h_3$ of Figure 6.8). Such a slice-line is called an *optimal slice-line*. Moreover, if an optimal slice-line is selected and the resulting two channel definition problems are optimally solved in the two subslices, then the combined solutions constitute an optimal solution to the original slice. Therefore, to get an optimal solution to the channel definition problem (corresponding to the entire layout), we need to enumerate only decision sequences (trial slice-lines) that have the potential of leading to optimal solutions. This drastically reduces the number of enumerated decision sequences. This is similar to the dynamic programming algorithm strategy, where the solution to the problem is arrived at as a result of a sequence of decisions (Horowitz and Sahni, 1984). The dynamic programming algorithm is based on the *optimality principle*, which we state next, in the context of the channel conversion problem.

## Optimality principle

*Let $[d_1, d_2, ..., d_i, ..., d_k]$ be a sequence of decisions with respect to the first slice-line, second slice-line, $\cdots$ $k^{th}$ slice-line. If $d_1$ to $d_k$ is an optimal sequence of k consecutive decisions, then $d_1$ to $d_i$ is an optimal sequence of decisions and $d_i$ to $d_k$ is also an optimal decision sequence with respect to the state (the subslices) resulting from the $d_1$ to $d_i$ decision sequence.*

The consequence of the above discussion is that the slice graph should be constructed in a depth first manner, starting at the root (representing the

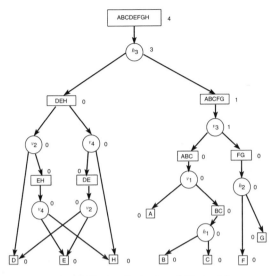

**Figure 6.9** The slice graph corresponding to the layout of Figure 6.8.

entire layout), and making a decision at each slice vertex as to the slice-lines that should be tried with that particular slice.

The slice graph will have a single source vertex (with no incoming edges) corresponding to the entire layout. The graph will have exactly $n$ sink vertices (with no outgoing edges), where $n$ is equal to the number of layout blocks (see Figure 6.9). The construction of the slice graph will be illustrated later with the help of an example.

Once the slice graph has been constructed, the algorithm proceeds with a breadth first traversal of the graph, from the sinks to the source. During the course of this traversal, the maximum bonus of each slice vertex is computed. The maximum bonus corresponding to a slice $s$ is defined recursively as follows (Cai and Otten, 1989):

$$Bonus(s) = \max_{l_s}\{Bonus(l_s) + Bonus(s_1) + Bonus(s_2)\} \qquad (6.1)$$

where $s_1$ and $s_2$ are the subslices resulting from cutting slice $s$ with slice-line $l_s$ (see Figure 6.7).

Once we reach the source vertex, the channels that lead to the maximum bonuses are identified by tracing back the slice graph from the source to the sinks. The steps of the algorithm are summarized in Figure 6.10.

**Example 6.1**  Suppose we are given the slicing floorplan of Figure 6.8. The floorplan has eight blocks identified with the letters A to H, and three cross-junctions $c_1$, $c_2$, and $c_3$. Let $c_i^h$ and $c_i^v$ denote the horizontal

**Algorithm** Channel_Conversion;
**Begin**
    1. Determine the conversion bonuses of all channel crossings;
    2. Determine the bonus of each slice-line;
    3. Construct the slicing graph;
    4. **For** each sink vertex $v$ **Do**
        $Bonus(v) \leftarrow 0$
        **EndFor**;
    5. Traverse the graph sinks-to-source, computing along the way
       the maximum bonus of each slice vertex;
    6. Traverse the graph source-to-sinks, selecting along the way
       the slice-lines that incurred maximum bonuses;
    7. Output the selected slice-lines;
**End**.

**Figure 6.10** Channel conversion algorithm.

and vertical conversions of cross-junction $c_i$, $i = 1, 2, 3$. $c_i$ is said to be vertically (horizontally) converted if the horizontal (vertical) channel is split into left and right (top and bottom) sub-channels. Assume that the three cross-junctions have the following bonuses:

$$Bonus(c_1^v) = 0; \quad Bonus(c_1^h) = 2;$$
$$Bonus(c_2^v) = 0; \quad Bonus(c_2^h) = 1;$$
$$Bonus(c_3^v) = 1; \quad Bonus(c_3^h) = 0;$$

We would like to identify the channel structure corresponding to a maximum bonus conversion of all three cross-junctions.

SOLUTION   The slicing structure of Figure 6.8 assumes that the crossings are initially converted in their preferred directions. The slicing graph corresponding to this floorplan is given in Figure 6.9, where the slice vertices are represented by boxes and the slice-lines by circles. The number next to each vertex is the maximum bonus corresponding to the slice vertex or slice-line. The sinks (vertices A to H) are assigned zero bonuses.

The graph is constructed as follows. Starting at the source vertex $ABCDEFGH$, we find that there is only one optimal slice-line $h_3$. Line $h_3$ converts the cross junctions $c_1$ and $c_2$ in their preferred directions. On the other hand, the line $v_{1,2} = v_1 \cup v_2$ is not optimal since it does not convert the cross-junction $c_1$ into its preferred direction. The subslices resulting from cutting the original floorplan with slice-line $h_3$ are [DEH] and [ABCFG]. For slice [DEH], there are two slice-lines of equal merit (both have a zero bonus). Therefore, both should be included in the slice graph. However, the other subslice [ABCFG] has only one optimal slice-line $v_3$, with a bonus equal to 1. The other slice-line, $v_1$, is not optimal, and therefore is not included in the graph. This process is continued until the entire graph is constructed (see Figure 6.9).

Now that the slice graph has been constructed, we first determine the bonuses of all slice-line vertices. The bonus of a horizontal (vertical) slice-line is equal to the sum of all horizontal (vertical) conversion bonuses of the cross-junctions traversed by that line. Therefore, for our example, the slice-line bonuses will be,

$$
\begin{aligned}
Bonus(h_1) &= Bonus(h_2) = 0; \\
Bonus(h_3) &= Bonus(c_1^h) + Bonus(c_2^h) = 3; \\
Bonus(v_1) &= Bonus(v_2) = Bonus(v_4) = 0; \\
Bonus(v_3) &= Bonus(c_3^v) = 1.
\end{aligned}
$$

The next step is to traverse the slice graph from the sinks to the source, and compute the maximum bonuses of the intermediate slice vertices. Proceeding this way, we get,

$$
\begin{aligned}
Bonus(EH) &= Bonus(v_2) + Bonus(E) + Bonus(H) = 0; \\
Bonus(DE) &= Bonus(v_4) + Bonus(D) + Bonus(E) = 0; \\
Bonus(DEH) &= \max \{ Bonus(v_2) + Bonus(D) + Bonus(EH); \\
&\qquad Bonus(v_4) + Bonus(DE) + Bonus(H) \} = 0.
\end{aligned}
$$

The maximum bonuses of the remaining slice vertices are computed in a similar way and are as follows,

$$
Bonus(BC) = 0, \ Bonus(FG) = 0, \ Bonus(ABC) = 0,
$$
$$
Bonus(ABCDFG) = 1, \text{ and } \ Bonus(ABCDEFGH) = 4.
$$

The last step of the algorithm is to proceed from the source to the sinks in order to determine the slice-lines that contributed to the maximum bonuses. For this example, $h_3$ is selected first. Then proceeding down to subslice DEH, we have two equally good choices, either line $v_3$ or line $v_4$. Suppose we randomly select slice-line $v_4$. Then, the remaining slice lines are $v_2$, $v_3$, $v_1$, $h_1$, and finally $h_2$. Hence, the initial slicing structure is the optimal one. The order constraint graph corresponding to these channels is given in Figure 6.11. Notice that this constraint graph is cycle free.

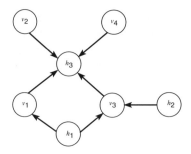

**Figure 6.11** Order constraint graph corresponding to the channels of Figure 6.8.

## Channel ordering

Once all nets have been assigned to individual channels, the final step is to assign the nets to individual tracks within every channel, i.e., to perform detailed routing. The channels are usually routed one at a time in a specific order. Channel ordering is an important intermediate step executed prior to detailed routing and after global routing. This step is needed to specify to the detailed router which channel to route first, which second, and which last. Obviously, it is assumed that all routing regions are channels.

Channel ordering is an important final step of global routing. The order in which channels should be routed is dictated by the fact that pin locations must be fixed before performing detailed routing of that channel. Therefore, of the two channels of a T-intersection, the crosspiece channel must be routed before the base channel (Ohtsuki, 1986). This is for the following two reasons:

1. To route the base channel, we need the pin information at the T-junction, i.e., the nets going through the junction. This necessitates that the crosspiece be routed before the base.

2. When routing the crosspiece channel, we may realize that we need to move blocks at the left (top) and/or right (bottom) of that channel to provide for extra tracks. This will change the pin positions within the base channel. This is another compelling reason to route the crosspiece channel before the base.

To order the channels, an order constraint graph (OCG) is built as follows. Each channel is represented by a vertex. There is an arc $(i, j)$ in the OCG if and only if channels $i$ and $j$ touch at a T-junction of which $i$ is the crosspiece and $j$ is the base (see Figure 6.4).

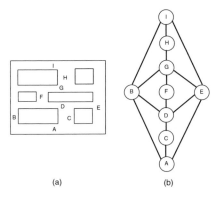

(a)                              (b)

**Figure 6.12** Channel connectivity graph. (a) A building-block layout. (b) Corresponding channel connectivity graph.

## 6.3.2 Routing regions representation

Once the routing regions have been defined, a routing graph is constructed. There are three general approaches to construct this graph.

1.  Use a channel connectivity graph $G = (V, E)$ where each channel is represented by a vertex. Each edge models the adjacency between the corresponding channels (see Figure 6.12). Vertices can be assigned weights to indicate the number of nets passing through the channel and/or the number of available tracks in that channel. Notice that the channel connectivity graph is the order constraint graph when arc directions are removed (compare Figures 6.4(b) and 6.12(b)).

2.  Use a bottleneck graph $G = (V, E)$ (Ohtsuki, 1986), where only switchboxes are modelled by vertices. There is an edge $(u, v) \in E$ if and only if the corresponding switchboxes are on opposite sides of the same vertical or horizontal channel. These routing channels are called bottlenecks, hence the name of the graph. The concepts of switchboxes and bottleneck regions are illustrated in Figure 6.13(a) where switchboxes are shaded. A building-block layout and its corresponding bottleneck graph are given in Figure 6.13. Notice that edges in this graph model horizontal and vertical channels.

3.  Use a grid graph $G = (V, E)$ where vertices model global cells and edges adjacencies between these cells (see Figure 6.14). For two-layer routing, each vertex is assigned two numbers indicating the number of available horizontal and vertical tracks.

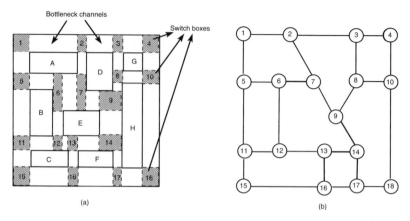

**Figure 6.13** Bottleneck graph. (a) A building-block layout. (b) Corresponding bottleneck graph.

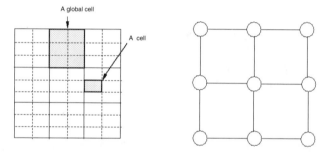

**Figure 6.14** Grid graph. (a) A two-dimensional grid. (b) Corresponding grid graph where each global cell (3 × 2 grid cells) is modelled by a vertex.

## 6.4 SEQUENTIAL GLOBAL ROUTING

Sequential global routing is the most widely used approach. This approach is graph based. Once the routing channels have been identified and the corresponding routing graph constructed, global routing proceeds as follows. For each net, we mark the vertices of the channel connectivity graph in which the net has pins. Hence, routing the net amounts to identifying a tree (preferably the shortest) covering those marked vertices.

If the net has pins in only two vertices, the problem reduces to finding the shortest path between the marked vertices. If the graph is a grid-graph, we can use Lee algorithm, which was explained in the previous chapter. For all three graph models, we can use Dijkstra's (1959) shortest path algorithm. Figure 6.15 gives a formal description of this algorithm.

**Algorithm** Shortest_Path($s, G$);
    (* $s$: a source vertex, and $G$ is a weighted graph *)
    (* $D_i$: estimated shortest distance from $s$ to node $i$; *)
    (* $d_{ij}$: weight of edge $(i, j)$ (or distance between nodes $i$ and $j$); *)
    (* $M$: set of permanently marked nodes. *)
**Begin**
    1. (* Initialization *)
    $M \leftarrow s$;
    $D_s \leftarrow \emptyset$;
    **ForEach** $j \in V(G)$ **Do** $D_j \leftarrow d_{sj}$ **EndFor**;
    2. (* Find the next closest node *)
    Find a node $i \notin M$ such that $D_i = \min_{j \notin M} D_j$;
    $M \leftarrow M \cup \{i\}$ :
    **If** M contains all nodes **Then** $STOP$; **EndIf;**
    3. (* Update markings *)
    **ForEach** $j \notin M$ **Do** $D_j \leftarrow \min_i(D_j; D_i + d_{ij})$ **EndFor**;
    **Goto** 2;
**End**.

**Figure 6.15** Dijkstra shortest path algorithm.

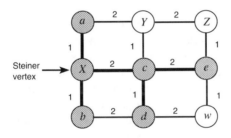

**Figure 6.16** Steiner tree corresponding to the net $M = \{a, b, c, d, e\}$.

However, in general, nets have three or more pins. Finding the shortest paths covering three or more nodes is known as the Steiner tree problem. This problem is of crucial importance to sequential routing and is the subject of the following subsection.

## 6.4.1 The Steiner tree problem

Let $M$ be the set of marked vertices. A tree connecting all vertices of $M$ as well as other vertices of $G$ that are not in $M$ is called a Steiner tree (refer to Figure 6.16). A minimum Steiner tree is a Steiner tree with minimum length.

The Steiner tree problem is NP-hard. Therefore, instead of finding a minimum Steiner tree, heuristics are used to identify as quickly as possible

**Algorithm** Steiner_Tree;
**Begin** $M \leftarrow$ set of marked nodes; (* nodes in which the net has pins *)
   $s \leftarrow$ select a node from $M$;
   $M \leftarrow M - \{s\}$;
   Apply Dijkstra_algorithm to find $\pi_{s,e}$, the shortest path from $s$ to some node $e$ of $M$;
   $M \leftarrow M - \{e\}$;
   $V \leftarrow \{s, e\}$; (* nodes of the Steiner tree *)

   **While** $M \neq \phi$ **Do**
   **Begin**
      $e \leftarrow next(M)$; (* get another node from $M$ *)
      Apply Dijkstra_algorithm to find $\pi_{e,x}$, the shortest path from $e$ to some node $x$ of $V$;
      $V(\pi_{e,x}) \leftarrow$ nodes covered by $\pi_{e}, x$;
      $V \leftarrow V \cup V(\pi_{e,x})$;
      (* remove marked nodes covered by the path $\pi_{e,x}$ *)
      $M \leftarrow M - M \cap V(\pi_{e,x})$;
   **EndWhile**
**End**.

**Figure 6.17** Steiner tree heuristic.

a tree of reasonable length not necessarily of minimum length. Most Steiner tree heuristics use a modification of minimum shortest path algorithm of Dijkstra or a variation of Lee's maze routing algorithm. Usually the heuristic proceeds in a greedy fashion as follows. First, one of the marked vertices is selected. Then the shortest path to any one of the remaining marked vertices is identified. Then, one of the remaining marked vertices is picked and a shortest path from that node to any of the nodes of the partial tree is identified. This process continues until all marked vertices have been processed. A formal description of this heuristic is given in Figure 6.17.

**Example 6.2**    Apply the algorithm of Figure 6.17 on the grid-graph of Figure 6.16 and determine a minimum weight Steiner tree corresponding to the net $\{a, b, c, d, e\}$. Assume that all horizontal edges have weights equal to 2 and all vertical edges have weights equal to 1.

SOLUTION    The vertices of the graph are $\{a, b, c, d, e, W, X, Y, Z\}$. Initially $V = \phi$ and the set of marked vertices $M = \{a, b, c, d, e\}$. Suppose vertex $a$ is selected first. The shortest path to any of the remaining marked vertices is $\pi_{a,b} = [a, X, b]$. Therefore, the sets $V$ and $M$ become, $V = \{a, X, b\}$, and $M = \{c, d, e\}$. Assume that vertex $c$ is selected next. The shortest path from $c$ to any of the vertices of $V$ is $\pi_{c,X} = [c, X]$. Therefore, the sets $V$ and $M$ become, $V = \{a, X, b, c\}$, and $M = \{d, e\}$. Assume that vertex $d$ is selected next. The shortest path from $d$ to any of the vertices of $V$ is $\pi_{d,c} = [d, c]$. Therefore, the

**Algorithm** Steiner_Approximation;
**Begin**
    $M \leftarrow$ set of marked nodes; (* nodes in which the net has pins *)
    Find the shortest paths between all pairs of nodes in $M$;
    $P \leftarrow$ sequence of paths sorted in descending order of their lengths;
    $V \leftarrow \phi$;
    $E \leftarrow \phi$;
    **While** $V \neq M$ **Do**;
        **Begin**
            $path \leftarrow next(P)$; (* get the next shortest path and remove it from $P$ *);
            **ForEach** $(i,\ j)\ \epsilon\ E(path)$ **Do**
                **If** $(i,\ j)$ does not create a cycle in the graph $G(V, E)$
                    **Then**
                        $V \leftarrow V\ \cup\ \{i, j\}$;
                        $E \leftarrow E\ \cup\ \{(i, j)\}$;
                **EndIf**
            **EndFor**
        **EndWhile**
**End**.

**Figure 6.18** Approximation heuristic of Steiner tree using a variation of Kruskal minimum spanning tree algorithm.

sets $V$ and $M$ become, $V = \{a, X, b, c, d\}$, and $M = \{e\}$. Finally vertex $e$ is selected last. The shortest path from $e$ to any of the vertices of $V$ is $\pi_{e,c} = [e, c]$. Therefore, the sets $V$ and $M$ become, $V = \{a, X, b, c, d, e\}$ and $M = \phi$. Hence, the Steiner tree identified by the algorithm has the following edges: $(a, X)$, $(X, b)$, $(X, c)$, $(c, d)$, and $(c, e)$. The weight of the tree is equal to 7.

Another heuristic that could be applied to find a sub-optimal Steiner tree is based on a variation of a minimum spanning tree algorithm. The heuristic can be summarized as follows. Let $M$ be the set of the nodes (vertices) in which the net has pins. We first find the shortest paths between all pairs of nodes in $M$. There are exactly $\binom{n}{2}$ such paths, where $n$ is equal to the number of nodes in $M$. These paths are sorted in increasing lengths, and are processed one at a time. The shortest path identifies the first branch of the Steiner tree. Each following path identifies another branch of the tree. The algorithm stops when all nodes have been covered and connected. This may very likely happen before processing all paths. A formal description of this heuristic is given in Figure 6.18.

If the graph is a grid graph, then Lee algorithm modified for multipin nets, as described in the previous chapter, can be used to route the net between the global cells (vertices). This algorithm will be described in the following subsection. All three heuristics exhibit the same time complexity and quality of solution.

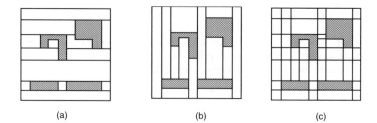

(a)                         (b)                         (c)

**Figure 6.19** Two-dimensional routing model. Shaded spaces indicate cells. The unshaded spaces are routing areas. (a) Horizontal routing areas. (b) Vertical routing areas. (c) Routing regions model.

## 6.4.2 Global routing by maze running

The first step in global routing is to model the routing regions. Figure 6.19 depicts a two-dimensional channel model. To simplify the explanation, we drop the requirement that all routing regions be channels.

Horizontal and vertical routing areas are defined by extending the horizontal and vertical edges of the placed cells up to the bounding frame. This is illustrated in Figure 6.19(a) and (b). Routing regions in the model are the intersections between horizontal and vertical routing areas. This is illustrated in Figure 6.19(c). Notice that this is a different approach to defining the routing regions from what we have seen in the previous section. This approach is straightforward. However, the routing regions are not guaranteed to be channels. This will require a switchbox router or a grid router for the following detailed routing step. Nevertheless, we shall adopt this last technique of routing region definition to illustrate the maze running approach to global routing. We shall also be referring to routing regions as channels.

Once the routing channels (regions) have been identified, the task now is to assign nets to them. To accomplish this, the channels are modelled by a weighted undirected graph called *channel connectivity graph*. Nodes of the graph correspond to channels and edges between nodes indicate that the corresponding channels are adjacent. For two-layer routing, each node is assigned two weights giving the horizontal capacity (width) and the vertical capacity (length) of the corresponding channel (see Figure 6.20) (Rothermel and Mlynski, 1983).

The sequential approach is the simplest and most widely used approach to global routing. This approach consists of picking one net at a time and finding an optimal (or sub-optimal) Steiner tree which covers all the pins of the net. Two general approaches are possible in this case: (1) the order dependent approach and (2) the order independent approach.

For order independent global routing, each net is routed independently of all other nets. Then congested passages are identified and the affected nets are rerouted, while penalizing paths going through such passages. This approach avoids net ordering and considerably reduces the complexity of the search space since the only obstacles are the cells. However, this might require a large number of iterations before a feasible global routing solution is found.

For order dependent global routing, first the nets are ordered according to some criteria. Then the nets are routed in the resulting order, while updating the available routing space after each net. The search is slightly more complex, since the number of obstacles has increased (cells and already routed nets). Furthermore, net ordering is crucial to the final outcome of the global routing step. Both approaches are somewhat similar, in that, they try to identify a Steiner tree for one net at a time. In the remainder of this section, we shall be describing the order dependent approach only.

Each side of a block is unambiguously attached to a unique channel. Hence, each pin is uniquely associated with a channel. Therefore, the nodes of the channel connectivity graph in which a net has pins are unambiguously determined. Then globally routing a net amounts to determining a tree that covers all those nodes in which the net has pins.

To illustrate the search process, we shall focus on the easy case when the net has pins in only two nodes. In that case, assigning a net to channels is accomplished by searching for a path in the channel-graph. A path from the channel which contains the source node to the one that contains the target node is searched. The search procedure is similar to the one used in Lee algorithm. For simplicity, we assume that the length of all edges in the channel-graph is one unit. As in Lee algorithm, starting from a node labelled $k$, all adjacent nodes are labelled $k + 1$. The labelling procedure continues until the target node is reached. The shortest path in the channel graph is found by a sequence of nodes with decreasing labels. Once the path is found, the net is assigned to the channels and for all nodes (channels) in the path the capacity weights $w$ and $l$ are decreased according to the width and length of the net to be routed.

Figure 6.20(a) gives a placement of cells with two nets, $A - A'$ and $B - B'$, to be connected. Figure 6.20(b) shows the corresponding channel-graph. Pin $A$ is adjacent to channel number 6 and pin $A'$ to channel number 8. The shortest path in the graph from node 6 to node 8 (6–7–8) is shown in Figure 6.20(d).

Observe that the weights of nodes in this part of the graph in Figure 6.20(d) are updated. The width and length of the channel are reduced by an amount equal to the space used by the wiring segment. If a path bends in a channel then both its vertical and horizontal weights are reduced. *If the*

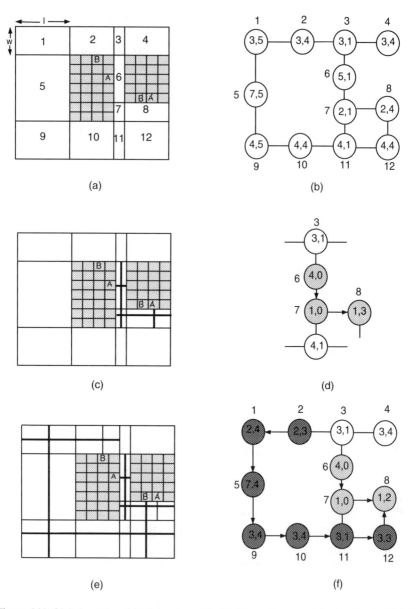

**Figure 6.20** Global routing. (a) Channel model. (b) Channel connectivity graph. (c) Global routing $A-A'$. (d) Actual channel graph with $A-A'$ routed. (e) Global routing $B-B'$. (f) Actual channel graph with $A-A'$ and $B-B'$ routed.

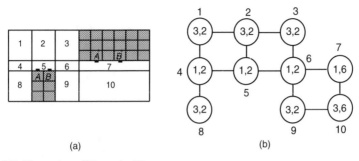

(a)     (b)

**Figure 6.21** Illustration of Example 6.3.

*path goes only vertically (horizontally) then only the weight corresponding to its length (width) is reduced.* This means that one entire row and one entire column of each channel (6, 7, and 8) are assigned to this net as shown in Figure 6.20(c). The next step is to assign channels to net $B - B'$. Pin $B$ is adjacent to channel 2 and pin $B'$ to channel 8. Let us look at two possible paths which connect node 2 to node 8. One is the path 2–3–6–7–8 and the other is 2–1–5–9–10–11–12–8. Channels 6 and 7 have a width of unity and have been assigned to net $A - A'$. Due to this assignment the horizontal capacities of channels 6 and 7 have been depleted to zero. The available shortest path therefore is 2–1–5–9–10–11–12–8. The updated weights in the channel-graph are shown in Figure 6.20(f). If the net has a third pin located in another vertex of the graph, then the expansion will continue at that third pin and will terminate when any of the three nodes 6, 7, or 8 is reached.

Another application of the above technique is to determine the required separation between cells in order to ensure routability of the chip. By routability is meant the certainty that *all* the nets of the circuit will be completely connected by the detailed router. Therefore, the global router is used only to determine the required separation between cells. When detailed routing is done for the entire chip, it is not necessary that the detailed router follows exactly the channels assigned to nets by the global router.

The modification to the above method to determine separation between cells would be to start with a zero separation, and this will represent the initial weights of the nodes. Next, every time a path is found, the weights of the corresponding nodes (channels) are increased. At the end of this procedure, the relative placement of the cells is maintained, but the minimum separation between the cells will be as given by the horizontal and vertical weights of the nodes of the channel-graph. Notice here that the estimations are somewhat pessimistic since they assume that nets do not share routing tracks.

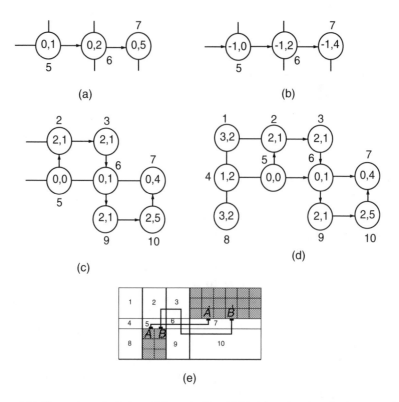

**Figure 6.22** Illustration of solution of Example 6.3. (a) Weights of nodes after loose routing $A-A'$ between nodes 5–6–7. (b) Weights of nodes after loose routing $B-B'$ between nodes 5–6–7. (c) Alternative path for net $B-B'$. (d) Updated channel-graph. (e) Connection of nets $A-A'$ and $B-B'$.

In order to avoid congestion of nets between channels, an upper limit for each node of the channel-graph can be set. The global router then will look for alternative paths in the channel-graph. The following example will further clarify this point.

**Example 6.3** A placement containing two cells is shown in Figure 6.21(a). The corresponding channel-graph is shown in Figure 6.21(b). The cells contain two nets $A-A'$ and $B-B'$. Determine if the circuit is routable.

SOLUTION The weights in the nodes indicate the width ($w$) and length ($l$) of the channel. Since nodes $A$ and $A'$ are adjacent to channels 5 and 7, a shortest path through the graph must be found between these nodes. The path obtained is 5–6–7. The weights of these nodes are updated as shown in Figure 6.22(a). Since the initial widths of these channels were unity, they are now both reduced to zero.

Now $B - B'$ also may use the same path. If this is done, the weights of the nodes corresponding to channels 5, 6, and 7 are further reduced as shown in 6.22(b). Since the updated widths of the channels are $-1$ (negative), the channel widths in the actual placement must be increased by one unit. This can be done by shifting the cells above the channel one unit upwards. Then both the nets $A - A'$ and $B - B'$ can be routed through channels 5, 6, and 7.

Assume that there is no provision to increase the initial separation between cells. Is the placement routable?

Observe that an alternative path from 5 to 7 exists, namely 5–2–3–6–9–10–7. Since all channels except 6 along this path have widths and heights greater than one, and the path from channels 3 to 9 through 6 uses only the length of channel 6 and whose current length is 2, the layout is routable. The weights of the updated nodes are shown in Figure 6.22(c). The updated channel-graph and wires through the channels on the given placement are shown in Figures 6.22(d) and (e) respectively.


## 6.5 INTEGER PROGRAMMING

Here, global routing is formulated as a 0–1 integer program. The layout is modelled as a grid-graph, where each node represents a grid cell (super cell). The boundary between any two adjacent grid cells $l$ and $k$ is supposed to have a capacity of $c_{l,k}$ tracks. This corresponds to a positive weight $c_{l,k}$ on the arc linking nodes $l$ and $k$ in the grid-graph. For each net $i$, we need to identify the different ways of routing the net. Suppose that for each net $i$, there are $n_i$ possible trees $t_1^i, t_2^i,..., t_{n_i}^i$, to route the net. Then, for each tree $t_j^i$, we associate a variable $x_{i,j}$ with the following meaning:

$$x_{ij} = \begin{cases} 1 & \text{if net } i \text{ is routed according to tree } t_j^i \\ 0 & otherwise \end{cases} \qquad (6.2)$$

For each net $i$, we associate one equation to enforce that only one tree will be selected for that net,

$$\sum_{j=1}^{n_i} x_{i,j} = 1 \qquad (6.3)$$

Therefore, for a grid graph with $M$ edges and $T$ trees, we can represent the routing trees of all nets as a 0–1 matrix $A_{M \times T} = [a_{i,p}]$ where,

$$T = \sum_{i=1}^{N} n_i \qquad (6.4)$$

where $N$ is the number of nets, and,

$$a_{i,p} = \begin{cases} 1 & \text{if edge } i \text{ belongs to tree } t_k^{\,l} \text{ and } p \text{ as defined in Equation 6.6.} \\ 0 & otherwise; \end{cases}$$

$$(6.5)$$

$$p = \sum_{m=1}^{l-1} n_m + k \qquad (6.6)$$

A second set of equations is required to ensure that the capacity of each arc (boundary) $i$, $1 \le i \le M$, is not exceeded , i.e.,

$$\sum_{k=1}^{N} \sum_{l=1}^{n_k} a_{i,p} \times x_{l,k} \le c_i \qquad (6.7)$$

Finally, if each tree $t_j^i$ is assigned a cost $g_{i,j}$, then a possible objective function to minimize is,

$$F = \sum_{i=1}^{N} \sum_{j=1}^{n_i} g_{i,j} \times x_{i,j} \qquad (6.8)$$

Therefore, a possible 0–1 integer programming formulation of global routing is,

$$\begin{cases} \sum_{i=1}^{N} \sum_{j=1}^{n_i} g_{i,j} x_{i,j} \leftarrow minimize \\ \text{subject to :} \\ \sum_{j=1}^{n_i} x_{i,j} = 1 & 1 \le i \le N \\ \sum_{k=1}^{N} \sum_{l=1}^{n_k} a_{i,p} x_{l,k} \le c_i & 1 \le i \le M, \text{ and } p \text{ as defined in 6.6} \\ x_{k,j} = 0, 1 & 1 \le k \le N, \text{ and } 1 \le j \le n_k \end{cases}$$

$$(6.9)$$

Note that if $g_{i,j} = g$ for all $i$ and $j$ and if we change the objective to the maximization of $F$, then this is equivalent to seeking a solution that achieves the maximum number of connections.

**Example 6.4**   Suppose you are given the gate-array layout of Figure 6.23(a). The layout has four cells and three nets. Assume that the capacity of each boundary is equal to two tracks. Formulate the corresponding global routing problem as a 0–1 integer program.

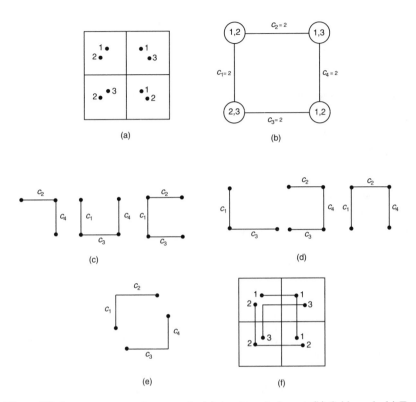

**Figure 6.23** Integer programming example. (a) A gate-array layout. (b) Grid-graph. (c) Trees of net 1. (d) Trees of net 2. (e) Trees of net 3. (f) Completed routing.

SOLUTION   First, a grid-graph is built (see Figure 6.23(b)). Each node of the graph is marked with the label of those nets which have pins in that node. There are three trees for net 1, three trees for net 2, and two trees for net 3. These are given in Figure 6.23(c), (d), and (e). The cross capacity of each boundary is equal to two tracks. There are four boundaries corresponding to four edges in the grid graph. The matrix $A$ will be as follows:

|   | $t_1^1$ | $t_2^1$ | $t_3^1$ | $t_1^2$ | $t_2^2$ | $t_3^2$ | $t_1^3$ | $t_2^3$ |
|---|---|---|---|---|---|---|---|---|
| 1 | 0 | 1 | 1 | 1 | 0 | 1 | 1 | 0 |
| 2 | 1 | 0 | 1 | 0 | 1 | 1 | 1 | 0 |
| 3 | 0 | 1 | 1 | 1 | 1 | 0 | 0 | 1 |
| 4 | 1 | 1 | 0 | 0 | 1 | 1 | 0 | 1 |

Assume that the cost of tree $t_i^j$ is equal to the length of that tree. In that case,

$$g_{1,1} = 2, \quad g_{1,2} = 3, \quad g_{1,3} = 3,$$
$$g_{2,1} = 2, \quad g_{2,2} = 3, \quad g_{2,3} = 3, \text{ and}$$
$$g_{3,1} = 2, \quad g_{3,2} = 2.$$

Therefore, the resulting 0–1 integer program is

$$
\begin{cases}
F = 2x_{1,1} + 3x_{1,2} + 3x_{1,3} + 2x_{2,1} + 3x_{2,2} + 3x_{2,3} + 2x_{3,1} + 2x_{3,2} \leftarrow minim \\
Subject\ to: \\
x_{1,1} + x_{1,2} + x_{1,3} = 1 \\
x_{2,1} + x_{2,2} + x_{2,3} = 1 \\
x_{3,1} + x_{3,2} = 1 \\
x_{1,2} + x_{1,3} + x_{2,1} + x_{2,3} + x_{3,1} \leq 2 \\
x_{1,1} + x_{1,3} + x_{2,2} + x_{2,3} + x_{3,1} \leq 2 \\
x_{1,2} + x_{1,3} + x_{2,1} + x_{2,2} + x_{3,2} \leq 2 \\
x_{1,1} + x_{1,2} + x_{2,2} + x_{2,3} + x_{3,2} \leq 2 \\
x_{i,j} = 0,1 \quad 1 \leq i \leq 3, 1 \leq j \leq 3
\end{cases}
$$

The integer programming formulation is elegant and finds a globally optimum assignment of the nets to routing regions. However, this approach suffers from the following problems:

1. We need to identify several Steiner trees for each net. This could be a very time consuming step. Moreover, how many such trees would be enough? Obviously, the number of trees should not be too large, as the size of the integer program is proportional to the number of such trees. Furthermore, it is not an easy task to enumerate, say, the best $n_i$ Steiner trees for each net $i$.
2. The trees should be selected so as to guarantee the feasibility of the problem. Hence, when the formulated integer program does not have a solution we cannot tell if it is because of a bad selection of the routing trees or if the problem is infeasible no matter how many trees are included.
3. There are too many $a_{i,j}$'s, leading to too many constraints.
4. There may be too many arcs, i.e., too many $c_i$'s, leading to too many constraints.
5. All constraints are integer constraints.

As we will see in a later section, hierarchical approaches come to the rescue, and can be used to solve some of the above problems. This problem decomposition will of course be at the price of not achieving global optimality!

## 6.6 GLOBAL ROUTING BY SIMULATED ANNEALING

The first reported application of simulated annealing to global routing appeared in 1983 (Vecchi and Kirkpatrick, 1983). The authors formulate the problem as an unconstrained integer program, where all edge capacities are equal to one. Only two terminal nets are considered. Furthermore, only routes with one bend are considered. The cost function used is a sum of the squares of the loads of all individual routing regions.

In this section, we will describe the simulated annealing formulation of global routing adopted in the TimberWolf package (Sechen and Sangiovanni-Vincentelli, 1986). TimberWolf is a package for the placement and global routing of standard-cell designs with two metal layers.

After an initial placement phase, TimberWolf proceeds with global routing. Global routing is solved in two stages. The objective of the first stage is to assign the nets to the horizontal routing channels so as to minimize the overall channel densities. At the end of the first stage, all nets that are switchable (assignable to an adjacent channel) are identified. The goal of the second stage is to attempt to reduce the overall channel densities by changing the channel assignment of the switchable nets.

After global routing, TimberWolf proceeds with a refinement of the placement by randomly interchanging neighbouring cells. After each interchange, both stages of the global router are invoked to reroute the nets affected by the interchange. It is only during the second stage of global routing (as well as the placement refinement phase) that simulated annealing is used.

### 6.6.1 The first stage

In this stage, an order independent sequential approach is taken. The nets are taken one at a time and routed.

Before describing the algorithm of this stage, we first need to define the required terminology (Sechen and Sangiovanni-Vincentelli, 1986).

A group of pins of a given cell that are internally connected are called a *pin cluster*. The pins of the same cluster are all equivalent. The $x$-coordinate of a pin cluster $P$ is equal to the average of the $x$-coordinates of its constituent pins, i.e.,

$$x(P) = \frac{1}{|P|} \times \sum_{i \in P} x(i) \qquad (6.10)$$

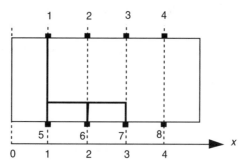

**Figure 6.24** Illustration of the pin cluster concept; $P = \{1, 5, 6, 7\}$ and $x(P) = \frac{1+1+2+3}{4} = \frac{7}{4}$.

These concepts are illustrated in Figure 6.24.

A portion of a net connecting two pin clusters, say $P_1$ and $P_2$, is called a *net segment*. If $P_1$ and $P_2$ belong to two different cells placed on the same row and both have pins on the top and bottom sides of the cells, then the net segment connecting both clusters is called a *switchable segment*. Hence, a switchable segment can be assigned to the routing channel above the cell row or the channel below it. The decision as to which channel each switchable segment should be assigned to is based on the cost function.

The cost function used is equal to the sum of the total channel densities, i.e.,

$$D = \sum_{\forall c} d(c) \qquad (6.11)$$

where $d(c)$ is the density of channel $c$, which is equal to the number of nets assigned to the channel.

The global routing algorithm of the first stage has four distinct steps which are executed for each net.

1. *Initialization:* All pin clusters of the net are identified, together with their $x$-coordinates. Then the pin clusters are sorted on their $x$-coordinates, from smallest to largest.

2. *Construction of a cluster graph:* In this step, a cluster graph is constructed, where nodes model the pin clusters and edges possible connections (net segments) between the corresponding pin clusters.

3. *Construction of a minimum spanning tree:* In this step, Kruskal algorithm is used to construct a minimum spanning tree of the cluster graph.

4. *Identification of all net segments:* In this final step of the first stage, individual pins within the pin clusters are selected to form the actual net segments. Also, if the net segment is switchable, then two pairs of pins are selected, one pair for the upper row, and one for the lower row.

A semi-formal description of the algorithm used in the first stage is given in Figure 6.25.

**Example 6.5**    Assume we are given the partial standard-cell design of Figure 6.26(a). Suppose that each pin cluster has exactly two pins available on opposite sides of the cell. Assume further that one of the nets is connecting exactly five pin clusters labelled $a$, $b$, $c$, $d$, and $e$ as illustrated in Figure 6.26(a). We would like to illustrate how the cluster graph is constructed for this net.

SOLUTION    Once all pin clusters have been identified, we determine the $x$-coordinate of each pin cluster and sort them on their $x$'s. This results in the following sorted sequence $clusters = [a, e, d, b, c]$. The sorted sequence will be processed one element at a time until the cluster graph is constructed (one edge at a time).

At the first execution of the outer repeat loop, $P_1 = a$, $a$ is removed from the sequence $clusters$, and the variable $TorB = 0$. $TorB$ is equal to 0, +1, −1, to indicate that the next pin is at the same row, the bottom row, or the top row respectively. The closest pin to the right of $a$ that is located at the top, bottom, or same row is pin cluster $d$. Therefore, $P_2 = d$, $row(d) = 2$, and $TorB$ is set equal to 1 ($P_2$ is at the bottom of $P_1$). Hence the edge $(a, d)$ is added to the cluster graph. The next closest pin cluster to the right of $a$ is $b$. However because $row(b) = 1 = row(a)$, the edge $(a, b)$ is not added. The reason is that there must exist at least one intermediate pin cluster between $a$ and $b$. Therefore, $b$ should be connected to its closest left neighbour not to $a$. Then, the next closest pin cluster to the right of $a$ is $c$. Since $row(c) = row(a) = 1$, edge $(a, c)$ is not considered. At this time, we exit from the inner repeat loop, $TorB$ is reset to 0, and the $head(clusters)$ returns pin cluster $e$, which is removed from $clusters$. The closest pin cluster to the right of $e$ and located in the same row or adjacent row is pin cluster $d$. Therefore, $P_2 = d$, $TorB = -1$, and the edge $(e, d)$ is added to the graph. No other pin in the same or adjacent row is to the right of $e$. Therefore, we exit from the inner repeat loop. Continuing in this way, edge $(d, b)$, then edge $(b, c)$ will be added to the cluster graph. At this moment, the sequence $clusters$ becomes empty. The cluster graph thus constructed is given in Figure 6.26 (b). The positive numbers indicated on the edges of the graph represent the lengths of the corresponding net segments. Notice that the graph is already in the form of a minimum spanning tree. As observed by the authors of this approach (Sechen and Sangiovanni-Vincentelli, 1986), the cluster graph will in most cases turn out to be a minimum spanning tree. Notice that, among the net segments selected, only segment $b-c$ is

**Algorithm** Stage_1_Global_routing;
**Begin**
  **For Each** net $n$ **Do**
    Identify the pin clusters of net $n$;
    **ForEach** pin cluster $P$ find $x(P)$;
    Sort the pin clusters of $n$ on their $x$-coordinates;
    **Repeat**
    (* Construct the cluster graph corresponding to net $n$ *)
      $TorB \leftarrow 0$;
      (* $TorB = 1, -1, 0$, if $row(P_2) = row(P_1) + 1, row(P_1) - 1$, or $row(P_1)$, respectively. *)
      **If** clusters=[ ] **Then Exit**;
      **EndIf**;
      $P1 \leftarrow Head(Clusters)$; (* Get the leftmost element of clusters *)
      **Repeat**
        **find** the closest pin cluster $P_2$ to the right of $P_1$
        **such that** $row(P_2) = row(P_1), row(P_1) + 1$, or $row(P_1) - 1$
        **If** there is no such $P_2$ **Then**   **Exit**; (* exit from the innermost repeat loop *)
        **Else**
          **Case**

| | |
|---|---|
| (1) $row(P_2) = row(P_1)$: | $TorB = 0$;<br>(* connect nodes $P_1$ and $P_2$ *)<br>add the edge $(P_1, P_2)$ to the cluster graph;<br>**Exit**; |
| (2) $row(P_2) = row(P_1) + 1$: | **If** $TorB \neq 1$ **Then**<br>**Begin**<br>  add the edge $(P_1, P_2)$ to the cluster graph;<br>  **If** $TorB = -1$ **Then**   **Exit**;<br>  **Else** $TorB = +1$; (* $P_1$ is on top *)<br>  **EndIf**;<br>**EndIf**; |
| (3) $row(P_2) = row(P_1) - 1$: | **If** $TorB \neq 1$ **Then**<br>**Begin**<br>  add the edge $(P_1, P_2)$ to the cluster graph;<br>  **If** $TorB = +1$ **Then**   **Exit**;<br>  **Else** $TorB = -1$; (* $P_2$ is on top *)<br>  **EndIf**;<br>**EndIf**; |

          **End Case**
        **EndIf**
      **Until** $doomsday$
    **Until** $doomsday$;
    Construct a minimum spanning tree (MST) on the nodes of the cluster graph;
    **ForEach** net segment in the MST **Do**
      (* Identify all switchable net segments *)
      **If** the segment connects pin clusters at the same row **Then**
        **If** the segment is switchable **Then**
          two pairs of pins are selected;
          (* one pair for the channel above and one for the channel below; *)
        **EndIf**;
      **EndIf**;
    **EndFor**;
  **EndFor**;
**End**.

**Figure 6.25** The first stage of TimberWolf global router.

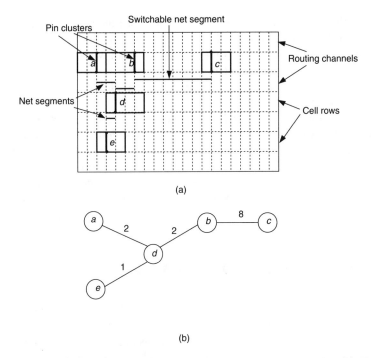

**(a)**

**(b)**

**Figure 6.26** Construction of the cluster graph. (a) A partial standard-cell design. (b) Cluster graph constructed during stage 1 of TimberWolf.

switchable. Therefore for this segment, two pairs of pins are selected, one if the segment is assigned to the channel above, and one if it is assigned to the channel below.

## 6.6.2 The second stage

In stage 2, the simulated annealing search technique is used to refine the global routing solution produced by the stage 1 sequential algorithm. Only switchable net segments are considered for rerouting. The simulated annealing algorithm has been given and explained in Chapters 2 and 3. In this subsection we shall describe only the cost function used, how new solutions are generated, and the cooling schedule.

The objective of this stage is to minimize the total channel density as expressed by Equation 6.11. The *generate* function used to obtain new solutions is as follows. First, a switchable segment is randomly selected from the pool of switchable net segments. Then, the channel assignment of the selected segment is switched from its current channel to the opposite one. If the switch reduces the value of the cost function, then the switch is accepted. If the new solution has the same cost as the previous one, then the switchable segment is assigned to the channel with the smaller density over the span of the net segment. The purpose of this decision is to facilitate the following step of detailed routing. New solutions with higher cost functions are not accepted.

As for the cooling schedule, the temperature is maintained equal to zero throughout the search. Hence only downhill moves are accepted. Furthermore, since $T = 0$, there is no need for the inner loop, nor a need to update the schedule parameter. The stopping criterion used is a function of the number of switchable segments. The search is stopped after the generation of $S = 30 \times N$ new states, where $N$ is the number of switchable net segments.

## 6.7 HIERACHICAL GLOBAL ROUTING

In order to reduce the complexity of the overall global routing problems, hierarchical approaches proceed with a hierarchical decomposition of the problem into subproblems. The subproblems are solved individually. Then the partial solutions are combined to produce a solution to the original global problem.

There have been several hierarchical formulations of the global routing problem. Among the first reported formulations was that of Burstein and Pelavin (1983). Several other hierarchical formulations were also reported (Ohtsuki, 1986; Lauther, 1987; Sadowska, 1986). The approaches proposed by Sadowska (1986) and Lauther (1987) are similar and were reported to generate superior solutions to all other hierarchical approaches. This approach to global routing is used in both the PROUD (Tsay et al., 1988) and BEAR (Dai, 1988) layout programs.

In this section we shall be describing the hierarchical formulation due to Sadowska (1986). It is a top-down approach. The top level is the entire layout, which is decomposed in a top-down fashion. At each level of the hierarchy, an attempt is made to minimize the cost of nets crossing cut lines of this current level of the hierarchy. The set of cut lines is derived from the design floorplan. The floorplan is not constrained to have a slicing structure.

This approach can be used with various design styles. We shall illustrate the approach for the sea-of-gates design style. In this design style, the cells are porous, i.e., through and over the cell routing is allowed. Therefore, it is realistic to assume that the routing resources are uniformly distributed across the layout surface. At the lowest level of the hierarchy,

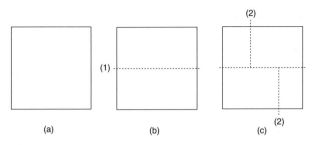

**Figure 6.27** Hierarchical partitioning based global routing.

the layout surface is divided into $R \times R$ grid regions. The capacity of each grid region boundary is equal to $C$ tracks. Hence, no more than $C$ nets (net segments) are allowed to cross a region boundary (see Figure 6.27).

Let $R_l$ be the number of grid regions of a given cut line $l$. When a cut line is applied, it can readily be divided into $M = \frac{R_l}{C}$ sections, each $C$ grid units long (see Figure 6.28). Suppose that there are $N$ nets, each with pins on both sides of a current cut line. Obviously we must have $N \leq M \times C$. Routing these nets amounts to assigning each net to a specific section of the cut line without exceeding the capacity of any of the sections. Such a global routing solution (at this level of hierarchy) is called a feasible solution. Let $w_{ij}$ be the cost of assigning net $i$ to section $j$ of the current cut line. Then a minimum global routing solution will be a feasible solution with minimum assignment cost. Hence, at each level of the hierarchy, and for each cut line at that level, global routing is formulated as a linear assignment problem as follows.

Let

$$x_{i,j} = \begin{cases} 1 & \text{if net } i \text{ is assigned to section } j \\ 0 & otherwise \end{cases}$$

Let us assume that each net is allowed to cross the cut line exactly once. Therefore,

$$\sum_{j=1}^{M} x_{ij} = 1, \quad 1 \leq i \leq N \qquad (6.12)$$

Furthermore, the assignment should not exceed the capacity of any of the $M$ sections of the cut line, i.e.,

$$\sum_{i=1}^{N} x_{ij} \leq C, \quad 1 \leq j \leq M \qquad (6.13)$$

Let $w_{ij}$ be the cost of assigning net $i$ to section $j$ of the current cut line, $1 \leq i \leq N$, and, $1 \leq j \leq M$. Then an optimum assignment will be one which minimizes the following cost function,

$$cost = \sum_{i=1}^{N} \sum_{j=1}^{M} w_{ij} x_{ij} \qquad (6.14)$$

This is a formulation of a classic network flow problem, whose minimum-cut maximum-flow solution can readily be obtained using available packages. Notice also that this formulation is very similar to the 0–1 integer programming formulation of Section 6.5. A possible choice of assignment costs is the sum of net lengths, where the lengths of individual nets are approximated by the semi-perimeter method. Another possible cost function would be to use a weighted sum of the net lengths, where critical nets are given higher weights. Still another objective function would be to minimize the number of wiring bends so as to favour assigning nets to their ideal sections (see Figure 6.28).

To each cut line corresponds a network flow problem. Once all of these problems are solved individually, the outcome is the specification for each region of a netlist in terms of pseudo-pins (locations on the region's boundaries at which nets cross) and cell terminals within that region. The detailed router will then take this netlist as input and determine the detailed routes within each region.

## 6.8 OTHER APPROACHES AND RECENT WORK

Recall that a crucial preparatory step to global routing is the determination of the routing areas. Numerous approaches have been suggested to perform such a task. Most of these approaches were described in this chapter (Cai and Otten, 1989; Ohtsuki, 1986).

In a recent paper by Cai and Wong (1991), an algorithm is given to define the routing areas as rectangular channels and switchboxes, such that the number of switchboxes is minimized. The same authors, in a more recent paper (1993), present an algorithm which decomposes the routing areas into rectangular channels while minimizing the number of L-shaped channels.

Another important problem encountered during global routing as well as detailed routing is that of finding a Steiner tree for each net. There are numerous works on this subject (Sarrafzadeh and Wong, 1992; Cong et al., 1992; Kahng and Robins, 1992).

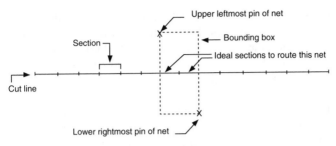

**Figure 6.28** Decomposition of a cut line into sections.

In one approach (Sarrafzadeh and Wong, 1992), a bottom-up hierarchical strategy to Steiner tree construction is described. At each level of the hierarchy a collection of partial Steiner trees is enumerated using a variation of minimum spanning tree algorithm. At each higher level of the hierarchy, the trees of the lower level are merged, while removing duplicate edges and avoiding the creation of cycles. The tree corresponding to the top level of the hierarchy is the desired Steiner tree for the current net.

Another approach uses a modification of minimum spanning tree and Steiner tree algorithms to grow a routing tree with bounded length and bounded maximum interconnect delay (Cong et al., 1992). In yet another recent approach (Kahng and Robins, 1992), an iterative heuristic is used to find optimal Steiner points. These points are found and added in a greedy fashion one at a time until no further reduction in the overall tree length is achieved. Experimental results indicate that the cost of the Steiner tree found is bounded by $\frac{4}{3}$ that of the optimal Steiner tree.

During floorplanning, location of pins on the boundary of blocks is not known. One of the main tasks of floorplan sizing is the determination of these pin locations. The pin assignment and global routing problems are heavily interdependent. Usually the quality of pin assignment solution is measured by the overall interconnection length and routing area, which require that global routing be performed. Hence, it seems that if both tasks are integrated we will achieve a solution of superior quality in terms of pin assignment cost as well as routability. One such approach has recently been reported, where global routing is used first to assign pins to particular sides of the blocks. Then in a second stage, channels are processed one at a time in order to assign pins to particular locations of the corresponding channel boundary (Cong, 1991).

With the advent of VLSI, the speed of circuits is becoming dominated by the interconnect delays. In order to avoid having long interconnect delays on the timing critical paths, there have been attempts to make the global routing step driven by the timing requirements of the circuit.

Minimizing the overall interconnection length will not necessarily lead to minimum interconnect delays. The number of wiring bends as well as other electrical characteristics of the nets are as important as the net length itself. In a recent approach (Cong *et al.*, 1992), global routing is performed while minimizing the total connection length as well as the longest interconnection delay. The Steiner tree heuristic suggested grows greedily a tree with bounded overall length and bounded maximum delay. Another timing driven global routing approach has been reported (Prasitjutrakul and Kubitz, 1992), where global routing is integrated with timing analysis. The routing process is guided by the interconnect delays. Whenever a connection is routed the slacks of the affected paths are updated at the corresponding path sinks. The cost function maximized is the slack of the longest path.

There are other approaches to global routing that were not described in this chapter. For example, in one approach global routing is formulated as a multicommodity network flow model (Shragowitz and Keal, 1987). The solution is obtained in two steps. In the first step, the unconstrained problem is solved. Then in the second step, an iterative procedure is applied, where at each iteration the connection that will result in a maximum decrease in overflow is rerouted.

In another approach (Clow, 1984), a line search algorithm was adopted to globally route general cell layouts, thus avoiding use of a grid. The algorithm is a generalization of the $A^*$ search algorithm widely use in artificial intelligence. A path is found between a source $s$ and a destination $d$ by greedily adding one line segment at a time, starting from the source $s$, until we reach the target $d$. This is achieved as follows. A graph with source $s$ and sink $d$ is constructed, adding one edge at a time, until a minimum path is established from $s$ to $d$. The edges correspond to line segments, and the vertices to line intersection points. Each edge is assigned a cost equal to the length of the corresponding line segment. The search for a shortest source-to-sink path is performed using the $A^*$ algorithm.

Recently, global routing was solved using neural networks (Shih *et al.*, 1991). The authors adopted a two-layer neural network. One layer is used to minimize net lengths and obtain an even distribution of the nets over the routing channels. The second layer is used to enforce the channel capacity constraints.

Traditionally, global routing has been used to elaborate a routing plan to be executed by the detailed router. However, another use of global routing is to assess floorplanning and placement solutions. During placement, global routing is used to guide the placement procedure into producing a routable placement. One of the early reported works in this respect describes a constructive global routing driven placement algorithm (Shragowitz *et al.*, 1988). The placement solution is constructed one slice at a time, starting from the left of the chip until all cells are placed. Whenever a group of cells is placed, global routing is performed to connect the recently placed cells to already placed ones and reserve the necessary routing resources. The simulated annealing approach described in this chapter also combines placement with global routing.

Routing in general can be a very time consuming process. Nevertheless, almost all of the proposed solution approaches lend themselves easily to parallel implementations. One of the latest reported parallel implementations of global routing describes two strategies for parallelizing this task (Rose, 1990). For the first strategy, several nets are routed in parallel. In the second strategy, several routing trees for each net are evaluated in parallel. The paper reports significant speed-up for both strategies.

There are very few books on the subject of physical design in general and global routing in particular. The book edited by Ohtsuki (1986) contains a chapter on the subject. Unfortunately the book is out of print. A more recent text which contains a small section on global routing is the book edited by Preas and Lorenzetti (1988). The most recent text, which gives a thorough and formal discussion of the problem of global routing is that of Lengauer (1990).

## 6.9 CONCLUSION

Global routing is a preparatory step to detailed routing. Usually, global routing is executed for the purpose of elaborating a routing plan for the detailed router.

Global routing is formulated differently for different design styles. For the gate-array design style the routing regions consist of fixed capacity horizontal and vertical channels. Since the array has a fixed size and fixed routing space, the objective of global routing in this case is twofold: (1) check the feasibility of detailed routing, and (2) elaborate a routing plan.

For the standard-cell design style the routing regions are horizontal channels with pins at their top and bottom boundaries. Global routing consists of assigning nets to these channels so as to minimize channel congestion and overall connection length. Here, the channels do not have pre-fixed capacities. Channels can always be made wider to achieve routability.

In building-block design style the cells are of various shapes and sizes. This leads to irregular routing regions. These routing regions may be decomposed into horizontal and vertical channels, and sometimes switch-boxes. The identification of these routing regions is a crucial first step to global routing. Here again, the routing regions do not have pre-fixed capacities. For both the standard-cell and building-block layout styles the objective of global routing is to minimize the required routing space and overall interconnection length while ensuring the success of the following detailed routing step. Therefore the cost function is a measure of the overall routing and chip area. Constraints could be a limit on the maximum number of tracks per channel and/or constraints on performance.

An important problem we are faced with in all design styles is the identification of the shortest set of routes to connect the pins of individual nets. This is known as the Steiner tree problem. There are several good heuristics for this problem, some of which were given in this chapter.

Global routing is a natural multi-commodity network flow problem. Several approaches have been reported in the literature to solve this problem. They fall into four general classes: (1) the sequential approach, (2) integer programming, (3) the random search techniques such as simulated annealing, and (4) hierarchical decomposition. In this chapter, we described all four approaches. Among these four approaches, the most powerful seems to be the integer programming approach when combined with hierarchical decomposition.

## 6.10 BIBLIOGRAPHIC NOTES

The sequential approaches, especially those relying on a maze running search, are the easiest and exhibit a high completion rate.

The integer programming formulation of global routing is elegant and eliminates the need to order the nets. Unfortunately it may lead to unacceptably large programmes. However, when combined with a hierarchical decomposition approach, this technique exhibits acceptable run times and solutions of superior quality to those obtained from other approaches.

The hierarchical approaches are very fast and usually lead to superior global routing solutions to all other approaches. Hence, this approach is the most suitable for large designs. Nets are processed simultaneously, thus leading to an order independent approach.

The stochastic iterative approaches usually exhibit the largest run time, since the search space of rip-up and rerouting can be quite large. Therefore, this approach cannot be used with large designs. However, they usually achieve high completion rate and good solution quality.

# EXERCISES

**Exercise 6.1**   Discuss the basic differences between detailed routing and global routing. Which one is the more difficult problem? Explain.

**Exercise 6.2**   What are the general uses of global routing? Explain.

**Exercise 6.3**   Briefly explain the two main global routing models.

**Exercise 6.4**   Discuss the global routing problem and objectives for various layout styles, namely, gate-array, sea-of-gates, standard-cell, and building-block design styles.

**Exercise 6.5**   Using the floorplan given in Figure 6.29:

(a) Construct the corresponding channel intersection graph.
(b) Describe a global routing algorithm that will use the channel intersection graph to perform the tasks of global routing.

**Exercise 6.6**   (*) For the figure of the previous exercise,

(a) Construct the corresponding order constraint graph.
(b) Suggest at least two techniques for removing cycles from the order constraint graph.

**Exercise 6.7**   Given the floorplan structure of Figure 6.29:

(a) Construct the corresponding channel connectivity graph.
(b) Assume that a net has five pins located on the following block sides: *right(B), bottom(E), left(G), top(C), and left(H)*. Find the channel assignment for the above net using the Steiner tree heuristic of Figure 6.17.
(c) Find the channel assignment for the above net using the Steiner tree heuristic of Figure 6.18.
(d) Compare the trees obtained in (c) and (d).

**Exercise 6.8**   Perform a time complexity analysis of the Steiner tree heuristic of Figure 6.17.

**Exercise 6.9**   Perform a time complexity analysis of the Steiner tree heuristic of Figure 6.18.

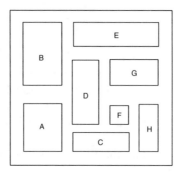

**Figure 6.29** A floorplan structure.

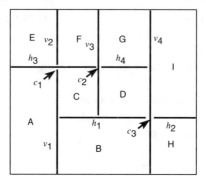

**Figure 6.30** A channel conversion example.

**Exercise 6.10** (*) In some cases (e.g., before floorplan sizing), we do not know beforehand as to which side of the blocks the pins belong. Give a solution approach that will find a side assignment for the pins as well as a channel assignment for the nets.

**Exercise 6.11** Suppose you are given the slicing structure of Figure 6.30, which has nine blocks labelled A to I and five slice-lines. Assume further that the conversion bonuses of the three crossings are:

$bonus(c_1^v) = 0$; $bonus(c_1^h) = 2$;
$bonus(c_2^v) = 1$; $bonus(c_2^h) = 0$;
$bonus(c_3^v) = 1$; $bonus(c_3^h) = 0$;

(a) Construct the slice graph corresponding to this slicing structure. Show all intermediate construction steps.
(b) Use the algorithm of Figure 6.10 to identify the channels which maximize the bonus conversion of all crossings.

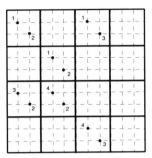

**Figure 6.31** A coarse grid example.

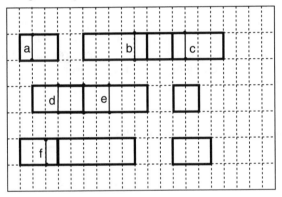

**Figure 6.32** A standard-cell placement example.

(c) Construct the order constraint graph corresponding to the channel structure resulting from (b).

(d) Use the order constraint graph of part (c) to indicate a possible conflict free channel ordering for the following detailed routing step.

**Exercise 6.12** Perform a time complexity analysis of the channel conversion algorithm of Figure 6.10.

**Exercise 6.13** Given the coarse grid of Figure 6.31:

(a) Formulate the corresponding mathematical programme to optimally route the four nets as indicated in that figure. Assume that the cross boundary capacity in the horizontal as well as the vertical direction of the coarse cells is equal to two.

(b) Solve the formulated integer programme using the mathematical programming package available to you.

**Exercise 6.14**   For the mathematical programming approach to global routing, give the size of the resulting mathematical program in terms of the number of nets and the number of trees per net.

**Exercise 6.15**   Using the standard-cell placement of Figure 6.32 apply the algorithm of Figure 6.25 to construct the cluster graph corresponding to the net $\{a, b, c, d, e, f\}$.

**Exercise 6.16**   Does the global router provide pessimistic or optimistic estimation of routing resources needed for detailed routing? Discuss.

**Exercise 6.17**   Assume a layout of cells where the '*power − index*' of each cell is known. The amount of current drawn by the cell is proportional to its power-index. It is required that the width of a routing segment belonging to a net be proportional to the current flowing through the net.

1. In the global routing phase that determines separation between cells, given the power-index for each cell, explain what modifications are to be made to determine channel dimensions.

2. In the detailed routing phase, if the current flowing through each net is assumed to be constant then the standard Lee algorithm can be applied. Explain how you will modify the Lee algorithm to route nets of different widths.

**Exercise 6.18**   (*) **Programming Exercise:** Implement a program which takes as input a slicing structure described as Polish expression and constructs the corresponding channel connectivity graph (CCG) similar to that of Figure 6.12.

**Exercise 6.19**   (*) **Programming Exercise:** Implement a program which takes as input a channel connectivity graph (CCG) and a net (set of channel labels) and marks the corresponding nodes in the CCG, then returns the set of marked nodes.

**Exercise 6.20**   (*) **Programming Exercise:** Implement a program which accepts as input a CCG and a subset from the set of vertices in $V(\text{CCG})$ and construct a Steiner tree on these vertices (output the edges of the tree).

**Exercise 6.21** (*) **Programming Exercise:** Implement a program which accepts as input a slicing structure and a netlist (net={channel labels}) and returns a channel assignment for each net. Assume that each channel has a weight equal to one, and that the objective minimized is the overall connection length.

# REFERENCES

Burstein, M. and R. Pelavin. Hierarchical wire routing. *IEEE Transactions on Computer-Aided Design*, CAD-2(4):233–234, 1983.

Cai, H. and R. J. H. M. Otten. Conflict-free channel definition in building-block layout. *IEEE Transactions on Computer-Aided Design*, CAD-8(9):981–988, September 1989.

Cai, Y. and D. F. Wong. Channel/switchbox definition for VLSI building-block layout. *IEEE Transactions on Computer-Aided Design*, CAD-10(12):1485–1493, September 1991.

Cai, Y. and D. F. Wong. On minimizing the number of L-shaped channels in building-block layout. *IEEE Transactions on Computer-Aided Design*, CAD-12(6):757–769, September 1993.

Clow, G. W. A global routing algorithm for general cells. *Proceedings of 21st Design Automation Conference*, pages 45–51, 1984.

Cong, J. Pin assignment with global routing for general cell design. *IEEE Transactions on Computer-Aided Design*, CAD-10(11):1401–1412, November 1991.

Cong, J., A. B. Kahng, G. Robins, G. M. Sarrafzadeh and C. K. Wong. Provably good performance-driven global routing. *IEEE Transactions on Computer-Aided Design*, CAD-11(6):739–752, June 1992.

Dai, W. M. BEAR: A macrocell layout system for VLSI circuits. *PhD Thesis, EECS Dept, University of California*, Berkeley, 1988.

Dijkstra, E. W. A note on two problems in connection with graphs. *Numer. Math.*, 1:269–271, October 1959.

Horowitz, E. and S. Sahni. *Fundamentals of Computer Algorithms*. Computer Science Press, Rockville, MD., 1984.

Kahng, A. B. and G. Robins. A new class of iterative Steiner tree heuristics with good performance. *IEEE Transactions on Computer-Aided Design*, CAD-11(7):893–902, July 1992.

Lauther, U. Top-down hierarchical global routing for channelless gate arrays based on linear assignment. *Proceedings of IFIP*, VLSI187, pages 109–120, 1987.

Lengauer, T. *Combinatorial Algorithms for Integrated Circuit Layout*. John Wiley & Sons, 1990.

Ohtsuki, T. *Layout Design and Verification (Advances in CAD of VLSI)*. North Holland, 1986.

Prasitjutrakul, S. and W. J. Kubitz. A performance-driven global router for custom VLSI chip design. *IEEE Transactions on Computer-Aided Design*, CAD-11(8):1044–1051, August 1992.

Preas, B. and M. Lorenzetti. *Physical Design Automation of VLSI Systems*. The Benjamin/Cummings Publishing Company, Inc. 1988.

Rose, J. Parallel global routing for standard cells. *IEEE Transactions on Computer-Aided Design*, CAD-9(10):1085–1095, October 1990.

Rothermel, H. and D. A. Mlynski. Automatic variable-width routing for VLSI. *IEEE Transactions on Computer-Aided Design*, CAD-2(4):272–284, October 1983.

Sadowska, M. M. Router planner for custom chip design. *Digest of Technical papers*, ICCAD, pages 246–249, 1986.

Sarrafzadeh, M. and C. K. Wong. Hierarchical Steiner tree construction in uniform orientations. *IEEE Transactions on Computer-Aided Design*, CAD-11(9):1095–1103, September 1992.

Sechen, C. and A. L. Sangiovanni-Vincentelli. Timberwolf 3.2: A new standard cell placement and global routing package. *Proceedings of 23rd Design Automation Conference*, pages 432–439, 1986.

Shih, P., K. Chang and W. Feng. Neural computation network for global routing. *Computer Aided Design*, 23(8):539–547, October 1991.

Shragowitz, E. and S. Keel. A global router based on a multicommodity flow model. *Integration, The VLSI Journal*, 5:3–16, 1987.

Shragowitz, E., J. Lee and S. Sahni. Placer-router for sea of a-gates design style. *Computer Aided Design*, 20(7):382–397, July 1988.

Tsay, R. S., E. S. Kuh, and C. P. Hsu. Proud: A sea-of-gates placement algorithm. *IEEE Design and Test of Computers*, pages 318–323, December 1988.

Vecchi, M. P. and S. Kirkpatrick. Global wiring by simulated annealing. *IEEE Transactions on Computer-Aided Design*, CAD-2(4):215–222, October 1983.

# CHANNEL ROUTING

## 7.1 INTRODUCTION

*Channel routing* is a special case of the routing problem where interconnections are made within a rectangular region having no obstructions. A majority of modern IC routing systems are based on channel routers. These systems apply a 'divide-and-conquer' strategy in which the layout routing problem is divided into channel routing problems which are solved separately. Channel routers are used in the design of custom chips as well as uniform structures such as gate-arrays and standard-cells. In layout design using gate-arrays, after the placement and global routing phases the channel router is invoked to perform final interconnection within individual wiring bays. Similarly channel routers are used to complete interconnection of standard-cell based designs. In custom layout of VLSI chips channel routers are used to complete interconnection between macro blocks.

Channel routing strategy is very popular because it is efficient and simple, and it guarantees 100 per cent completion if channel width is adjustable. The actual router implementation is technology dependent. Different technologies introduce different instances of the problem. In this chapter we present the classical channel routing problem. The classical model (also known as the Manhattan model) consists of a rectangular space between two parallel rows of pins (terminals). The locations of these pins are fixed and aligned with vertical grid lines. Two layers are available for routing one exclusively used for horizontal wires and the other for vertical wires. Horizontal wire segments called *trunks* run along tracks and vertical

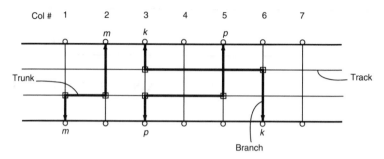

**Figure 7.1** Channel, terminals, trunks and branches.

wire segments called *branches* connect trunks to terminals as shown in Figure 7.1.

## 7.2 PROBLEM DEFINITION

The channel to be routed is defined by a rectangular region with two rows of terminals along its top and bottom sides. A number between 0 and $N$ is assigned to each terminal. These numbers are labels of grid points located at the top and bottom of the rectangle as shown in Figure 7.1. They represent the netlist. Terminals having the same label $i$ $(1 \leq i \leq N)$ must be connected by net $i$. Zeros indicate that no connection has to be made to the corresponding point. The netlist is usually represented by two vectors TOP and BOT. TOP($k$) and BOT($k$) represent the grid points on the top and bottom sides of the channel in column $k$, respectively.

The task of the channel router is to specify for each net a set of horizontal and vertical wire segments that interconnect the net, and whose end points are located on the terminals or tracks. In case of a standard-cell design methodology, the objective is to use a minimum number of tracks to complete routing. Therefore, the width of the channel (i.e., the number of tracks required) is to be determined by the router. For gate-array design methodology the objective is to finish routing using a specified number of tracks.

Generally two or three layers are available for routing. We shall restrict our discussion to two-layer routing (also known as H–V routing). Poly is used on one layer and metal1 on the second. Connections between segments of poly and metal1 are made with contact cuts. In two-layer routing, all horizontal wires are laid out on tracks on one layer and all vertical wires on the other. If two horizontal segments belonging to different nets do *not* overlap, then they may be assigned to the same track. If they overlap then they *must* be assigned to different tracks. Referring to

Figure 7.1, the horizontal segment corresponding to net $m$ does not overlap with the segment of net $p$ but the intervals of segments $k$ and $p$ overlap. Thus, there are horizontal constraints on nets whether or not they can be assigned to the same tracks.

Also, any two nets must not overlap at a vertical column. If we assume that there is only one horizontal segment per net, then it is clear that the trunk of a net connected to the upper terminal at a given column must be placed above the trunk of another net connected to the lower terminal at that column. In Figure 7.1, the terminal corresponding to net $k$ appears on the top of the column (column 3) and the terminal corresponding to net $p$ on the bottom. Clearly the trunk of net $k$ must be above the trunk of net $p$. Therefore, we also have vertical constraints among nets. Briefly no two nets can overlap at contact, branch or track.

### 7.2.1 Constraint graphs

For any instance of the channel routing problem we can associate two constraint graphs, one to model the horizontal constraints and the other to model the vertical constraints. For both graphs, every net is represented by a vertex.

### Horizontal constraint graph

The horizontal constraint graph denoted by $HCG(V, E)$ is an undirected graph where a vertex $i \in V$ represents net $i$ and edge $(i, j) \in E$ if the horizontal segments of net $i$ and net $j$ overlap.

### Vertical constraint graph

The vertical constraints in the channel routing problem can be represented by a directed graph $VCG(V, E)$, where each node $i \in V$ corresponds to net $i$, and each vertical column introduces an edge $(i, j) \in E$ if and only if net $i$ has a pin on the top and net $j$ on the bottom of the channel in the same column. That is, for any two nets with pins at the same column on opposite sides of the channel, there will be an edge between their corresponding vertices in the VCG.

Therefore, if there is a cycle in the VCG, the routing requirement cannot be realized without dividing some nets. We will now illustrate the construction of the above graphs with the help of an example.

**Example 7.1** For the channel routing problem shown in Figure 7.2, construct the HCG and VCG.

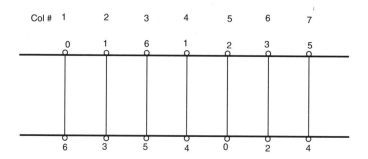

**Figure 7.2** Netlist for Example 7.1

SOLUTION   The netlist can be represented by two vectors TOP and BOT given by TOP=[0,1,6,1,2,3,5] and BOT=[6,3,5,4,0,2,4]. To determine if two horizontal segments of nets overlap we define a set $S(i)$, where $S(i)$ is the set of nets whose horizontal segments intersect column $i$. The number of elements in each set is called the local density. Since horizontal segments (trunks) of distinct nets must not overlap, the horizontal segments of two nets in any set $S(i)$ must *not* be placed in the same horizontal track.

For the above channel routing problem the values of $S(i)$ are:

$$S(1) = \{6\} \qquad S(2) = \{1,3,6\} \qquad S(3) = \{1,3,5,6\}$$
$$S(4) = \{1,3,4,5\} \qquad S(5) = \{2,3,4,5\} \qquad S(6) = \{2,3,4,5\}$$
$$S(7) = \{4,5\}$$

Since elements of $S(i)$ represent trunks of those nets that must not be on the same horizontal track, we can eliminate those sets which are already subsets of other sets. For example, $S(1) = \{6\}$, and $S(2) = \{1,3,6\}$ are subsets of $S(3) = \{1,3,5,6\}$. Therefore they need not be considered. The remaining sets $S(i)$ after elimination are called **maximal sets**. For this example, the maximal sets are:

$$S(3) = \{1,3,5,6\} \qquad S(4) = \{1,3,4,5\} \qquad S(5) = \{2,3,4,5\}$$

The HCG (also known as the interval graph) is now constructed by placing an edge between vertices $i$ and $j$ if both $i$ and $j$ belong to a set $S(k)$, for some $k$. For example, $S(3) = \{1,3,5,6\}$. Therefore edges are placed between vertices (1,3), (1,5), (1,6), (3,5), (3,6), and (5,6). The complete HCG is shown in Figure 7.3(a). An alternative representation of the HCG is the zone representation. A zone representation is a graphical representation of the maximal sets $S(i)$. Each set $S(i)$ is

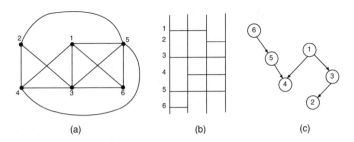

**Figure 7.3** (a) Horizontal constraint graph. (b) Zone representation. (c) Vertical constraint graph.

| Column | $S(i)$ | Zone |
|---|---|---|
| 1 | $\{6\}$ | |
| 2 | $\{1,3,6\}$ | 1 |
| 3 | $\{1,3,5,6\}$ | |
| 4 | $\{1,3,4,5\}$ | 2 |
| 5 | $\{2,3,4,5\}$ | |
| 6 | $\{2,3,4,5\}$ | 3 |
| 7 | $\{4,5\}$ | |

**Table 7.1 Zone table for channel routing of Example 7.1.**

represented by a column and the elements of the maximal sets $S(i)$ are represented by line segments as shown in Figure 7.3(b). Note that Figure 7.3(b) has three columns corresponding to the three maximal sets $S(3)$, $S(4)$, and $S(5)$. The first column has four horizontal line segments at 1, 3, 5, and 6, corresponding to the elements of the maximal set $S(3)$. The second and the third columns correspond to maximal sets $S(4)$ and $S(5)$ respectively. In terms of the interval graph a zone is defined by a maximal clique, and the clique number is the density. The zone-table for channel routing problem of Example 7.1 is given in Table 7.1.

The VCG is simpler to construct. For every column $k$ of the channel not containing a zero in either TOP($k$) or BOT($k$) a directed edge is drawn from vertex TOP($k$) to vertex BOT($k$). For example, in the given netlist, TOP(2) = 1 and BOT(2) = 3. Therefore the VCG will have an edge from vertex 1 to vertex 3. Similarly there is an edge from vertex 6 to vertex 5, and so on. The complete VCG is shown in Figure 7.3(c).

## 7.3 COST FUNCTION AND CONSTRAINTS

One common objective of most layout design systems is to reduce the total overall area required by the chip. In the channel routing problem the length of the channel is fixed. The objective then is to assign the required segments to the given layers so as to connect electrically all the nets with *minimum* number of tracks. Unnecessary contact cuts and vias are also highly undesirable. They cause an increase in area, decrease in yield and reliability and poor electro-magnetic characteristics. Therefore reducing the number of vias must be an important objective of any routing heuristic.

The number of layers available for routing is constant. In most existing VLSI technologies, the number of available layers is either two or three. For three-layer routing, in addition to poly and metal (known as metal1), a second layer of metal (metal2) is available. Several routing models exist for three-layer channel routing. One of them known as VHV routing uses layer-1 for vertical segments originating from the upper boundary of the channel, layer-3 for vertical segments originating from the lower boundary, and layer-2 for all horizontal segments (Chen and Liu, 1984; Vijayan *et al.*, 1989). In such a model there are obviously no vertical constraints. If layer-1 and layer-3 are assigned for horizontal runs (tracks) and the middle layer for vertical runs, the resulting model is called HVH routing. Each of the two models (HVH and VHV) has its strengths and weaknesses. VHV is good when the VCG has long chains, and HVH is good when VCG has a large number of incomparable nodes (two nodes are *incomparable* when there is no path from any of the two nodes to the other). The solution to VHV model is trivial while studies have shown that the HVH model uses fewer tracks and is more economical (Cong *et al.*, 1988). A mixed HVH–VHV algorithm for three-layer channel routing was reported by Pitchumani and Zhang (1987).

As mentioned earlier, in this chapter we present heuristics to solve the two-layer H–V routing problem. We try to minimize the first objective, that is, the number of routing tracks.

## 7.4 APPROACHES TO CHANNEL ROUTING

Most of the solution techniques of the channel routing problem are based on the left-edge algorithm with several extensions and variations of this technique. In this algorithm tracks are processed one at a time. In this section we present just the basic left-edge algorithm. Then, the dogleg algorithm proposed by Deutch (1976) which performs splitting of nets, is described. Finally another technique that uses merging of nets proposed by

Yoshimura and Kuh is explained (1982). All the above techniques aim at reducing the total number of horizontal tracks required to perform channel routing.

### 7.4.1 The basic left-edge algorithm

The original left-edge channel routing algorithm was proposed by Hashimoto and Stevens (1971). It attempts to maximize the placement of horizontal segments in each track. Segments of nets to be connected are sorted in the increasing order of their *left* end points from the left-edge of the channel, hence the name. The basic algorithm imposes the restriction that each net consists of a single trunk, and that trunks (horizontal segments) are routed on one layer and branches (vertical segments) on the other. In the absence of vertical constraints the algorithm produces a solution with minimum number of tracks given by $\max_i |S(i)|$, which is also the lower bound on the number of tracks.

The procedure for assigning trunks to segments is as follows. First the edges are sorted on their leftmost end-points as explained above. The algorithm then selects the trunk corresponding to the first net and places it on the lowermost track, the net is deleted from the list. The algorithm then scans through the remaining list for the first net that does not overlap with the placed net. If it finds one, then it assigns it to the same track. The process is repeated until no more nets can be placed in the first track. The algorithm starts again using the remaining unplaced nets in the list and fills track 2, and so on until all the nets in the list are placed. The algorithm is also known as the *unconstrained left-edge algorithm* and is illustrated in Figure 7.4.

**Example 7.2**    For the netlist shown in Figure 7.2 (Example 7.1), use the left-edge algorithm to assign nets to tracks.

SOLUTION    The trunks, sorted in the order of their left end points, are 6, 1, 3, 5, 4 and 2. This is illustrated in Figure 7.5. Let us ignore the vertical constraints for the moment. Using the left-edge algorithm we try to assign the sorted segments to tracks. The first segment chosen from the above sorted list is 6 and is placed in track 1. The next segment in sequence is 1. But since we have an edge (1,6) in HCG (see Figure 7.3(a)) it cannot be placed in the same track as 6. So also is the case with the trunks of nets 3 and 5 which are after net 1. The next net in sequence is 4 and since there is no edge (6,4) in HCG, 4 is assigned to the same track. The last element in the sorted list is 2, and although there is no edge (6,2) in HCG, we do have (4,2), therefore 2 is not assigned to track 1.

**ALGORITHM** Unconstrained_ChannelRouter;
**Begin**
1.  Sort all nets on their leftmost end positions;
2.  Select the net with the lowest left position;
    Place it on the lower most available track;
    Delete net from list;
3.  Continue scanning the list and select from it nets
    that do no overlap with the nets assigned to this track;
    Assign the nets to the current track and delete from list;
4.  If list $\neq \phi$ **Then Goto** 2;
5.  **Exit**
**End.**

**Figure 7.4** Unconstrained left-edge algorithm.

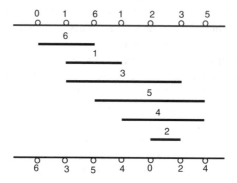

**Figure 7.5** Sorted horizontal segments of Example 7.2.

The set of remaining sorted nets contains 1, 3, 5, and 2. Now the same procedure is repeated to place the remaining segments in track 2, and then in track 3, and so on. The final solution is shown in Figure 7.6.

In the absence of vertical constraints the above solution is acceptable. But we do have vertical constraints, and as mentioned earlier, ignoring them will create short-circuit between nets. We leave it to the reader to verify that the above solution is not acceptable if two-layer routing is used where horizontal segments are assigned to one layer and vertical segments to the other.

A more elaborate algorithm which takes into account the vertical constraints is the *constrained left-edge algorithm* reported by Perskey *et al.* (1976). As in the previous case, horizontal segments are placed on tracks from the lower left corner of the routing region. The algorithm will place a horizontal segment of a net only if it does not have any descendants in the vertical constraint graph. The algorithm is commonly known as the *constrained left-edge algorithm* and is illustrated in Figure 7.7.

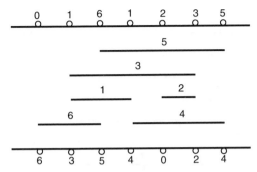

**Figure 7.6** Optimal ordering for netlist of Example 7.2 ignoring vertical constraints.

**ALGORITHM** Constrained_ChannelRouter;
**Begin**
1. Sort all nets on their leftmost end positions;
2. Select the next net $n$ with the lowest left-end position;
    **If** $n$ has no descendents in VCG
      **Then**  **Begin**
               Place $n$ on the lowermost available track;
               Delete $n$ from the sorted list;
               Delete $n$ from VCG
           **End**
      **Else**  **Goto**  2
    **EndIf**
3. Continue scanning the sorted list and from it select
    those nets which do not overlap with nets assigned
    to this track and have no descendents in VCG;
    Remove all selected nets from the list;
4. **If** list $\neq \phi$ **Then Goto**  2;
5. **Exit**;
**End.**

**Figure 7.7** Constrained left-edge algorithm.

**Example 7.3**  Obtain a solution to the channel routing problem of Figure 7.2 taking both the horizontal and vertical constraints into account.

SOLUTION  The same procedure as above is now repeated but taking into consideration the vertical constraints. In this case, a segment corresponding to a net can be placed in a track only if the nets corresponding to its descendants have already been assigned.

Referring to the vertical constraint graph of Figure 7.3(c), we see that only nodes 4 and 2 have no descendents. Now scanning the sorted list we ignore nets 6, 1, 3, and 5 because they all have descendents and the corresponding nets have not been assigned. The first candidate is

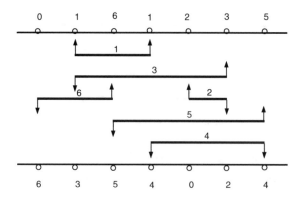

**Figure 7.8** Final correct solution of Example 7.3.

net 4. Therefore net 4 is assigned to track 1 and is deleted from the sorted list as well as from the VCG. Continuing the scanning of the sorted list, we reach net 2, which cannot be assigned to track 1 because of horizontal constraint (Figure 7.3(a)).

The nets remaining in the list are 6, 1, 3, 5, and 2. We now search for candidates that can go into track 2. Scanning the sorted list, we ignore nets 6, 1, and 3 since these have descendants in the VCG. The next net, which is 5, is chosen and assigned to track 2. Net 2, the next in sequence cannot be assigned to the same track as net 5 because of horizontal constraint. The above procedure is continued, and the final solution is shown in Figure 7.8.

## 7.4.2 Dogleg algorithm

The algorithm mentioned in the previous section will fail if there are cycles in the VCG. Consider the channel routing problem and its VCG shown in Figure 7.9(a) and (b) respectively. In this example the constrained left-edge algorithm will not be able to route net 1 or net 2, since each one is the descendant of the other in the VCG. Figure 7.9(c) shows that a solution to this problem is possible only if we allow horizontal segments of the net to be split.

In many instances, even if the VCG contains no cycles, it is desirable to allow splitting of horizontal tracks in order to reduce the channel density. Consider the example routed in Figure 7.10(a). The optimal solution shown uses three tracks if splitting of horizontal segments is not allowed. However the same example can be realized with only two tracks by horizontal splitting, as shown in Figure 7.10(b).

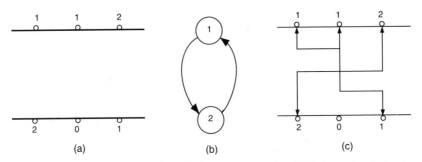

**Figure 7.9** (a) Routing problem. (b) Vertical constraint graph. (c) Solution using doglegging.

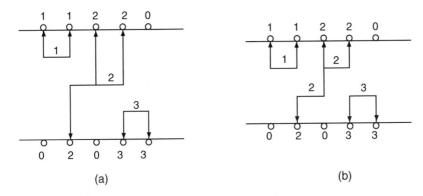

**Figure 7.10** Example illustrating doglegging. VCG has no cycles. (a) Problem and solution without doglegging uses three tracks. (b) Solution with doglegging uses two tracks.

The splitting of horizontal segments of a net is called *doglegging*. This is used not only to avoid vertical conflicts but also to minimize the number of horizontal tracks. In the case of doglegging we assume that the horizontal splitting of a net is allowed at terminal positions only and no additional vertical tracks are allowed.

## The algorithm

This algorithm was proposed by Deutch (1976) to avoid vertical constraint loops and to decrease the density of the channel. It helps in reducing the number of horizontal tracks, particularly for channels with multipin signal nets.

Deutch's algorithm takes each multipin net and breaks it into individual horizontal segments. A break occurs only in columns that contain a pin for that net. Figure 7.11(a) illustrates the horizontal segment

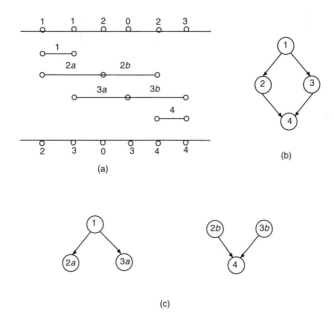

(a)

(b)

(c)

**Figure 7.11**   (a) Channel routing problem. (b)VCG without splitting of nets. (c) VCG with nets split.

definition. Figure 7.11(b) illustrates the VCG without splitting of nets and Figure 7.11(c) shows the new VCG when nets are split.

Using the new vertical constraint graph the dogleg algorithm is similar to the constrained left-edge algorithm. Horizontal segments are sorted in increasing order of their left end points. The first segment in the list that has no descendants in the vertical constraint graph is placed in the channel. The node corresponding to this section of the net is removed from the vertical constraint graph. Then, the next net in the list that does not overlap with the first segment and has no descendants is placed. This process continues for elements in the list from left to right until all segments have been completed.

**Example 7.4**   For the channel routing problem shown in Figure 7.11(a) find a solution using the dogleg algorithm described in this section.

SOLUTION   The sorted list of net segments is [1,2a,3a,2b,3b,4]. The set of nets $S(i)$ whose horizontal segments cross column $i$ are given by

$S(1) = \{1, 2a\}$      $S(2) = \{1, 2a, 3a\}$      $S(3) = \{2a, 2b, 3a\}$
$S(4) = \{2b, 3a, 3b\}$      $S(5) = \{2b, 3b, 4\}$      $S(6) = \{3b, 4\}$

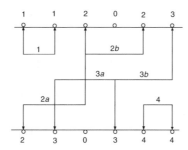

**Figure 7.12** Solution of Example 7.4

Referring to Figure 7.11(c) the net segments that do not have descendants in the VCG are $2a$, $3a$, and 4. Scanning the sorted list of net segments we find that the first net segment that does not have descendants is $2a$. This segment is placed in the lowermost track. Continuing the scanning of the sorted list, the next segment in sequence is $3a$, but due to horizontal constraint (see $S(2)$ above) it cannot be placed in the same track as segment $2a$. Net 4, the next candidate in sequence, does not have any horizontal constraint with segment $2a$, and therefore is placed in the same horizontal track. The placed nets are deleted from the sorted list and the corresponding nodes are deleted from the VCG. The above procedure is repeated by placing the remaining segments in the next track, and so on. The final solution is shown in Figure 7.12.

The original dogleg algorithm suggested by Deutch has one difference from the procedure explained above. That is, the segments are chosen as suggested above for the first track, and horizontal segments are placed from left to right in that track. Then the algorithm switches to the top track in the channel and places horizontal segments in it, from right to left; then the second track from bottom is considered, and so on, until all segments are placed. The above technique can be easily incorporated into the algorithm above by selecting a horizontal segment that has no descendants for the placement in the bottom tracks, and segments that have no ancestors for placement in the top tracks. This symmetric alternating is claimed to produce routing with smaller total vertical length than routing all nets from the bottom to the top of the channel.

### 7.4.3 Yoshimura and Kuh algorithm

If there is a path $n_1$-$n_2$-$n_3$-$\cdots$-$n_k$ in the vertical constraint graph, then obviously no two nets among $\{n_1, n_2, n_3, \cdots, n_k\}$ can be placed on the same track. Therefore, if the longest path length in terms of the number of nodes on the path is $k$, at least $k$ horizontal tracks are necessary to realize the interconnections.

In this section we present two algorithms proposed by Yoshimura and Kuh (1982). The first algorithm uses the VCG and the zone representation of HCG and attempts to minimize the longest path in the VCG. This is done by merging nodes of VCG (which correspond to nets), so that the longest path length after merging is minimized as much as possible. Obviously, this merging is performed for the purpose of minimizing the channel density.

The second algorithm proposed by Yoshimura and Kuh achieves longest path minimization through matching techniques on a bipartite graph. Both techniques report better results than the dogleg algorithm. Before describing these algorithms we first introduce the required terminology.

## Definitions

Let $i$ and $j$ be the nets for which,

1. there exists no horizontal overlap in the zone representation, and
2. there is no directed path between node $i$ and node $j$ in the vertical constraint graph,

(i.e., net $i$ and net $j$ can be placed on the same horizontal track). Then, the operation 'merging of net $i$ and net $j$' results in the following.

1. It modifies the vertical constraint graph by shrinking node $i$ and node $j$ into a single node $i \cdot j$.
2. It updates the zone representation by replacing net $i$ and net $j$ by net $i \cdot j$ which occupies the consecutive zones including those of net $i$ and net $j$.

**Example 7.5** Consider the netlist given below.

0 1 4 5 1 6 7 0 4 9 0 0
2 3 5 3 5 2 6 8 9 8 7 9

The zone table for this example is given in Table 7.2. The VCG and the zone representation are shown in Figure 7.13. Consider nets 6 and 9. Since there is no horizontal overlap in the zone representation (both nets 6 and 9 do not appear in the same $S(i)$), and no vertical conflict (they are on separate vertical paths), nets 6 and 9 are candidates for merging. The merge operation explained above can then be applied.

| Column | $S(i)$ | Zone |
|---|---|---|
| 1 | $\{2\}$ | |
| 2 | $\{1,2,3\}$ | |
| 3 | $\{1,2,3,4,5\}$ | 1 |
| 4 | $\{1,2,3,4,5\}$ | |
| 5 | $\{1,2,4,5\}$ | |
| 6 | $\{2,4,6\}$ | 2 |
| 7 | $\{4,6,7\}$ | 3 |
| 8 | $\{4,7,8\}$ | |
| 9 | $\{4,7,8,9\}$ | |
| 10 | $\{7,8,9\}$ | 4 |
| 11 | $\{7,9\}$ | |
| 12 | $\{9\}$ | |

**Table 7.2 Zone table for channel routing problem of Example 7.5.**

The updated VCG and the zone representation are shown in Figure 7.14. Due to merging, both nets 6 and 9 will be placed in the *same* horizontal track. However the position of the track is not yet decided.

## The algorithm

It can be easily proved that if the original vertical constraint graph has no cycles then the updated vertical constraint graph with merged nodes does not have cycles either (see Exercise 7.9). Since it is assumed that there is no cycle in the initial VCG, we can repeat the operation *merging of nets* without generating any cycle in the graph. The algorithm shown in Figure 7.15 merges nets systematically according to the zone representation.

**Example 7.6** Apply the algorithm of merging of nets to the channel routing problem of Example 7.5 and obtain the routed solution.

SOLUTION
**Zone 1:** Refer to Figure 7.13. The set of nets that terminates at zone 1 is $L = \{1,3,5\}$ and the set of nets which begins at zone 2 is $R = \{6\}$. The merge operation can merge nets (1,6), (3,6) or (5,6). Verify that only the merging of nets 5 and 6 causes minimum increase in the length of the longest path. Therefore nets 5 and 6 are merged to form a merged net $5 \cdot 6$ which ends at zone 3. The updated set $L$ is $L = \{1,3,5\} - \{5\} = \{1,3\}$ (see Figure 7.16(a) and (b)).

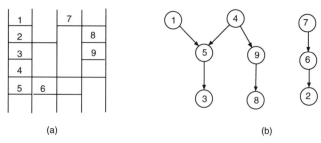

(a)                                     (b)

**Figure 7.13** Zone representation and VCG for Example 7.5. Note that transitive edges (e.g., from 1 to 3) are not shown.

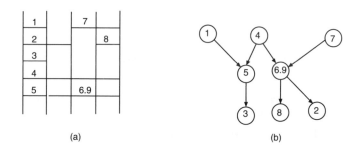

(a)                                     (b)

**Figure 7.14** Modified zone representation and VCG with merged nodes for Example 7.5.

**Algorithm** merge1$(z_s, z_t)$;
**Begin**
1. L= {}; $z_s$ =leftmost zone; $z_t$ =rightmost zone.
2. **For** $z = z_s$ to $z_t$ **Do**
   **Begin**
3.     $L = L + \{$nets which terminate at zone $z\}$;
4.     $R = \{$nets which begin at zone $z + 1\}$;
5.     merge $L$ and $R$ so as to minimize the increase of the
       longest path in the vertical constraint graph;
6.     $L = L - \{n_1, n_2, \cdots, n_j\}$, where$\{n_1, n_2, \cdots, n_j\}$ are the nets merged at Step 5;
   **EndFor**;
**End.**

**Figure 7.15** Algorithm #1 to merge nets.

**Zone 2:** In the next iteration, nets that terminate at zone 2 are added to $L$. Note that net 2 ends at zone 2. The updated set $L$ is $\{1, 2, 3\}$. Only one net begins at zone 3, that is net 7, therefore $R = \{7\}$. In the VCG (see Figure 7.16(b)), since nets 2 and 3 are along the same path as net 7, the only candidate that can be merged with net 7 is net 1. The new

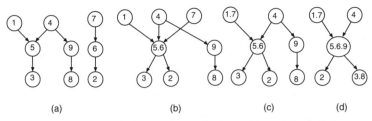

(a)   (b)   (c)   (d)

**Figure 7.16** Updated vertical constraint graphs for problem of Example 7.6.

merged net is $1 \cdot 7$. Next (see Figure 7.16(c)), $L$ is updated by adding nets that end at zone 3 (net $5 \cdot 6$) and removing the nets that are merged (net 1), $L = \{1, 2, 3\} - \{1\} + 5 \cdot 6 = \{2, 3, 5 \cdot 6\}$. The set of nets which begin at zone 4 is $R = \{8, 9\}$.

**Zone 3:** Observe that the merged net $5 \cdot 6$ ends at zone 3. In this step the merged net $5 \cdot 6$ is merged with net 9, and net 3 is merged with net 8 to form the merged nets $5 \cdot 6 \cdot 9$ and $3 \cdot 8$.

The above procedure continues until the last zone is reached. Figure 7.16 illustrates how the vertical constraint graph is updated by the algorithm. Thus, applying the algorithm of Figure 7.15, first, net 5 and net 6 are merged. Then net 1 and net 7, and finally nets $5 \cdot 6$ with net 9 and net 3 with net 8 are merged. The final graph is shown in Figure 7.16(d).

In the next step we apply the left-edge algorithm and assign horizontal tracks to the nodes of the graph. The list of nets sorted on their left-edges is $[2, 3 \cdot 8, 1 \cdot 7, 5 \cdot 6 \cdot 9, 4]$. The final routed channel is shown in Figure 7.17.

## Minimizing the longest path

The key step of the algorithm in Figure 7.15 is Step 5 where two sets of nets are selected and merged. Let us define a few terms needed for the explanation of the merging process.

First the VCG is modified by adding two artificial nodes $s$ (source) and $t$ (sink), and arcs from $s$ to previously ancestor-free nodes and from previously descendant-free nodes to $t$ (see Figure 7.18).

Let $P = \{n_1, n_2, \cdots, n_p\}$ and $Q = \{m_1, m_2, \cdots, m_q\}$ $(p > q)$ be the two sets of candidate nets for merging, where elements of $P$ are on a separate vertical path from that of $Q$.

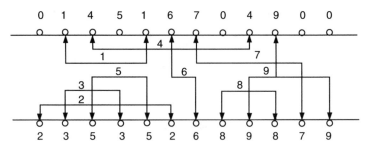

**Figure 7.17** Channel routing solution of Example 7.6.

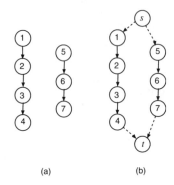

**Figure 7.18** (a) Vertical constraint graph. (b) Modified vertical constraint graph.

Let $u(n)$, $n \in P \cup Q$, be the length of the longest path from $s$ to $n$. Let $d(n)$, $n \in P \cup Q$, be the length of the longest path from $n$ to $t$.

When two nodes $n \in P$ and $m \in Q$, that are on two separate paths from $s$ to $t$ are merged, the length of the longest path will either increase or will remain the same. The exact increase denoted by $h(n, m)$ is given by,

$$h(n, m) = h^{\max}(n, m) - \max\{u(n) + d(n), u(m) + d(m)\} \qquad (7.1)$$

where $h^{\max}(n, m) = \max\{u(n), u(m)\} + \max\{d(n), d(m)\}$. The proof is left as an exercise (see Exercise 7.10).

The purpose here is to minimize the length of the longest path after the merge. However, it will be too time consuming to find an exact minimum merge. An heuristic merging algorithm is proposed by Yoshimura and Kuh (1982). In this heuristic, first a node $m \in Q$ which lies on the longest path before the merger is chosen; furthermore, it is furthest away from either $s$ or $t$. Next, a node $n \in P$ is chosen such that the increase of the longest path after merger is minimum. If there is a tie then the selected $n$ is such that

**Algorithm**   merge2 $(P, Q)$;
**Begin**
   **While** $Q \neq 0$ **Do**
   **Begin**
      among $Q$, find $m*$ which maximizes $f(m)$;
      among $P$, find $n*$ which minimizes $g(n, m*)$, and which
      is neither ancestor nor successor of $m*$;
      merge $n*$ and $m*$;
      remove $n*$ and $m*$ from $P$ and $Q$, respectively;
   **Endwhile**;
**End**.

**Figure 7.19** Algorithm #2 to merge nets.

$u(n) + d(n)$ is nearly maximum and that the condition $\frac{u(m)}{d(m)} = \frac{u(n)}{d(n)}$ is nearly satisfied. This is in order to incur the minimum increase in path length. The above heuristic is implemented by introducing the following:

1. for every $m \in Q$;
$$f(m) = C_\infty \times \{u(m) + d(m)\} + \max\{u(m), d(m)\}, \quad C_\infty \gg 1$$
2. for every $n \in P$, and every $m \in Q$;
$$g(n, m) = C_\infty \times h(n, m) - \{\sqrt{u(m) \times u(n)} + \sqrt{d(m) \times d(n)}\}$$

A formal description of the merging algorithm corresponding to Step 5 of Figure 7.15 is given in Figure 7.19. Next we illustrate the algorithm with an example.

**Table 7.3 Tabulated** $f(m)$ **for Example 7.7.**

|  | $P$ | | | $Q$ | |
|---|---|---|---|---|---|
| $m$ | 1 | 3 | 4 | 6 | 7 |
| $u(m)$ | 1 | 3 | 4 | 2 | 3 |
| $d(m)$ | 4 | 2 | 1 | 2 | 1 |
| $f(m)$ | | | | 202 | 203 |

**Example 7.7**   Given $Q = \{6, 7\}$ and $P = \{1, 3, 4\}$, for the VCG shown in Figure 7.18, using the technique explained above, find a pair of nodes to be merged.

SOLUTION

$u(1) = 1, \quad u(3) = 3, \quad u(4) = 4, \quad u(6) = 2, \quad u(7) = 3$
$d(1) = 4, \quad d(3) = 2, \quad d(4) = 1, \quad d(6) = 2, \quad d(7) = 1$

Let us pick $C_\infty = 50$. Then
$f(6) = C_\infty \times 4 + \max\{2, 2\} = 202$
$f(7) = C_\infty \times 4 + \max\{3, 1\} = 203$

| $n$ | 1 | 3 | 4 |
|---|---|---|---|
| $h(n,7)$ | 2 | 0 | 0 |
| $g(n,7)$ | 96.27 | -4.41 | -4.46 |

**Table 7.4  Tabulated values of $h(n,7)$ and $g(n,7)$ for Example 7.7.**

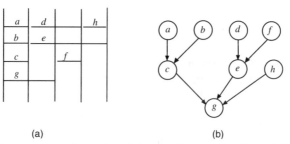

(a)                                    (b)

**Figure 7.20** Zone representation and vertical constraint graph for Example 7.8.

The values of $u(m)$, $d(m)$ and $f(m)$ are tabulated in Table 7.3. Since $f(7) > f(6)$, node 7 is chosen from $Q$.

Next we evaluate $h(n, m^*)$ where $m^* = 7$.

$$h(1,7) = \max\{u(1), u(7)\} + \max\{d(1), d(7)\} -$$
$$\max\{u(1) + d(1), u(7) + d(7)\}$$
$$h(1,7) = 3 + 4 - \max\{5, 4\} = 2$$

Similarly

$$h(3,7) = 3 + 2 - \max\{5, 4\} = 0 \text{ and}$$
$$h(4,7) = 4 + 1 - \max\{5, 4\} = 0$$

Next, we evaluate $g(n, 7)$,

$$g(n, m) = 50 \times h(n, m) - \{\sqrt{u(m) \times u(n)} + \sqrt{d(m) \times d(n)}\}$$
$$g(1, 7) = 50 \times h(1, 7) - \{\sqrt{u(1) \times u(7)} + \sqrt{d(1) \times d(7)}\}$$
$$g(1, 7) = 96.268$$

The results of computation of $h(n, m)$ and $g(n, m)$ are shown in Table 7.4. Since $g(4, 7)$ is the smallest, we can merge node 4 with node 7.

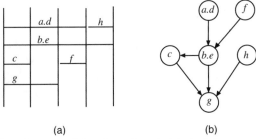

(a)                              (b)

**Figure 7.21** Merging of nets $a$ and $d$, and nets $b$, and $e$ blocks futher merging.

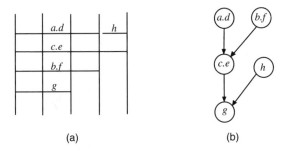

(a)                              (b)

**Figure 7.22** Merging of nets $a$ and $d$, $c$ and $e$, and $b$ and $f$.

## An improved algorithm based on matching

In the algorithm of the previous section it is possible that a merging of nets may block subsequent mergings. To avoid this type of situation as much as possible and in order to make the algorithm more flexible, Yoshimura and Kuh introduced another algorithm. In this algorithm a bipartite graph $G_h$ is constructed where a node represents a net and an edge between net $a$ and net $b$ indicates that nets $a$ and $b$ can be merged. A merging is expressed by a matching on the graph which is updated dynamically (Yoshimura and Kuh, 1982).

In this section two examples will be presented, one to illustrate how a merging can block further mergings, and then an example to illustrate the idea of merging based on matching.

**Example 7.8** Given the problem instance of Figure 7.20. Let us assume that at zone 1 the algorithm merges net $a$ with net $d$, and net $b$ with net $e$ respectively (if we follow the merging algorithm of the last section these mergings will not occur, but they are assumed only for illustration). The vertical constraint graph and zone representations are modified as shown in Figure 7.21. The merged vertical constraint graph

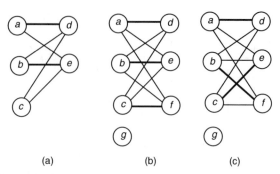

(a)              (b)              (c)

**Figure 7.23** Bipartite graphs $G_h$ for Example 7.9.

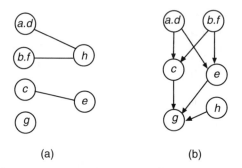

(a)                        (b)

**Figure 7.24** (a) Updated graph $G_h$. (b) Updated VCG after processing zone 3.

indicates that net $f$ cannot be merged with either net $c$ or net $g$ because a cycle would be created. However if we merge net $a$ with $d$ and net $c$ with $e$, then net $f$ can be merged with net $b$ as illustrated in Figure 7.22. Therefore the final solution is order dependent.

**Example** 7.9   In this example we introduce a bipartite graph where a node represents a net and an edge between net $a$ and net $b$ signifies that net $a$ and net $b$ can be merged. A merging is expressed by a matching on the graph. The idea is explained using the example of Figure 7.20. In that example we see that net $d$ as well as net $e$ can be merged with any of the three nets $a$, $b$ or $c$ in zone 1. The algorithm constructs a bipartite graph $G_h$ as illustrated in Figure 7.23(a) and a temporary merging is feasible but neither the vertical constraint graph nor the zone representation is updated. Next we move to zone 2, where net $g$ terminates and net $f$ begins. So we add $g$ to the left and $f$ to the right of $G_h$ as shown in Figure 7.23(b). Since the VCG in Figure 7.20(b) indicates that net $f$ can be merged with either net $a$, net $b$ or net $c$, three edges are added and the matching is also updated as shown by the heavier lines in Figure 7.23(b). Of course there is no guarantee that the merging which corresponds to the updated matching satisfies the vertical constraints (horizontal constraints are satisfied automatically), so the algorithm checks the constraints and modifies the matching as shown in Figure 7.23(c).

At zone 3, net $d$ and net $f$ terminate. This means that, in processing zone 3, node $d$ and node $f$ should be moved to the left side in graph $G_h$ and merged with their partner nets $a$ and $b$, respectively, as shown in Figure 7.24(a). Net $c$ and net $e$ have not been merged yet, since $e$ has not terminated. The vertical constraint graph is also updated as shown in Figure 7.24(b). A matching is next sought for the updated $G_h$. The above procedure continues until all zones have been processed (see Exercise 7.12).

## 7.4.4 Greedy channel router

The channel routers discussed in the earlier sections route the channel one track at a time. They are based on the left-edge algorithm or its variations. In this section we describe the working of a greedy heuristic for channel routing known as the '*greedy channel router*' (Rivest and Fiduccia, 1982).

The algorithm routes the channel column by column starting from the left. In each column the router applies a sequence of greedy but intelligent heuristics to maximize the number of tracks available in the next column. The router does not use horizontal or vertical constraints. All decisions are made locally at the column. The algorithm handles routing problems even with cycles in the VCG. Routing is always completed, sometimes with additional columns at the end of the channel. Unlike other channel routers, in this technique a net can occupy two different tracks until the heuristic decides to merge them. Also, instead of doglegging only at terminal locations, the greedy router allows any horizontal connection to change tracks.

Before we explain this technique we present some definitions and terms used in the explanation of the procedure.

## Definitions

*Initial-channel-width*: This is the number of tracks initially available to the greedy router. Generally this number is the local density. As routing proceeds, a new track is added only if routing cannot continue at a given column. When a track is added, the current assignment of nets to tracks to the left of the current column is preserved, and routing does *not* restart, but continues from the current column with additional tracks. Clearly, different *initial-channel-widths* will give different results. In some cases, the router may have to use area beyond the channel. This area can be minimized by iterating the entire procedure with a higher value of *initial-channel-width*. According to Rivest and Fiduccia (1982) best results are usually obtained with initial channel width equal to just a little less than the final channel width.

*Minimum-jog-length*: A jog is a vertical wire placed on a net to bring it closer to the channel side where the next pin of the net is located. The router will make no *jogs* shorter than $k$, the *minimum-jog-length*. Usually this value is taken to be $\frac{w}{4}$, where $w$ is the best channel width available. This parameter affects the number of vias and tracks in the final routing. A high value of $k$ reduces the number of vias, while a lower value reduces the number of routing tracks.

*Steady-net-constant*: This is a window size (given in number of columns). This parameter determines the number of times a multipin net changes tracks, and is used to classify nets as rising, falling, or steady.

*Rising, falling, and steady nets*: When routing at a given column, the router classifies each net which has a pin to its right as either *rising, falling, or steady*. A net is $rising$ if its next pin (after the current column) will be on the top of the channel, and the net has no pin on the bottom of the channel within a window size equal to the *steady-net-constant*. $Falling$ nets are defined similarly. Finally, a net is $steady$ if it is neither a rising net nor a falling net.

*Split net and collapsible net*: If more than one track is occupied by one net, the net is called a *split net*. A split net is eventually collapsed to a single track (or zero tracks if the last pin of the net is passed) by making an appropriate jog. Therefore, a split net is also called a collapsible net.

*Spillover area*: This is the additional area (columns) used at the end of the track to collapse split nets to complete routing.

The greedy router takes as input the specific channel routing problem and three non-negative integers, *initial-channel-width, minimum-jog-length*, and *steady-net constant*. The router scans the channel column-by-column from left to right. It uses six steps in routing each column. The wiring at a given column is completed before proceeding to the next. We will now explain the general idea and the various steps involved.

**Step 1. Make feasible top and bottom connections in minimal manner:** In this step, each pin connection at the current column is brought to an empty track or to a track occupied by the same net, whichever uses the least amount of vertical wire. If the channel is fully occupied, then bringing a new net is deferred until Step 5. If two nets (one from the top and the other

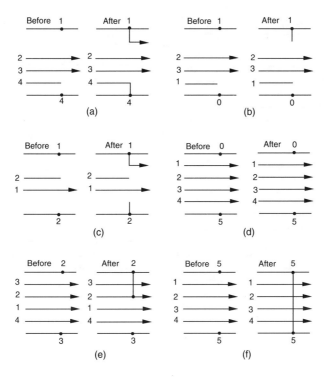

**Figure 7.25** (a)–(f) Step 1 of greedy channel router.

from the bottom) have a conflict (overlap), then the one that uses the least wire is connected, the other is again deferred to Step 5. If there are no empty tracks and both the pins (on top and bottom) belong to the same net, and this net is not placed on any track, then a vertical straight-through connection is possible. The various possibilities that are processed in this step are shown in Figure 7.25(a)–(f). Each figure has two parts, one part depicting the channel status *before* the step and the other *after* the processing of the step. In Figure 7.25(a) pin '1' is brought to an empty track and pin '4' is brought to track 4 which is occupied by net '4'. Pin '1' in Figure 7.25(b) and pins '1' and '2' in Figure 7.25(c) are brought to an empty track and *not* to a track occupied by the net, since the former uses a lesser amount of wire. In Figure 7.25(d) there is not enough room to connect net '5', therefore nothing is done in this step, and addition of a track is deferred until Step 5. In Figure 7.25(e) two nets, one from top and the other from bottom create a conflict due to overlap. Therefore pin '2' is connected to net '2' while connection of net '3' is deferred until Step 5. Finally, in Figure 7.25(f) we see that there are no empty tracks, but a straight-through connection can be made.

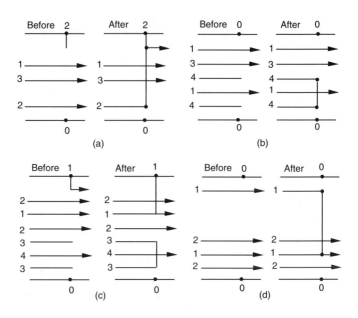

**Figure 7.26** (a)–(d) Step 2 of greedy channel router.

**Step 2. Free up as many tracks as possible by collapsing split nets:** In this step, a column tries to free up as many tracks as possible by making vertical connecting jogs that 'collapse' nets currently occupying more than one track. This step may also complete a connection by connecting a pin to the track that its net currently occupies. Step 1, as seen above, may stop at an available empty track (see Figure 7.26(a)).

This is an important step in the algorithm since it makes more tracks available to the nets arriving to the right of the channel. A collapsing segment (a piece of vertical wire that connects two adjacent tracks occupied by the same net) is used. Each collapsing segment has a weight of 1 or 2, depending on whether or not the net continues to the right beyond the current column. This weight represents the number of tracks freed as a result of the collapse. The best collapsing segment is one which frees most tracks, and does not overlap with segments of other nets or the routing of Step 1. This segment is found by a combinatorial search (Figure 7.26(a)–(b)). If there is a tie, the segment that leaves the outermost uncollapsed net as far as possible from the channel side is chosen (Figure 7.26(c)). The idea here is to keep the free area as close to the sides as possible. If there are still

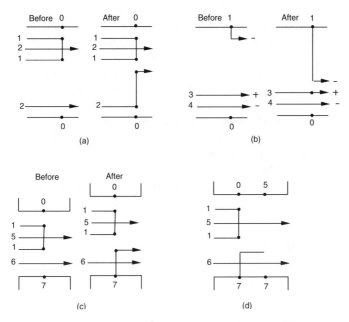

**Figure 7.27** Greedy channel router. (a) Step 3. (b) Step 4. (c) Step 5. (d) Step 6.

ties, then the segment (pattern) that maximizes the amount of vertical wire is used (see Figure 7.26(d)).

**Step 3. Add jogs to reduce the range of split nets:** In this step, the range of tracks assigned to the net of each uncollapsed split net is reduced by adding vertical jogs. These jogs have the effect of moving the net from the highest track to an empty one as far down as possible, and the one in the lowest level to another empty track which is as far up as possible. Of course, no jogs must be smaller than the minimum-jog-length. Furthermore, the jog must be compatible with vertical wiring already placed in the column by previous steps (see Figure 7.27(a)).

**Step 4. Add jogs to raise rising nets and lower falling nets:** This step tries to move all nets closer to the side they prefer, based on the side of the next pin of the net. If the next pin is on the top of the channel, the greedy router tries to move the net to a higher track. Moving the track is done by jog addition.

If such jogs are permissible and if the length of such jogs is greater than the minimum jog length, then they are added to move the net to an empty track which is as close as possible to its target side. This is shown in Figure 7.27(b); '−' (+) indicates falling (rising) net.

**Step 5. Widen channel if needed to make previously infeasible top or bottom connections:** If nets in the current column failed to enter the channel in Step 1, then new tracks are added and the nets are brought to these tracks. Such new tracks are placed as near the centre of the channel as possible if they do not conflict with existing wiring (see Figure 7.27(c)).

**Step 6. Extend to next column:** In this step, the list of tracks occupied by the unsplit nets that ended at the current column are deleted. And, the tracks occupied by both the unfinished nets and the split nets are extended to the next column (see Figure 7.27(d)).

This completes the description of the greedy router. The router will always complete routing successfully, although, to do so, sometimes it may use a few additional columns beyond the right end of the channel.

**Example 7.10**  Apply the greedy algorithm of Rivest and Fiduccia explained in Section 7.4.4 to route the netlist given below.

0  1  2  5  7  1  6  0  2  9  0  0
4  3  5  3  5  4  7  1  3  1  6  9

Let the *initial-channel-width* be 6 tracks, and the *minimum-jog-length* allowed be equal to 1.

SOLUTION  Let the tracks be numbered from $T_0$ to $T_5$ as shown in Figure 7.28. We shall explain the application of the above heuristic at each column starting from the leftmost.

**Column 1:** Connect pin 4 to $T_5$ and extend it to the next column.

**Column 2:** Connect pin 1 to $T_0$ and pin 3 to $T_4$ and extend tracks $T_0$, $T_4$, and $T_5$ to the next column.

**Column 3:** Connect pin 2 to $T_1$ and pin 5 to $T_3$. Jog net 5 from $T_3$ to $T_2$ since net 5 is a rising net. Extend tracks $T_0$, $T_1$, $T_2$, $T_4$, and $T_5$ to the next column.

**Column 4:** Connect pin 5 to $T_2$ and pin 3 to $T_4$. Jog net 5 from $T_2$ to $T_3$ since net 5 now is a falling net. Extend tracks $T_0$, $T_1$, $T_3$, $T_4$, and $T_5$.

**Column 5:** Connect pin 7 to $T_2$ and pin 5 to $T_3$. Extend tracks $T_0$, $T_1$, $T_2$, $T_4$, and $T_5$ to the next column.

**Column 6:** Connect pin 1 to $T_0$ and pin 4 to $T_5$. Jog net 1 from $T_0$ to $T_3$ since net 1 is a falling net. Extend tracks $T_1$, $T_2$, $T_3$, and $T_4$.

**Column 7:** Connect pin 6 to $T_0$ and pin 7 to $T_5$. Merge tracks $T_2$ and $T_5$. Extend tracks $T_0$, $T_1$, $T_3$, and $T_4$.

**Column 8:** Connect pin 1 to $T_5$. Jog net 5 from $T_0$ to $T_2$ and jog net 1 from $T_5$ to $T_3$. Extend tracks $T_1$, $T_2$, $T_3$, and $T_4$ to the next column.

**Column 9:** Connect pin 2 to $T_1$, and pin 3 to $T_5$. Merge tracks $T_4$ and $T_5$ and extend tracks $T_2$ and $T_3$.

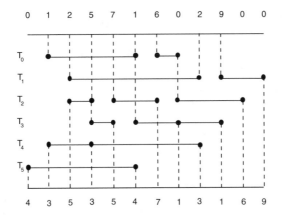

**Figure 7.28** Solution to greedy router problem of Example 7.10.

**Column 10:** Connect pin 9 to $T_0$ and pin 1 to $T_5$. Jog net 9 from $T_0$ to $T_1$, merge Tracks $T_3$ and $T_5$ and then extend tracks $T_1$ and $T_2$.

**Column 11:** Connect pin 6 to $T_5$, merge Tracks $T_2$ and $T_5$, and extend tracks $T_0$ and $T_1$.

**Column 12:** Connect pin 9 to $T_5$ and merge tracks $T_1$ and $T_5$.

The complete solution obtained by the application of the greedy router is shown in Figure 7.28.

## 7.4.5 Switchbox routing

When rectangular cells are placed on the layout floor, normally two kinds of routing regions are created. These routing regions are called *channels* and *switchboxes*. Channel rectangular regions limit their interconnection endpoints or *terminals* to one pair of parallel sides. Switchboxes are generalizations of channels and allow terminals on all four sides of the region. In the detailed routing stage, channel routers and switchbox routers are required to complete the connection. The channel router is a special router designed for routing in an area with no inside obstructions and with terminals placed on two opposite sides. In earlier sections we studied the heuristics used for channel routing. In this section we discuss the switchbox routing problem which, as will be seen, is more difficult than the channel routing problem.

## Problem definition

A switchbox is a rectangular region with no inside obstructions, and with terminals lying on all four sides. The terminals are grouped into a collection $S$ of disjoint sets called *nets*.

To identify which terminals are to be connected, each terminal is labelled with a net identification number $k$, $1 \leq k \leq |S|$. Formally a switchbox is defined as a region $R = \{0, 1, \cdots, m\} \times \{0, 1, \cdots, n\}$ where $m$ and $n$ are positive integers. Each pair $(i, j)$ in $R$ is a grid point (Luk, 1985). The $i$th column is a set of grid points $COL(i) = \{(i, j) \mid j \in \{0, 1, \cdots n\}\}$, $1 \leq i \leq m$. The $j$th row or track is a set of points $ROW(j) = \{(i, j) \mid i \in \{0, 1, \cdots m\}\}$, $1 \leq j \leq n$. The zeroth and $m$th columns are the left and right boundaries of the switchbox respectively. Similarly, the zeroth and $n$th rows are the top and bottom boundaries of the switchbox. The connectivity and location of each terminal is represented as $LEFT(i) = k$, $RIGHT(i) = k$, $TOP(i) = k$, and $BOT(i) = k$, depending on which side of the switchbox it lies on, where $i$ stands for the coordinate of the terminal along the edge and $k$ for its identification number.

Lee-type algorithms are not suitable for solving this problem. Lee-type routers do not check ahead to avoid unnecessary blocking of other terminals.

The goal of the switchbox router is to connect electrically the terminals in each individual net. Connections run horizontally or vertically along rows and columns along grid lines. As in previous cases, only a single wire is allowed to occupy each row and each column segment. The wires are allowed to cross. An example of a switchbox is shown in Figure 7.29 for which we have

$$R = \{0, 1, 2, \cdots, 7, 8, 9\} \times \{0, 1, 2, \cdots, 15, 16, 17\}$$

$$
\begin{aligned}
TOP &= (1, 2, \cdots, 7, 8) & &= [8, 7, 1, 2, 6, 1, 5, 3] \\
BOT &= (1, 2, \cdots, 7, 8) & &= [10, 12, 1, 10, 3, 9, 5, 11] \\
LEFT &= (1, 2, \cdots, 14, 15, 16) & &= [0, 3, 10, 0, 0, 0, 2, 0, 11, 1, 0, 0, 13, 6, 0, 4] \\
RIGHT &= (1, 2, \cdots, 14, 15, 16) & &= [10, 5, 9, 2, 12, 5, 8, 11, 7, 5, 7, 3, 13, 6, 3, 4]
\end{aligned}
$$

## Switchbox routing algorithm

In this section we present an efficient switchbox routing algorithm due to Luk (1985). This algorithm is a modification and extension of the greedy routing heuristic of Rivest and Fiduccia discussed in Section 7.4.4. Some

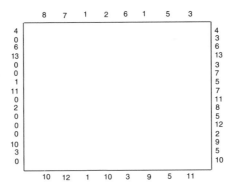

**Figure 7.29** Example of a switchbox.

operations of the greedy heuristic that are not vital to its operation are relaxed and modified to overcome the additional constraints of switchbox routing. The additional constraints are:

1. the matching of the terminals on the LEFT and RIGHT of the routing region;
2. bringing in left-edge terminals directly into the routing region as horizontal tracks at the column;
3. instead of jogging to the next top and bottom terminals as in Step 4 of the greedy router, the horizontal tracks must be jogged keeping in mind the target row, which is a row where a right edge terminal is located. This jogging is to ensure matching the nets with their right edge terminals.

## Jogging strategies

The main modification to the greedy channel router of Rivest and Fiduccia (1982) is in the jogging schemes applied to accommodate the additional switchbox constraints discussed above. The various jogging schemes are defined as follows.

1. $(Jog_R)$. For nets that have terminals on the RIGHT, this jog is performed until the net occupies a track that matches one of the right edge terminal positions.
2. $(Jog_{T/B})$. For nets that only have terminals on TOP and BOT, this jog is similar to the one in the greedy channel router.
3. $(Jog_{T/B}; Jog_R)$. In this jogging scheme, first $(Jog_{T/B})$ is performed on every net and then a switch is made to perform $(Jog_R)$ at the column where the last top and/or bottom terminals appear.

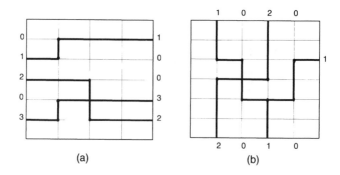

**Figure 7.30** Jogging schemes. (a) ($Jog_R$). Nets 1,2, and 3 jog to right side target terminal. (b) ($Jog_{T/B}$; $Jog_R$). ($Jog_{T/B}$) performed until column 3 and then ($Jog_R$).

Examples of the above jogging schemes are illustrated in Figure 7.30

The general structure of the switchbox routing algorithm is given in Figure 7.31. We will now briefly describe the various steps of the algorithm.

The algorithm begins by assigning one of the four edges of the switchbox as the LEFT edge. Then the direction of scanning is determined. This is done in Step 0. The quality of the final solution depends on the direction of the scan. A good heuristic based on augmented channel density distribution is proposed by Luk (1985).

Once the scan direction is decided, the LEFT edge terminals are brought into the first column. Then, for each column the next four steps are repeated.

In Step 1 the nets TOP($i$) and BOT($i$) are brought into empty rows. In the second step split nets are joined as much as possible to increase the number of free tracks. Step 3 is comprised of jogging. In Step 3a, as in the case of the greedy channel routing algorithm, trunks of each split net, which have no terminals on the right are brought closer by jogging. And in Step 3b, for those nets which have terminals on the right we use the combination of jogging strategies discussed above.

This procedure is called SWJOG. It divides the routing region into a left $p$-portion and a right $p$-portion. The jogging strategy to be applied depends upon the location of the column (in left or right $p$-portion) where the decision is to be made. The value of $p$ is between 0 and 1. Below we now enumerate the rules for SWJOG.

1. For nets that do not have right side terminals, always perform $Jog_{T/B}$.

2. For nets that have a right side terminal and whose rightmost top/bottom terminal is on the right $p$-portion of the routing region, perform $Jog_R$ for that net.

**Algorithm** SwitchboxRouter;
  **Begin**
    0. Determine Scan Direction;
       Bring in LEFT terminals into column.
    **For** $i = 1$ **To** $m - 1$ **Do**
    1. **If** empty tracks exist **Then**
       bring TOP($i$) and BOT($i$) into empty rows;
       **EndIf**;
    2. Join split nets as much as possible;
    3a. **ForEach** net with no right terminal **Do**
        Bring split nets closer by jogging;
       **EndForEach**;
    3b. **ForEach** net with right terminal **Do** SWJOG; **EndForEach**;
    4. When close to right edge fanout to targets;
    5. **If** Step 1 failed **Then** increase number of rows; **EndIf**;
       Update columns 1 to $i$;
    **EndFor**;
    6. **While** split nets exist **Do**;
       Increase number of columns by 1;
       Join split nets as much as possible;
       **EndWhile**;
  **End.**

**Figure 7.31** Switchbox routing algorithm.

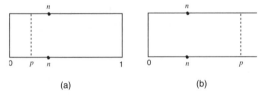

(a)            (b)

**Figure 7.32** For nets that have a terminal in RIGHT and whose rightmost top/bottom terminal is: (a) on right $p$-portion of routing region perform $(Jog_R)$; (b) on left $p$-portion of routing region perform $(Jog_{T/B}; Jog_R)$.

3. For nets that have a right side terminal and whose rightmost top/bottom terminal is on the left $p$-portion of the routing region, perform $(Jog_{T/B}; Jog_R)$, that is, $(Jog_{T/B})$ before the last top/bottom terminal and $(Jog_R)$ at and after the last top/bottom terminal.

4. The value of $p$ may vary between 0 and 1. If $p = 0$ perform $(Jog_R)$. Obviously, if $p = 1$ perform $(Jog_{T/B}; Jog_R)$. A typical value of $p$ is 0.5 (see Figure 7.32).

In the implementation, a distance dependent threshold scheme is used to avoid excessive jogging. A net is allowed to jog to its target row only if it can be brought to or beyond half-way between the initial position and final target position.

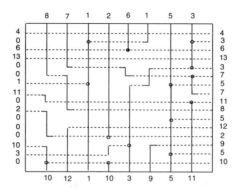

**Figure 7.33** Solution to switchbox routing of Figure 7.29.

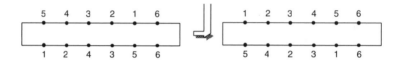

**Figure 7.34** Equivalent channel routing problems.

In Step 4 for nets that occupy more than one location on the RIGHT, when they get closer to the right edge, these nets are made to fan-out to their final terminal locations. Step 5 consists of increasing the number of rows if Step 1 failed. And in Step 6, if split nets exist then the number of columns is incremented and split nets are joined as much as possible. The complete routed solution of Figure 7.29 is shown in Figure 7.33.

The time efficiency of the switchbox router is the same as the greedy channel router. The router can be modified to route a region with terminals fixed on any three sides.

# 7.5 OTHER APPROACHES AND RECENT WORK

In this section we look at some related work in the area of channel and switchbox routing. Recent algorithms for both channel and switchbox routing, and techniques for multilayer and over the cell channel routing are discussed.

We begin this section with a channel routing approach proposed by Chaudry and Robinson (1991). Their approach is based on sorting and assumes that wires, in addition to running horizontally and vertically, can also run at 45° and 135°. The technique uses *bubble-sort* for routing two-pin nets, and can be easily extended to handle multiterminal nets.

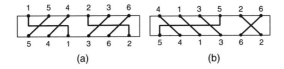

(a)                    (b)

**Figure 7.35** Examples of adjacent permutations.

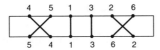

**Figure 7.36** $X$ routing used in *swap-router*.

## Channel routing by sorting

Without loss of generality the nets in TOP can be assumed to be numbered in sequence $1, 2, \cdots, n$.

**Example 7.11**   For example the channel routing problem TOP = [5,4,3, 2,1,6] and BOT = [1,2,4,3,5,6] can also be specified as TOP = [1,2,3,4,5,6] and BOT = [5,4,2,3,1,6], where the terminal labels in TOP are reordered to be in sequence and corresponding changes are made to the labels in BOT (see Figure 7.34). The problem can also be specified as [5,4,2,3,1,6].

Then, the nets in BOT are a permutation of the sequence in TOP. Two permutations $p_i$ and $p_{i+1}$ are said to be adjacent if the routing problem obtained by assigning $p_i$ to the lower side and $p_{i+1}$ to the upper side of the channel can be routed in one track. Possible adjacent permutations and the associated routing are shown in Figure 7.35. The solution to the channel routing problem is represented as a series of permutations $\{p_i\}$, $i = 1, 2, 3, \cdots, w$ such that $p_1$ is the given permutation (BOT) and $p_w = (1, 2, \cdots, n)$ (TOP), and $p_i$ is adjacent to $p_{i+1}$, for $0 \le i \le w$. The channel routing problem then amounts to finding a series of intermediate adjacent permutations $\{p_i\}$ such that the number of permutations $w$ is minimized. We now present the basic idea behind two routers, namely *Swap-router* and *Sort-router*. These routers are based on permutations and sorting.

### Swap-router

In swap-router, two nets that have adjacent terminals in the wrong order are interchanged. These nets can be connected using $X$ routing as shown in Figure 7.36. Note that it is assumed that connections can run at 45° and

135°. A series of adjacent permutations can be built using only $X$ routing. This corresponds to factoring the permutation as a product of transpositions. Routing is done from bottom to top. If $(a_1, a_2, \cdots, a_n)$ is the bottom permutation, we compare $a_i$ and $a_{i+1}$ for $i = 1, 3, 5, \cdots, n$; and swap the terminals if $a_i > a_{i+1}$. In the next step the process is repeated for $i = 2, 4, 6, \cdots, n$. The above two steps are repeated until all the terminals are in the correct order. Since two nets cross only once if their terminals are not in order the routing obtained by this swap-router is a minimal crossing solution.

Bounds on the channel width are obtained in terms of *span* number. The span of a terminal in a permutation is the difference of the terminal number and its position in the permutation. For example, in the permutation (5,4,1,3,6,2), number 1 has a span of –2, number 2 has a span of –4, number 3 has a span of –1, and so on. The span number tells us how far the number is from its correct position. Since in each step a net moves by at most one column, a net with span of $y$ will require $y$ steps. It can therefore be concluded that

$$number\ of\ steps \geq \max(|span_i|), \quad 1 \leq i \leq n \qquad (7.2)$$

Clearly, channel width can be reduced by removing the restriction of moving only one column at each step.

**Example 7.12**   Determine the number of tracks required to route the channel instance specified by [5,4,6,2,1,3] using the swap-router.

SOLUTION   For the problem under consideration, TOP = [1,2,3,4,5,6], BOT = [5,4,6,2,1,3]. We begin from bottom. In the first pass we compare $a_i$ and $a_{i+1}$ for $i = 1, 3, 5, \cdots$. This leads to a permutation (4,5,2,6,1,3). In the next pass, we repeat the above, but for $i = 2, 4, 6, \cdots$. The resulting permutation is (4,2,5,1,6,3). In the third pass, again we apply swapping for $i = 1, 3, 5, \cdots$. and the resulting permutation is (2,4,1,5,3,6). The last two permutations are (2,1,4,3,5,6) and (1,2,3,4,5,6) respectively. The routed solution is illustrated in Figure 7.37.

## Sort-router

From the previous discussion it is clear that any sorting algorithm based on exchanges can be easily converted to a channel router. An algorithm based on *bubble-sort* is presented by Chaudry and Robinson (1991). The steps of *bubble-sort* swap a pair of numbers only once if they are in the wrong

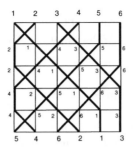

**Figure 7.37** Swap-router solution of Example 7.12.

order. Therefore, as in the case of *swap-router*, the *sort-router* always produces a minimal crossing solution. Since in one pass of the *bubble-sort* at least one number moves to its final place, it would require at most $n$ steps to sort the $n$ numbers. Thus the channel width will be $\leq n$, where $n$ is the number of nets. Here again 45° routing is allowed. We will illustrate this process with an example.

**Example 7.13** Apply the sort-router based on bubble-sort to the channel routing problem where TOP = [1,2,3,4,5,6], and BOT = [5,4,6, 2,1,3].

SOLUTION The problem instance to be routed is (5,4,6,2,1,3). We now have to sort these numbers. Each intermediate step of sorting will produce an adjacent permutation and will require one track. The numbers in bubble-sort can be either scanned from left to right or vice versa; and the number of passes required to complete sorting will depend on the direction of scan.

If the numbers are scanned from left to right we call this a *right-step*, and if they are scanned from right to left we call this a *left-step*. The number of steps required to route varies depending on the direction of scan. The intermediate permutations for both the *right-step* and *left-step* are shown below as *R-step* and *L-step* respectively.

$$R\text{-}step = \begin{vmatrix} 1 & 2 & 3 & 4 & 5 & 6 \\ 2 & 1 & 3 & 4 & 5 & 6 \\ 4 & 2 & 1 & 3 & 5 & 6 \\ 4 & 5 & 2 & 1 & 3 & 6 \\ 5 & 4 & 6 & 2 & 1 & 3 \end{vmatrix} \qquad L\text{-}step = \begin{vmatrix} 1 & 2 & 3 & 4 & 5 & 6 \\ 1 & 2 & 3 & 5 & 4 & 6 \\ 1 & 2 & 5 & 4 & 6 & 3 \\ 1 & 5 & 4 & 6 & 2 & 3 \\ 5 & 4 & 6 & 2 & 1 & 3 \end{vmatrix}$$

Coincidentally, in this example, the number of intermediate permutations for both the right step and the left step are the same. The channel routing solution of the above problem for both scan directions are given in Figure 7.38.

Details of choice of optimal step-type, and extensions to multiterminal nets, and multilayer routing are available in literature (Chaudry and Robinson, 1991).

## Over-the-cell channel routing

Another extension to the classical channel routing problem is *over-the-cell* channel routing. This method is employed when there are at least two layers in the routing channel, and one routing layer over the channel. Certain nets can be partially or totally routed on one side over the channel using the single available layer. Then, the remaining net segments are chosen for routing. Therefore, a common approach to over-the-cell channel routing is to divide the problem into three steps, namely, (1) routing over the channel, (2) choosing the net segments, and (3) routing within the channels. The third step can be accomplished easily using one of the conventional techniques discussed in this chapter. Cong and Liu (1990) showed that the first step can be formulated as the problem of finding a maximum independent set of a circle graph and can be solved optimally in quadratic time. In this step a row of terminals are routed on one side of the channel using a single routing layer. The result is that the number of hyperterminals are minimized. Cong and Liu called this problem *multiterminal single-layer one-sided routing problem* (MSOP). The second step is formulated as the problem of finding a minimum density spanning forest of a graph. An efficient heuristic which produces satisfactory results is proposed. A channel routing problem and its over-the-cell solution are illustrated in Figure 7.39.

## Techniques for multilayer channel routing

New techniques for routing multilayer channels and their implementation in a multilayer channel router called Chameleon were presented by Braun *et al.* (1988). These new techniques handle a variety of technology constraints. For example, (a) the line width and line spacings can be specified independently for each layer, (b) contact stacking can be forbidden or allowed, etc. Chameleon consists of two stages, a partitioner and a detailed router. The task of the partitioner is to divide the problem into two- and three-layer subproblems such that the global channel area is minimized.

A new approach to three-layer channel routing problem based on the idea of transforming a two-layer solution into a three-layer one was presented by Cong *et al.* (1988). Their transformation consists of several steps which can be formulated as two-processor scheduling, maze routing,

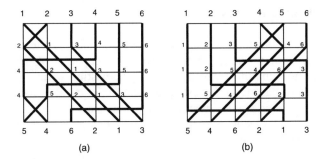

**Figure 7.38** Sort router solution of Example 7.13. (a) Left to right scanning. (b) Right to left scanning.

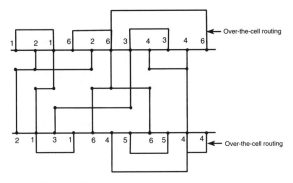

**Figure 7.39** Over-the-cell channel routing.

and the shortest path problem respectively. Since the above problems are well understood and have polynomial time complexity, three-layer channel routing can be solved optimally. HVH model for three-layer routing is used. Most of the above techniques can also be extended to four layers.

A channel routing heuristic which assigns wires track by track in a greedy way was proposed by Ho *et al.* (1991). The data structure and strategy used is simple and can be generalized to obtain a class of channel routing heuristics. This algorithm has a backtracking capability that increases the chance of completing the routing with a minimum number of tracks. In addition the concept discussed can be applied to switchbox routing, and three-layer and multilayer channel routing.

A new robust channel router with very simple heuristics was proposed by Yoeli (1991). It supports a wide variety of routing models and handles two-layer channels and three-layer channels (both HVH and VHV). Both these routers, that is, the greedy channel router and the robust channel router, achieve excellent results. For example, the Deutsch difficult example is routed in 19 tracks (which is its local density).

## Other switchbox routers

In the area of switchbox routing, Hamachi and Ousterhout (1984) presented a router called Detour which is capable of routing both switchboxes and channels which contain pre-existing wiring as obstacles. Their router is based on Rivest and Fiduccia's greedy channel router. Detour routes over single layer obstacles such as wiring, and jogs nets around multilayer obstacles such as contacts. Detour was developed as a part of the Magic layout system (Ousterhout *et al.*, 1984).

Cohoon and Heck (1988) presented a new fast router called BEAVER. This is a computational geometry-based-tool that uses heuristics which produce a switchbox solution that minimizes both via usage and wire length. It also maximizes the use of preferred routing layer. It consists of four tools that are run successively: (a) a corner router, (b) a line sweep router, (c) a thread router, and (d) a layerer. The corner router is used to make single bend terminal-to-terminal connections. The line sweep router is used for making straight-line connections, single-bend connections, and two-bend connections. The thread router makes connections of arbitrary form and the layerer completes switchbox routing by layering wires that have been assigned a location but are yet to be assigned to a layer. The quality of solution produced by BEAVER (in terms of wire length) is better than or comparable to previously reported solutions.

Finally we refer to switchbox routing algorithm called PACKER (Gerez and Herrmann, 1989). PACKER proceeds in two steps. In the first step, connectivity of each net is established without taking the other nets into account. This of course will result in conflicts and short circuits. In the second step these conflicts are removed iteratively using what is known as *connectivity preserving local transformations* (CPLTs). CPLTs perform the task of reshaping a segment without disconnecting it from the net.

## 7.6 CONCLUSION

In this chapter, we presented the problem of channel routing. Graph theoretic approaches to solve the channel routing problem were presented. The most commonly used technique, known as the left-edge algorithm was discussed in detail. Following this, the algorithm which performs splitting of nets, known as doglegging was described (Deutch, 1976). We also presented two algorithms due to Yoshimura and Kuh (1982). Their first algorithm uses merging of nodes so that the longest path length after merging is minimized as much as possible. The second algorithm proposed by Yoshimura and Kuh achieves longest path minimization through matching techniques on a bipartite graph. Both techniques report better results than the dogleg algorithm. Further, we discussed the working of a greedy router, which, unlike the above mentioned methods, routes the channel column by column. A modification of the greedy heuristic to route switchboxes, called the 'switchbox router' was also presented.

Interesting heuristics that employ sorting and swapping to route channels were presented in the section dedicated to 'other approaches'. Over-the-cell channel routing, techniques for multilayer channel routing, and other popular switchbox routers were briefly mentioned.

## 7.7 BIBLIOGRAPHIC NOTES

The original unconstrained channel routing algorithm was proposed by Hashimoto and Stevens (1971). A constrained left-edge algorithm was reported by Perskey et al. (1976). The dogleg routing algorithm technique is due to Deutch (1976). The technique of merging nets to reduce the total number of tracks used to accomplish final routing is due to Yoshimura and Kuh (1982). The greedy channel routing algorithm was proposed by Rivest and Fiduccia (1982), and was modified by Luk (1985) to solve the switchbox routing problem. Channel routing based on sorting as discussed in this chapter is by Chaudry and Robinson (1991). For details on over-the-cell channel routing the reader is referred to the work of Cong and Liu (1990).

## EXERCISES

**Exercise 7.1** Describe the basic channel routing algorithm. Explain why designers prefer to use channel and switchbox routers over maze or line-probe routers.

**Exercise 7.2** Given the following instance of the channel routing problem:

TOP = [2,1,5,1,2,3,6] and
BOT = [5,3,6,4,0,2,4]

1. Determine the maximal sets and find a lower bound on the channel width.
2. Draw the HCG and VCG.
3. Apply the algorithm given in Figure 7.7 to route the channel.

**Exercise 7.3** Find the time complexity of the algorithm given in Figure 7.7.

**Exercise 7.4** Suggest an improvement to the algorithm given in Figure 7.7 with respect to time complexity [*Hint:* Reduce time complexity of Steps 1 and 2].

**Exercise 7.5** Compare and discuss the routing algorithms given in Figure 7.7 and obtained in Exercise 7.4 in terms of quality of their solutions (qualitative discussion) and time complexity.

**Exercise 7.6** For the netlist given below find an assignment of nets to tracks to complete routing. Use doglegging to break cycles.

TOP = [1,4,2,0,2,3,4,5] and
BOT = [2,0,3,3,1,4,5,5].

**Exercise 7.7** Apply the basic left-edge algorithm to the channel routing problem of Example 7.5. Compare and comment on the result.

**Exercise 7.8** In a channel routing problem with vertical constraints, the length of the VCG can be reduced by merging nodes corresponding to nets that can be placed in the same track. Find a solution that uses the least number of tracks to the channel routing problem given below using the merging technique suggested by Yoshimura and Kuh.

TOP = [0,1,0,2,1,5,5,6,8,0,1, 9, 8,10,11,12] and
BOT = [2,3,1,4,4,3,6,3,1,9,9,11,10,11,12,12]

**Exercise 7.9** Prove that if the original VCG has no cycles then merging of nodes using Algorithm # 1 of Figure 7.15 does not generate cycles.

**Exercise 7.10** Prove that the increase in length of the longest path due to merging of two nodes in the vertical constraint graph is given by

$h(n, m) = h^{\max}(n, m) - \max\{u(n) + d(n), u(m) + d(m)\}$, where
$h^{\max}(n, m) = \max\{u(n), u(m)\} + \max\{d(n), d(m)\}$.
For definitions of $u$, and $d$ see Section 7.4.3.

**Exercise 7.11 Programming Exercise:** Design and implement a program to determine the vertical constraint graph of a channel. The program must be able to remove loops from the graph by inserting doglegs.

**Exercise 7.12**  Complete Example 7.9.

**Exercise 7.13**  Show that the function $\sqrt{u(m) \times u(n)} + \sqrt{d(m) \times d(n)}$ used in the cost function $g(n, m)$ minimized by the Yoshimura–Kuh algorithm is maximum when $\frac{u(m)}{u(n)} = \frac{d(m)}{d(n)}$.

**Exercise 7.14**  Solve the channel routing problem given in Exercise 7.6 using

1. the greedy channel router;
2. the swap and sort router.

**Exercise 7.15**  Route the switchbox given below by applying the technique discussed in Section 7.4.5.

$R = \{0, \cdots, 7\} \times \{0, \cdots, 5\}$

RIGHT $= (1, \cdots, 4) = [4,7,1,2]$
LEFT $= (1, \cdots, 4) = [3,6,5,1]$
BOT $= (1, \cdots, 6) = [0,3,2,8,8,2]$
TOP $= (1, \cdots, 6) = [5,6,2,4,7,5]$

# REFERENCES

Braun, D., J. L. Burns, F. Romeo, A. Sangiovanni-Vincentelli, K. Mayaram. Techniques for multilayer channel routing. *IEEE Transactions on Computer-Aided Design of Integrated Circuits and Systems*, 7(6):698–712, June 1988.

Chaudry, K. and P. Robinson. Channel routing by sorting. *IEEE Transactions on Computer-Aided Design of Integrated Circuits and Systems*, 10(6), June 1991.

Chen, Y. K. and M. Liu. Three layer channel routing. *IEEE Transactions on Computer-Aided Design of Integrated Circuits and Systems*, CAD-3, April 1984.

Cohoon, J. P. and P. L. Heck. BEAVER: a computational-geometry-based tool for switchbox routing. *IEEE Transactions on Computer-Aided Design of Integrated Circuits and Systems*, 7(6):684–697, June 1988.

Cong, J. and C. L. Liu. Over-the-cell channel routing. *IEEE Transactions on Computer-Aided Design of Integrated Circuits and Systems*, 9(4), April 1990.

Cong, J., D. F. Wong and C. L. Liu. A new approach to three- or four-layer channel routing. *IEEE Transactions on Computer-Aided Design of Integrated Circuits and Systems*, 7(10):1094–1104, October 1988.

Deutch, D. N. A dogleg channel router. *Proceedings of 13th Design Automation Conference*, pages 425–433, 1976.

Gerez, S. H. and O. E. Herrmann. Switchbox routing by stepwise reshaping. *IEEE Transactions on Computer-Aided Design of Integrated Circuits and Systems*, 8(12):1350–1361, December 1989.

Hamachi, G. T. and J. K. Ousterhout. A switchbox router with obstacle avoidance. *Proceedings of 21st Design Automation Conference*, pages 173–179, 1984.

Hashimoto, A. and J. Stevens. Wire routing by optimizing channel assignment within large apertures. *Proceedings of 8th Design Automation Conference*, pages 155–169, 1971.

Ho, T., S. S. Iyengar and S. Zheng. A general greedy channel routing algorithm. *IEEE Transactions on Computer-Aided Design of Integrated Circuits and Systems*, 10(2):204–211, February 1991.

Luk, W. K. A greedy switchbox router. *Integration, The VLSI Journal*, 3:129–149, 1985.

Ousterhout, J. K., G. T. Hamachi, R. N. Mayo, W. S. Scott and G. S. Taylor. Magic: A VLSI layout system. *Proceedings of 21st Design Automation Conference*, pages 152–159, 1984.

Perskey, A., D. Deutch and D. Schweikert. A system for the directed automatic design of LSI circuits. *Proceedings of 13th Design Automation Conference*, June 1976.

Pitchumani, V. and Q. Zhang. A mixed HVH-VHV algorithm for three-layer channel routing. *IEEE Transactions on Computer-Aided Design of Integrated Circuits and Systems*, CAD-6(4):497–502, July 1987.

Rivest, R. L. and C. M. Fiduccia. A greedy channel router. *Proceedings of 19th Design Automation Conference*, pages 418–424, 1982.

Vijayan, G., H. H. Chen and C. K. Wong. On VHV routing in channels with irregular boundaries. *IEEE Transactions on Computer-Aided Design of Integrated Circuits and Systems*, 8(2):146–152, February 1989.

Yoeli, U. A robust channel router. *IEEE Transactions on Computer-Aided Design of Integrated Circuits and Systems*, 10(2):211–219, February 1991.

Yoshimura, T. and E. S. Kuh. Efficient algorithms for channel routing. *IEEE Transactions on Computer-Aided Design of Integrated Circuits and Systems*, CAD-1:25–35, Jan. 1982.

# EIGHT

# LAYOUT GENERATION

## 8.1 INTRODUCTION

A compilation is usually implemented as a sequence of transformations $(SL, L_1)$, $(L_1, L_2)$, ... , $(L_k, TL)$ from a source language $SL$ to a target language $TL$. Each language $L_i$ is called an intermediate language. Intermediate languages are introduced for the purpose of decomposing the task of compiling from the source language to the target language. In **silicon compilation** $SL$ is a digital system specification and $TL$ is VLSI layout specification.

The input source to a silicon compiler, which is a representation of the digital system to be synthesized, may be either *behavioural, structural* or *physical*. The behavioural level representation is the most abstract and hides a large amount of detail. Lower in the hierarchy is the structural level representation, followed by the physical level.

### 8.1.1 Behavioural level

The behavioural level model of a digital system describes the way the system and its components interact with their environment, that is, the mapping from inputs to outputs *without* any reference to structure. The behaviour of a logic circuit can be defined in terms of Boolean functions. For example the behavioural description of a circuit that produces the *sum* output in a full-adder can be given as

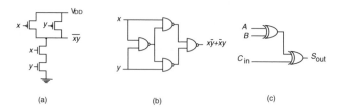

(a)                                    (b)                                 (c)

**Figure 8.1** (a) CMOS 2-input NAND gate. (b) XOR gates implemented using four NAND gates. (c) Sum output of a full-adder using two XOR gates.

$$S_{out} = A \oplus B \oplus C_{in} \tag{8.1}$$

where $\oplus$ indicates Exclusive-Or (XOR) operation, $A$, $B$, are the inputs and $C_{in}$ is the carry-in.

Behavioural descriptions are technology independent and hide implementation specific information. Higher levels of behavioural descriptions are possible. For instance the arithmetic add operation may be written as

$$C = A + B \tag{8.2}$$

In the above expression no indication of the size of the adder (number of bits) or its implementation is given. The adder may be implemented as a carry look-ahead adder or a ripple carry adder. Besides that, no information about $C$ is known (it may be a bus or a register). The behaviour of a digital system can also be expressed as an algorithm in a high level programming language such as Pascal (Trickey, 1987) or Fortran (Tanaka *et al.*, 1989).

## 8.1.2 Structural level

The abstraction level below the behavioural level is the structural level. A structural level specification primarily contains information about the *components* used and their *interconnection* (netlist, for example). In addition, structural level description *may* also contain information about the technology used, dimensions of devices, their electronic characteristic, input/output capacitances, etc.

Take the example of $S_{out}$, the *sum* output of the full-adder (Boolean Equation 8.1). Let us assume that the XOR gate is implemented using four 2-input NAND gates and that the NAND gate is implemented in CMOS technology. These details about implementation and technology can be included in the structural model. The circuit diagram of the 2-input NAND gate is given in Figure 8.1(a). The XOR gate implementation that uses four NAND gates is given in Figure 8.1(b), and the circuit to implement $S_{out}$ is

FULL_ADDER OUTPUT CIRCUIT
.SUBCKT NAND 1 2 4
*Transistors in a NAND circuit
** D G S B Transistor-Type
M1 3 1 4 3 P
M2 3 2 4 3 P
M3 4 1 5 0 N
M4 5 2 0 0 N
VDD 3 0 DC 5V
.MODEL P PMOS
.MODEL N NMOS
.ENDS NAND

.SUBCKT XOR 11 12 25
*NAND gates in XOR circuit
NAND1 11 12 22 NAND
NAND2 22 11 23 NAND
NAND3 22 12 24 NAND
NAND4 23 24 25 NAND
.ENDS XOR

*XOR gates in FULL_ADDER
XOR1 11 12 25 XOR
XOR2 25 13 31 XOR
CAP 31 0 1.0PF
.END

**Figure 8.2** SPICE model of circuit in Figure 8.1(c).

given in Figure 8.1(c). The complete structural model of the CMOS circuit in SPICE format is given in Figure 8.2. Observe that the SPICE model describes components and their interconnectivity. In addition to this, it contains information about the output capacitance. In SPICE, electrical characteristics of MOS devices can also be included (Nagel, 1975).

### 8.1.3 Physical level

The physical level of representation defines how the particular circuit is to be constructed to yield a specific structure and hence the behaviour. The lowest level of physical specification is the representation required by the various steps of the fabrication process.

MOS technology offers four to five layers of conducting materials in which wires are run to build circuits. They are diffusion (*p*-type or *n*-type depending on the type of transistor to be fabricated), polysilicon (also known as *poly*), and metal (one or two layers). When poly crosses over diffusion a transistor is formed as depicted in Figure 8.3. An electrical property of a MOS transistor that enables its use as a switch is that current

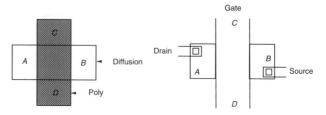

**Figure 8.3** Layout of a MOS transistor.

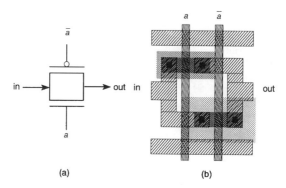

(a)                                    (b)

**Figure 8.4** (a) Transmission gate circuit. (b) Layout of CMOS transmission gate.

can pass along the diffusion wire from $A$ to $B$ only if the voltage on the poly wire $C-D$ is high. A low voltage on poly does not allow current to pass (this is the case for $n$MOS, for $p$MOS the voltage levels are reversed). A set of rules commonly known as '*design rules*' specify precisely the requirements of sizes of rectangles and restrictions on their spacing. A MOS layout of any digital system can be constructed using the above mentioned layers. As an example, the complete layout of a CMOS transmission gate is shown in Figure 8.4.

The layout can also be represented using a geometry language known as CIF (Caltech Intermediate Form) (Mead and Conway, 1980). This is a language used to describe the mask geometries of the chip. The CIF description of the CMOS transmission gate is given in Figure 8.5. In CIF, coordinates and dimensions of all rectangles that make the masks are accurately defined. In CIF, a box is created by a statement of the form

B <width> <height> <x-center> <y-center>

Where 'B' is a keyword abbreviation for box, <width> and <height> represent the size of the box, and (<x-center> <y-center>) are the

```
DS 1 25 2;
9 trans;
L CWP;
    B 348 180 138 -126;
L CMF;
    B 480 60 60 198;
    B 72 96 -96 72;
    B 228 60 138 90;
    B 120 72 -120 -12;
    B 72 48 -96 -72;
    B 72 36 216 42;
    B 120 72 240 -12;
    B 228 60 -18 -126;
    B 72 108 216 -102;
    B 480 60 60 -234;
L CPG;
    B 36 540 -18 -18;
    B 36 540 138 -18;
L CAA;
    B 228 60 -18 90;
    B 228 60 138 -126;
L CCA;
    B 24 24 -96 90;
    B 24 24 60 90;
    B 24 24 60 -126;
    B 24 24 216 -126;
L CSP;
    B 276 108 -18 90;
L CSN;
    B 276 108 138 -126;
DF;
C 1;
End
```

**Figure 8.5** CIF description file for the CMOS transmission gate of Figure 8.4.

coordinates of the centre of the box. The specification of the layer assignment of the box is stated as follows.

L <layer specification>

Layer specification is a code for the name of one of the layers used in the layout. For example, for CMOS technology the layer designations together with their meanings are: CWG (for well), CWP (for $p$-well), CAA (for active area), CSG (for select), CSP (for $p$ select), CSN (for $n$ select), CPG (for poly), CCP (for contact to poly), CCA (for contact to active), CMF (for metal-1), CVA (for via), CMS (for metal-2), and COG (for glass). The specification of a cell is introduced by the statement

DS <symbol-number> <scale>

where DS is a keyword meaning 'definition start', <symbol-number> is an integer that serves as the name of the cell, and <scale> is a pair of integers 'a' and 'b', such that all dimensions and coordinates of boxes are multiplied by the ratio $\frac{a}{b}$ (used for scaling designs). The end of a specification is marked by the statement DF (definition finish). The call of a cell is achieved by a statement of the form

C <symbol-number> <list of transformation>

where list of transformation may contain commands to rotate, translate or mirror the cell (Mead and Conway, 1980).

The traditional manner of creating the physical layouts of custom VLSI chip requires a human to interact with a graphics-based hierarchical layout editing program. For large designs, this approach is time consuming and error prone. An automated physical design system would take the behavioural or structural level description of systems and automatically synthesize VLSI layouts. That is, it would translate automatically a structural/behavioural description to a physical level representation.

In the previous chapters we presented algorithms and heuristics for automatic partitioning, placement and routing of functional modules of a digital system. These modules are layouts of Boolean logic functions and flip-flops, or layouts of macro cells such as ROMs, Programmable Logic Arrays (PLAs), multiplexers, etc. The layouts of such cells may be handcrafted (drawn using an intelligent interactive layout editor such as Magic (see Chapter 9)), or may be generated automatically (Ousterhout *et al.*, 1984).

A large portion of routine circuitry can be automatically produced by CAD software programs. Programs that take the structural or behavioural descriptions and produce layouts are called '*silicon compilers*.' Programs that automatically produce VLSI layouts of cells whose functions are known from input descriptions (or parameters) are called *layout generators*. These programs are generally technology dependent, and are suitable for the creation of layouts which have a well defined structure (e.g., ROMs, PLAs, multiplexers, etc.). These generators require a number of leaf cells. They also depend on the layout style adopted. The final structure of the layout is predefined.

In this chapter we shall restrict ourselves to MOS technology. We present techniques to automatically generate layouts of standard-cells where the input description is of an AND/OR MOS circuit (Uehara and VanCleemput, 1981). Another systematic method for performing chip layout called gate matrix is also presented (Lopez and Law, 1980).

Following this, we discuss the synthesis of multi-output two level AND/OR circuits as PLAs.

## 8.2 LAYOUT GENERATION

Generators are tools that produce high quality VLSI layouts from transistor level netlist descriptions or from Boolean functions. These tools capture the expertise of a designer. Layouts can be generated in terms of modules such as ROMs/RAMs, PLAs, etc., or as standard-cells, or as special structures. Layout generators provide the advantage of producing error free layouts with quick turn-around time. They use a systematic procedure to map netlist to physical level layouts that have high performance and minimum area.

Cell generators can generate layouts confining to certain constraints e.g., fixed height. Before going into a study of some popular structures in which layouts can be designed, we present the advantages of implementing a function using a functional cell over the implementation which uses universal gates.

Logic functions can be implemented by means of circuits consisting of one or more universal gates such as NAND or NOR, or by means of a single functional cell. However, synthesizing cells at the transistor level usually leads to more efficient implementations. For example consider the function

$$F = x\overline{y} + \overline{x}y \qquad (8.3)$$

An implementation of the function which uses only NAND gates is shown in Figure 8.1(b). In CMOS, this implementation requires 16 transistors and a large number of interconnects. Frequently, it is possible to implement a complex logical function more efficiently with a single functional cell, instead of using several primitive logic circuits such as NAND or NOR gates. The alternative implementation of the XOR function that takes advantage of the functional cell which realizes the function $\overline{x + yz}$ is shown in Figure 8.6(a). The CMOS circuit of the XOR function is shown in Figure 8.6(b). This alternative implementation results in better performance, reduced delay and smaller size than the design of Figure 8.1(b). It uses only ten transistors (Uehara and VanCleemput, 1981).

In this chapter we restrict ourselves to AND/OR networks. In CMOS, such networks can be realized by means of series/parallel connections of transistors, where the $p$MOS and $n$MOS sides of the circuit are dual of each other as shown in Figure 8.6(b). That is, if the $n$MOS side of the network is a series/parallel circuit, then the $p$MOS side will be a parallel/series circuit. The number of $p$MOS transistors is equal to the number of $n$MOS

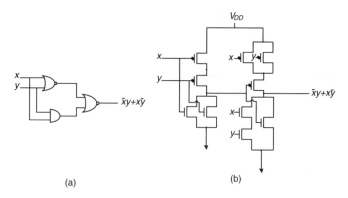

**Figure 8.6** (a) Alternative implementation of XOR function. (b) CMOS circuit of function in (a).

transistors and for every AND/OR element this number is also equal to the number of inputs to that element. This restriction in topology is commonly used by designers of CMOS circuits.

In the following paragraphs we will briefly introduce some layout styles whose details are discussed later in this chapter. We begin with fixed height layouts of functional cells.

## 8.2.1 Standard-cells

The function to be synthesized can be mapped to a regular structure which consists of an array of MOS transistors as shown in Figure 8.7. This array consists of a row of $p$MOS transistors and a row of $n$MOS transistors corresponding to the $p$MOS and $n$MOS sides of the circuit. The transistors are vertically aligned, that is, a $p$-transistor and an $n$-transistor that receive the same input are placed one above the other. The height of the functional cell may be assumed to be fixed as in 'standard-cell layout style'. The devices are interconnected depending on the function to be realized to yield the desired layout. More details on standard-cell layout generation are presented in Section 8.3.

An extension of this method is the gate matrix layout style discussed below.

## 8.2.2 Gate-matrix methodology

This methodology was first proposed by Lopez and Law (1980). It uses an orderly structure, which is a matrix composed of intersecting rows and columns. Columns are equally spaced, and are implemented in polysilicon as parallel wires running vertically. All transistors having a common input

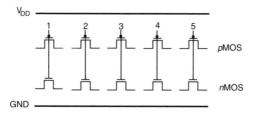

**Figure 8.7** Array of CMOS transistors for standard-cell layout implementation.

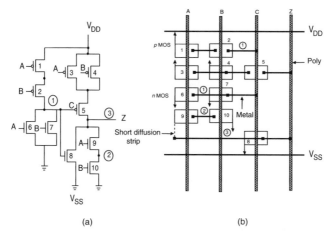

| (a) | (b) |

**Figure 8.8** (a) A static CMOS circuit. (b) Corresponding gate matrix layout.

are constructed along the same column. Thus, columns serve a dual purpose: (a) they are gates of many transistors which lie on the line, and (b) they also serve as a common connect among the transistors. The sources and drains of transistors are connected by horizontal segments of metal which are placed in the rows of the matrix. In addition, short vertical diffusion strips are sometimes necessary to connect the drains and sources of transistors to metal lines on different rows. Connections to Vdd/Vss are in a second metal layer (metal-2).

An example circuit and its corresponding gate matrix layout are shown in Figure 8.8. Lines A, B, C, and Z in Figure 8.8(b) correspond to inputs (A,B), internal node (C) and output (Z). Transistors 1,3,6, and 9 which have the same gate signal (A) are placed in the same column.

## 8.2.3 Programmable logic array

Irregular combinational functions can be mapped on to regular structures using PLAs. One very general way to implement a combinational logic function of $n$-inputs and $m$-outputs is to use a ROM of size $2^n \times m$ bits. The

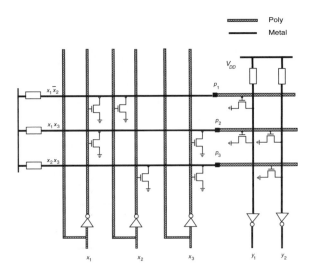

**Figure 8.9** PLA implementing two functions with three product terms.

$n$ inputs form the address of the memory and the $m$ outputs are the data contained in that address. Since it is often the case that only a small fraction of the $2^n$ minterms are required for a canonical sum-of-products (SOP) implementation, a large area is wasted by using a ROM. A PLA structure has all the generalities of a memory for the implementation of a combinational logic function. However, any specific PLA structure need contain a row of circuit elements only for product terms that are actually required to implement a given logic function. Since it does not contain entries for all possible minterms, it is usually far more compact than a ROM of the same function. To achieve further reduction in area, the various output functions can be jointly minimized before the PLA layout pattern is defined. The example below illustrates the overall structure of a PLA.

**Example 8.1**  Consider the two functions given below:

$$y_1 = x_1\overline{x_2} + x_1x_3 = p_1 + p_2$$
$$y_2 = x_1x_3 + x_2x_3 = p_2 + p_3$$

Both functions are expressed in the canonical sum-of-product form and can be implemented using a two level AND/OR circuit.

A PLA is a very regular structure consisting of an AND plane and an OR plane. The function of the AND plane is to produce the product terms, and the OR plane the sums of the product terms.

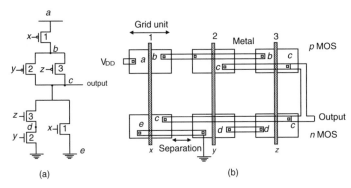

**Figure 8.10** (a) Circuit that implements the function $\overline{x+yz}$. (b) Corresponding CMOS standard-cell layout.

The $n$MOS circuit of a PLA that implements the above functions $y_1$ and $y_2$ is given in Figure 8.9. The three rows in the AND plane implement the three unique product terms ($p_1, p_2$, and $p_3$). The function of the AND plane is actually implemented using NOR logic which receives inverted inputs. The first row of the AND plane implements the product term $x_1\overline{x_2}$ since the output $p_1$ is high when $\overline{x_1}$ and $x_2$ are both low.

We have thus far seen three regular structures of layouts on to which combinational functions can be easily mapped, two for CMOS (standard-cell and gate matrix) and one for $n$MOS (PLAs). Of course PLAs can also be implemented in CMOS. In the following sections we give more details of the above three structures and then present algorithms for reducing the area of generated layouts.

## 8.3 STANDARD-CELL GENERATION

For the restricted CMOS environment discussed earlier in this chapter, where the $p$MOS network is the dual of the $n$MOS network, the problem of layout generation as standard-cells can be stated as follows. Given a circuit specification, produce a functionally correct, area efficient and design rule error free VLSI layout that represents the circuit. The AND/OR circuit may be specified as a logic level netlist. It may also be specified at the transistor level in a format similar to the one shown in Figure 8.2. The layout produced may be in CIF format (Figure 8.5).

To understand the problem of layout generation and its optimization in detail, consider the circuit of Figure 8.10(a). This network is the realization

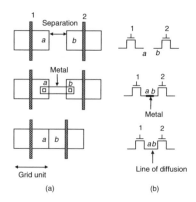

**Figure 8.11** Abutment of adjacent devices reduces width.

of the SOP specification of the function $\overline{x + yz}$ by means of a series/parallel connections of transistors. Since the number of $p$MOS transistors is the same as the number of $n$MOS transistors (the $p$MOS network is the dual of the $n$MOS network) the entire circuit can be constructed by assigning the devices of the circuit to the devices of the array (as shown in Figure 8.7) and interconnecting the corresponding sources and drains. One possible solution is shown in Figure 8.10(b). In this layout, the placement of devices of the circuit is in serial order of the labels on their gates. Any random order of devices would also yield a workable but not necessarily an efficient solution.

Before we proceed further, for the sake of explanation only, let us designate the diffusion area on the left side of the gate as the drain, and that to the right of the gate as the source, and the label of the device as the input number at its gate.

Now referring to Figure 8.10(b) we see that the source of device 1 (node $b$) is connected to the drain of device 2 with a metal strip and the two devices must be separated as dictated by the design rules. A physical separation whose value is dictated by the design rules is required if there is no connection between physically adjacent diffusion rectangles or if they are connected by metal. However if the source of a device is at the same potential as the drain of the device adjacent to it, then both devices can be connected by *abutting* their diffusion areas, therefore, reducing the separation to zero. This abutment does not violate any design rules. It will reduce the width of the array by a distance of one separation. This point is further illustrated in Figure 8.11.

It must be mentioned here that connecting sources/drains of adjacent devices with metal runs can improve speed but will require vias that affect yield. If speed is not as important as area, then the devices can be abutted as explained above.

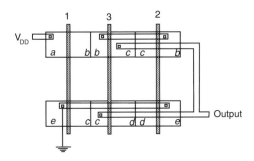

**Figure 8.12** Resultant layout of circuit in Figure 8.10.

## Area of the layout

Clearly, the area of the functional cell can be calculated as follows:

area = width × height
where, height = constant
width = basic grid size × (number of inputs + number of separations).

Since the number of inputs and the height of the layout are fixed, the problem of reducing the layout area then becomes the problem of reducing the number of separations. Separations can be reduced by ordering the devices in a row such that the physically adjacent devices can be connected by a diffusion area. Referring to the circuit of Figure 8.10(a) the order 1,3,2 of devices for both $p$MOS and $n$MOS devices will allow the sources and drains of adjacent devices 1,3 and 3,2 to be abutted (connected by diffusion). The resultant layout is illustrated in Figure 8.12. In the following paragraphs we present a graph-theoretic model of the circuit and an algorithm for judicious pairing of sources and drains to reduce the overall area of the layout.

## 8.3.1 Optimization of standard-cell layout

The optimization procedure explained in this section uses the following graph model (Uehara and VanCleemput, 1981). A $p$-side graph and an $n$-side graph are used to model the $p$MOS side and the $n$MOS side of a circuit, respectively. The $p$-side graph is defined as follows:

1. Every gate-drain and gate-source potential is represented by a vertex. Drains (or sources) that are at the same potential are represented by the same vertex.

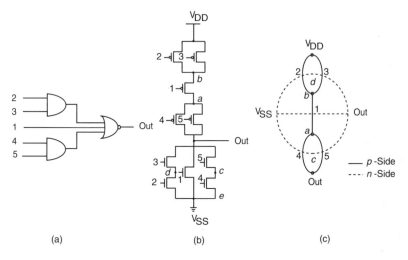

**Figure 8.13** *n-p*MOS graph construction.

**Figure 8.14** (a) Logic diagram. (b) Corresponding circuit diagram. (c) Graph.

2. Every transistor is represented by an edge connecting the drain and source vertices of that transistor (see Figure 8.13).

The *n*-side graph can be defined in a similar way. An example of such a graph for the circuit of Figure 8.14(b) is shown in Figure 8.14(c). Because of the restriction on the CMOS circuits under consideration, both the *n*-side and *p*-side graphs are series-parallel graphs. Edges correspond to transistors in both graphs and they are connected in a series/parallel manner according to the series/parallel connections of transistors in the circuit. For example, in Figure 8.14(c), edges labelled 2 and 3 are in parallel in the *p*-side graph, and are in series in the *n*-side graph. This corresponds to the *p*MOS transistors 2 and 3 being in parallel and the *n*MOS transistors 2 and 3 being in series. The names of input signals are used to label those edges. The *p*-side graph and the *n*-side graph are dual of each other. The following property of this graph model is of interest for the optimal layout of CMOS circuits.

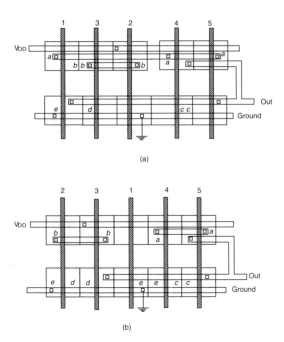

(a)

(b)

**Figure 8.15** (a) Layout generated from Euler path only on $n$-side. Path $< 1, 3, 2, 4, 5 >$ is an Euler path of $n$-side graph but not of $p$-side. (b) Layout generated from Euler path $< 2, 3, 1, 4, 5 >$. This is a Euler path in both $p$-side graph and $n$-side graph.

*If two edges $x$ and $y$ in the graph are adjacent, then it is possible to place the corresponding transistors in physically adjacent positions of the same row, and hence connect them by a diffusion area. In order to minimize the number of separation areas, it is necessary to find a set of paths of minimum-size, which corresponds to chains of transistors in the row. As indicated earlier, such a set will result in minimal area layout* (Uehara and VanCleemput, 1981).

A Euler path is a path with no repeated edges that contains all the edges of the graph. If there exists a Euler path then all gates can be chained by diffusion areas. If there is no Euler path, then the graph can be decomposed into several subgraphs which have Euler paths. In the latter case, each Euler path corresponds to a chain of transistors that is separated from the next chain by a separation area.

In order to reduce the size of an array, it is necessary to find a pair of paths in the dual graphs with the same sequence of labels. This is because $p$-type and $n$-type gates corresponding to the same input signal have the same horizontal position in the CMOS array. For example the path $< 1, 3, 2, 4, 5 >$ of the $n$-side graph in Figure 8.14(c) produces a chain of gates on the $n$MOS side, as shown in Figure 8.15(a). There is however no

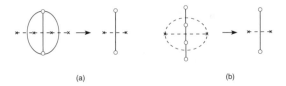

(a)                                    (b)

**Figure 8.16** Reduction of odd number of edges. (a) Odd number of edges in parallel(series) corresponding to $p$MOS($n$MOS) graph reduced to a single edge. (b) Odd number of edges in series(parallel) corresponding to $p$MOS($n$MOS) graph reduced to a single edge.

corresponding Euler path in the $p$-side graph. Therefore, a separation exists between gates 2 and 4 as shown. On the other hand, a path $< 2, 3, 1, 4, 5 >$ is a Euler path in both the $p$-side and the $n$-side graph of Figure 8.14(c). Therefore, all gates can be chained together by diffusion areas without any separation as shown in Figure 8.15(b).

The general procedure proposed by Uehara and VanCleemput (1981) consists of three steps which are itemized below:

1. Enumerate all possible decompositions of the graph to find the minimum number of Euler paths that cover the graph.
2. Chain the gates by means of a diffusion area according to the order of the edges in each Euler path.
3. If more than two Euler paths are necessary to cover the graph model, then provide a separation between each pair of chains.

The problem of finding an optimal layout of a Boolean function using the restricted CMOS design style is then reduced to the decomposition of the corresponding graph into a minimum number of Euler paths (Uehara and VanCleemput, 1981). Next we present a possible approach to determine minimum number of Euler paths. The approach relies on the following reduction technique which consists of reducing every odd number of series or parallel edges into a single one. This is repeated until no further reduction is possible (see Figure 8.16).

**Theorem 1**  If there is a Euler path in the reduced graph, then there exists a Euler path in the original graph.
PROOF  It is possible to reconstruct a Euler path in the original graph by replacing each edge of the Euler path in the reduced graph by a sequence of the original odd number of edges.

**Theorem 2**  If the number of inputs to every AND/OR element is odd, then
(1) the corresponding graph has a single Euler path;

(2) there exists a graph such that the sequence of edges on the Euler path corresponds to the order of inputs on a planar representation of the logic diagram.

PROOF

(1) The CMOS implementation of an AND/OR function has a number of series/parallel transistors that is equal to the number of variables of that function. Since the number of edges in series or in parallel is always odd, the graph model can be reduced to a single edge which is a Euler path itself. So there exists a Euler path on the original path according to Theorem 1.

(2) It is possible to construct the graph as follows:

(a) start with an edge corresponding to the CMOS circuit's output,

(b) select an edge corresponding to the output of a gate and replace it by the series-parallel graph for that gate,

(c) reorganize the sequence of new edges on the Euler path being constructed such that it corresponds to the order of the inputs on the planar representation of the logic diagram. Such a rearrangement of edges in the Euler path is always possible when the number of inputs to an AND/OR element and, hence, the number of edges in series or in parallel is odd.

It should be noted that this algorithm assumes that every gate has an odd number of inputs. This is obviously not the case for most AND/OR networks in actual practice.

## The heuristic algorithm

The heuristic algorithm takes advantage of Theorem 2. Additional inputs called 'pseudo' inputs are introduced and the original problem is modified so that every gate in a logic diagram has an odd number of inputs. It is guaranteed by Theorem 2 that there exists a Euler path for this modified problem. This Euler path contains edges corresponding to the original inputs and also edges corresponding to the new pseudo inputs, which are possible separation areas. The topology of the circuit should be selected such that the number of separation areas is minimized. The heuristic algorithm consists of the following steps:

1. To every gate with an even number of inputs a pseudo input is added.

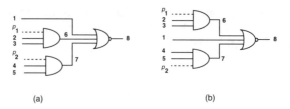

(a)                                             (b)

**Figure 8.17** (a) Adding pseudo inputs to AND/OR circuit. (b) Moving pseudo inputs to the upper and lower ends of the figure.

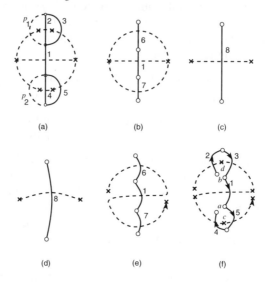

**Figure 8.18** (a) The graphs corresponding to the original circuit with pseudo edges added. (b) Merging edges $p_1$, 2 and 3, into a single edge 6. (c) Merging edges 6, 1 and 7 into a single edge 8. (d) Single edge representing existence of a single Euler path. (e) Euler paths after one expansion of the reduced graph. (f) Euler path indicating the order of devices in the row which consists of edges 2, 3, 1, 5, and 4.

2. Add this new input to the gate such that the planar representation of the logic diagram shows a minimum interlace of pseudo and real inputs. The reason is that a pseudo input at the top or at the bottom of the logic diagram does not contribute to the separation areas.

3. Obtain the CMOS circuit corresponding to the original gate level logic circuit.

4. Construct the graph such that the sequence of edges corresponds to the vertical order of inputs on the planar logic diagram.

5. Chain together the gates by means of diffusion areas, as indicated by the sequence of edges on the Euler path. 'Pseudo' edges indicate separation areas.

6. The final circuit topology can be derived by deleting pseudo edges that are in parallel with other edges and by contracting pseudo edges which are in series with other edges (Uehara and VanCleemput, 1981).

The procedure is applied to the circuit of Figure 8.14(a). Figure 8.17(a) shows the addition of pseudo inputs.

Figure 8.17(b) is the same circuit with pseudo inputs moved to the top and bottom of the logic diagram. The application of the reduction algorithm to the circuit is shown in Figure 8.18(a)–(c) and the reconstruction of the graph to find the desired sequence is shown in Figure 8.18(d)–(f).

## 8.4 OPTIMIZATION OF GATE-MATRIX LAYOUT

In Section 8.2 we introduced the structure of a gate matrix. In this section we will look at the problem of optimizing its layout. In formulating the problem, we consider only the $n$-transistor circuit, since the $p$-part of the circuit can be determined once the $n$-part is known. Also, we ignore connections to power supply and ground since they can be placed in metal-2 layer. The optimization problem and the solution explained in this section are formulated by Wing *et al.* (1985). We begin with some definitions.

### Terminology and notations

Let

$T = \{t_i | t_i$ is a transistor$\}$.

$G = \{g_i | g_i$ is a distinct transistor gate$\}$.

$N = \{n_i\}$ be the set of nets.

$C = \{c_i\}$ be the set of columns of the gate matrix.

$R = \{r_i\}$ be the set of rows of the gate matrix.

$f : G \to C$ be the function that assigns the gates to the columns. We call the function $f$ the *gate assignment function*.

$h : N \to R$ be the function that assigns the nets to the rows. We call the function $h$ the *net assignment function*.

The gates of the transistors are labelled with distinct names (except that two or more gates connected to the same node will have the same name). A poly strip is assigned to each distinct transistor gate and to each output terminal of the circuit. The poly strips are referred to as *gate lines*. In a circuit, the transistors are connected to one another at their source, drain and gate. A net is assigned to each node of the circuit and the set of nets is

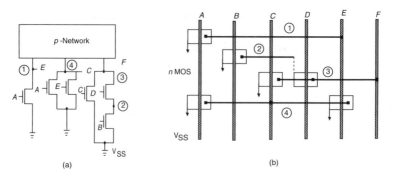

**Figure 8.19** (a) CMOS circuit of Example 8.2. (b) Corresponding gate-matrix layout.

denoted by $N = \{n_i\}$. Every transistor (channel) connects two nodes (nets) or a node and ground. Each net will be realized by a segment of horizontal metal line which is connected to a drain, source, or gate of the transistor of the node.

For each pair of functions $(f, h)$, there is a layout $L(f, h)$. Each layout $L(f, h)$ will result in a gate matrix with a certain number of rows. The number of columns is fixed. The gate matrix layout optimization problem can be stated as follows: Given a set of transistors $T$ together with a set of gates $G$ and a set of nets $N$, find a pair of functions $f : G \to C$ and $h : N \to R$ such that in the layout $L(f, h)$ has a minimum number of rows and is *realizable*.

We first consider the problem of reducing the number of rows in the gate-matrix layout. We then determine the conditions which make the layout realizable.

## The abstract model

Since each transistor is assigned a gate line, it is convenient to regard each net as being a set of gate lines. An abstract representation of this relation between gates and nets is as follows. Each gate $g_j$ will be represented by a vertical line and each net $n_i$ by a horizontal line segment. Associated with each net $n_i$ is a *net-gates set* $X(n_i) = \{g_j | n_i \text{ is connected to the source, drain or gate of a transistor assigned to } g_j\}$. Also, associated with each gate $g_j$ is a *gate-nets set* $Y(g_j) = \{n_i | g_j \in X(n_i)\}$, which lists all the nets connected to gate $j$.

**Example 8.2**   For the circuit shown in Figure 8.19 find the *net-gates set* $X(n_i)$ and the *gate-nets set* $Y(g_j)$.

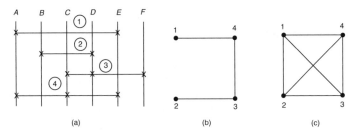

**Figure 8.20** (a) Abstract representation of gate-matrix layout of Figure 8.19. (b) The connection graph $H$. (c) The interval graph $I(L)$.

SOLUTION   The circuit has four nets labelled 1 to 4. There are five transistors whose gates are labelled A, B, C, D, and E. The output terminal node is F. The gate matrix will therefore require six columns and at most four rows. The *net-gates sets* $X(n_i)$ for the four nets are as follows:

$$X(1) = \{A, E\}: \quad X(2) = \{B, D\};$$
$$X(3) = \{C, D, F\}; \quad X(4) = \{A, C, E\}.$$

Similarly, the *gate-nets sets* $Y(g_j)$ for all $g_j$ are:

$$Y(A) = \{1, 4\}; \quad Y(B) = \{2\}; \quad Y(C) = \{3, 4\};$$
$$Y(D) = \{3, 2\}; \quad Y(E) = \{1, 4\}; \quad Y(F) = \{3\}.$$

A gate-matrix layout can be represented by either the set $X(n_i)$, $n_i \in N$, or the set $Y(g_j)$, $g_j \in G$. Given $X(n_i)$, the problem is to find a pair of functions $f : G \to C$ and $h : N \to R$ such that the number of rows in a realizable layout is minimum.

## Graph representation

The gate-nets sets defined above can be represented by a *connection graph* $H = (V, E)$ where each net $n_i$ is represented by a vertex $v_i$, and the edge set $E = \{< v_i, v_j > |n_i$ and $n_j \in Y(g_k)$ for some gate $g_k\}$. The connection graph can be derived from the set $Y(g_k), g_k \in G$, and it completely describes how the nets are connected to the gates.

In the abstract representation of the layout (see Figure 8.20), the nets can be regarded as intervals which overlap one another. We define an *interval graph* associated with the layout $L(f, h)$ as $I(L) = (V, B)$ where $V$ is the same vertex set as in the connection graph $H$ and the edge set $B = \{< v_i, v_j > |n_i$ and $n_j$ overlap in $L\}$. Figure 8.20(a) shows the abstract representation of the circuit ($L$) given in Figure 8.19(b). The connection graph and the interval graph are given in Figure 8.20(b) and (c) respectively.

Now we are in a position to explain the gate matrix optimization problem. As stated earlier, the number of columns in a gate matrix layout is fixed. Therefore the problem reduces to assigning nets to rows such that the total number of rows required is minimized. The problem is similar to the unconstrained channel routing problem discussed in Chapter 7. The difference however is that unlike the channel routing problem, the order of columns can be changed to reduce the length of nets connecting the columns. An optimal order of columns is one which assigns nets to rows such that the total number of rows required in $L(f, h)$ is minimal.

The problem can be divided into two subproblems: (1) find an optimal column ordering, and (2) assign the net segments to rows. The second subproblem can be solved using the left-edge algorithm presented in Chapter 7 (Hashimoto and Stevens, 1971). A possible solution to the problem of column ordering was given by Ohtsuki *et al.* (1979). In the following paragraphs we will describe the solution to the above two subproblems for the circuit of Figure 8.19.

To solve the above problem, we define a matrix $A = [a_{ij}]$ whose rows are the vertices (nets) of the connection graph $H$ and whose columns are the sets of all dominant cliques of $H$. A dominant clique is a clique that is not contained in another clique. Element $a_{ij} = 1$ if and only if vertex $i$ is contained in dominant clique $j$; otherwise $a_{ij} = 0$. For the example circuit under consideration (see Figure 8.20) we have three dominant cliques in $H$: {1,4}, {2,3}, and {3,4}. Therefore,

$$
A = \begin{array}{c} \\ 1 \\ 2 \\ 3 \\ 4 \end{array}
\begin{array}{ccc}
1 & 2 & 3 \\
1 & 0 & 0 \\
0 & 1 & 0 \\
0 & 1 & 1 \\
1 & 0 & 1
\end{array}
$$

The above matrix is known as *vertex versus dominant clique* matrix (Fulkerson and Gross, 1965).

A connection graph is an interval graph if and only if its vertex versus dominant clique matrix has the property that the columns can be so arranged that in each row, the ones appear in consecutive positions. In general, the connection graph may or may not be an interval graph and the matrix $A$ may or may not have the consecutive ones property. $A$ can be made to have consecutive ones by replacing the zeros by ones (called fill-ins) between the leftmost and rightmost ones in each row. The resultant matrix then corresponds to an interval graph whose largest dominant clique has a size equal to the largest number of ones in any column. (The size of the clique is equal to the number of vertices in the clique.)

Our first subproblem can now be stated as follows: *Given the vertex versus dominant clique matrix of a connection graph find a permutation of the columns such that, when zeros are replaced by ones in some rows to satisfy the consecutive ones property, the largest number of ones in any column is minimized.* As will become clear later, minimizing the number of ones in any column will result in reducing the net lengths. The above matrix $A$ is reproduced below with fill-ins (represented by **x**). Note that column 2 has the largest number of ones, which is equal to three.

$$
\begin{array}{c c c c}
 & 1 & 2 & 3 \\
1 & 1 & 0 & 0 \\
2 & 0 & 1 & 0 \\
3 & 0 & 1 & 1 \\
4 & 1 & \mathbf{x} & 1
\end{array}
$$

The computational complexity of the above problem is $Q!$ where $Q$ is the number of dominant cliques. A fast heuristic that generates many solutions from which the best can be picked is given by Wing et al. (1985). The strategy used is to reconstruct matrix $A$ column by column in such a way, that as each column is added, the number of fill-ins needed to satisfy the consecutive ones property is minimized (Wing et al., 1985). This is accomplished by adding columns in such a way that (1) the number of new edges added by the new clique is minimized and (2) the column is added to the right or left of already placed columns that will result in minimum number of fill-ins.

Let us denote the matrix $A$ constructed by the above procedure as $A^*$. One possible $A^*$ matrix is given below. We shall use this to illustrate the gate matrix optimization problem.

$$
\begin{array}{c c c c l}
 & 3 & 1 & 2 & \textit{Old column} \\
 & 1 & 2 & 3 & \textit{New column} \\
 & & & & \\
1 & 0 & 0 & 1 & \\
2 & 1 & 0 & 0 & \\
A^* = 3 & 1 & 1 & 0 & \\
4 & 0 & 1 & 1 &
\end{array}
$$

## Determination of gate assignment function $f$

Matrix $A^*$ in our example does not require any fill-ins since it satisfies the consecutive ones property. The matrix $A^*$ that has the consecutive ones property also corresponds to an interval graph $I(L)$ where $L$ is the layout to

| Gate | $\cap[l_i, r_i]$ | $[L_k, R_k]$ | $f$ |
|------|------------------|--------------|-----|
| $A$ | $l_1, r_1 \cap l_4, r_4$ | $[3,3]$ | 6 |
| $B$ | $l_2, r_2$ | $[1,1]$ | 1 |
| $C$ | $l_3, r_3 \cap l_4, r_4$ | $[2,2]$ | 4 |
| $D$ | $l_2, r_2 \cap l_3, r_3$ | $[1,1]$ | 2 |
| $E$ | $l_1, r_1 \cap l_4, r_4$ | $[3,3]$ | 5 |
| $F$ | $l_3, r_3$ | $[1,2]$ | 3 |

**Table 8.1 Intervals specifying relative positions of gates and the gates assignment function $f$.**

| | Net $n_i$ | $h(n_i) = Row$ |
|---|-----------|----------------|
| **Table 8.2 Table for net assignment function $h$.** | 2 | 1 |
| | 4 | 1 |
| | 3 | 2 |
| | 1 | 2 |

be determined. The technique to find $f$ presented in this chapter was given by Wing *et al.* (1985) and uses the results of Ohtsuki *et al.* (1979). The procedure is summarized below.

Let $l_i$ and $r_i$ be the numbers of the leftmost and the rightmost columns in matrix $A^*$ in which row $i$ has 1's. We define the *interval* of row (net) $i$ as $[l_i, r_i]$. Then, gate $g_k$ in the layout $L$ must lie within the interval $[L_k, R_k]$ where,

$$[L_k, R_k] = \bigcap_{n_i \in Y(g_k)} [l_i, r_i] \tag{8.4}$$

The intervals $[L_k, R_k]$ specify the relative positions of the gates. The assignment function $f$ uses these intervals to determine the relative positions of the columns. This is performed as follows. The intervals are topologically sorted in ascending order where $[L_i, R_i] \leq [L_j, R_j]$ if and only if $(L_i < L_j)$ or $((L_i = L_j)$ and $(R_i \leq R_j))$. For the matrix $A^*$ given above, the corresponding intervals together with the assignment function are given in Table 8.1. For example, referring to Table 8.1, since the interval of C is $[2,2]$ and that of E is $[3,3]$, column C is assigned to the left of column E.

## Determination of net assignment function $h$

Having determined the gate assignments, we now move on to determine the function $h$, that is, the assignment of nets to rows. This function is determined by applying the left-edge algorithm (Hashimoto and Stevens, 1971). Sorting the rows of $A^*$ on their left edge results in the order $[2,3,4,1]$. Now scanning the sorted list, we assign net 2 to row 1. Due to horizontal

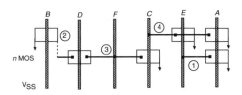

**Figure 8.21** Final layout of circuit in Figure 8.19.

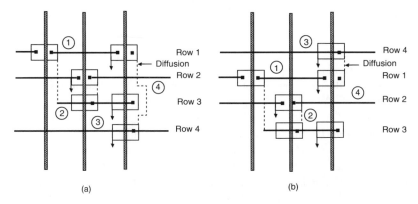

(a)                                      (b)

**Figure 8.22** (a) Gate-matrix layout with net $D(1,4)$ unrealizable. (b) Permutation of rows to make the layout realizable.

conflict, the next net, that is 3, is skipped. Net 4 however can be assigned to row 1. Then, rescanning the list, net 3 is assigned to row 2. Finally net 1 is also assigned to row 2 since it does not horizontally conflict with net 3 (see Table 8.2). The final layout $L(f, h)$ is shown in Figure 8.21.

## Realizable layouts

In the above examples we saw that vertical diffusion runs are sometimes required to connect sources and drains of transistors in different rows. For a given layout $L(f, h)$, for each pair of nets $(n, m)$ such that $h(n) \neq h(m)$ we define the set of diffusion runs $D(n, m) = \{g_k | h(n) \neq h(m)$ and there exists a vertical diffusion run between net $n$ and net $m$ near gate line $k\}$. $D(n, m)$ is said to be realizable if each of its vertical diffusion runs can be placed such that it does not overlap with any transistor in the same column on a row between row $h(n)$ and row $h(m)$. For example in Figure 8.22, $D(1,2)$ is realizable but $D(1,4)$ is not. Thus, each function $h : N \to R$ induces a set of diffusion runs. The function $h$ is *realizable* if every $D(n, m)$ it induces is realizable. A layout $L(f, h)$ is realizable if $h$ is realizable.

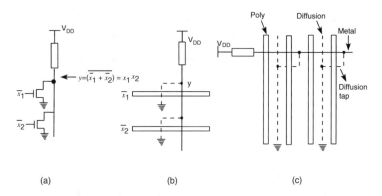

**Figure 8.23** (a) NOR implementation of AND function. (b) Stick diagram representation. (c) AND plane of PLA for the product term $x_1 x_2$.

The solution to the gate matrix layout optimization problem can therefore be divided into two stages. In the first stage the problem of realizability is ignored and a layout which requires minimum number of rows is found. If the layout found is realizable then we are done. If the layout is not realizable then we attempt to find a permutation of the rows to obtain a realizable function $h$. If a satisfactory permutation is not available then the last resort is to increase the spacing between some pairs of poly strips to accommodate vertical diffusion runs which will not overlap any transistor.

## 8.5 PROGRAMMABLE LOGIC ARRAYS

In this section we will discuss the problem of minimizing the layout of a PLA. We begin with a discussion of the structure and method of construction of PLAs.

As mentioned earlier, a PLA consists of two planes, the AND plane that implements the unique product terms of all the functions to be implemented by that PLA, and the OR plane which sums the product terms generated by the AND plane. Wires that carry product terms or simply *terms* (the outputs of the AND plane) are called *term* wires. These term wires run horizontally in the AND plane. In nMOS implementation the term wires are implemented in metal, with the pullup transistors on the left. The inputs are fed to the AND plane using poly wires that run vertically. Between each pair of vertical poly wires in the AND plane runs a diffusion wire connected to ground. At certain places, a diffusions *tap* is connected to a term wire; it crosses a poly wire, and connects to the adjacent diffusion ground wire as shown in Figure 8.23. The effect of a diffusion tap (crossing

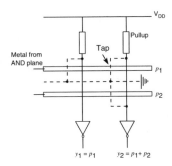

**Figure 8.24** Structure of OR plane. $p_1$ and $p_2$ are fed by the AND plane. $y_1 = p_1$, and $y_2 = p_1 + p_2$.

a poly and connecting the output to ground) is that a transistor is formed. Since the transistor formed connects the output to ground, if the vertical poly wire is on, that is, the input is high, any current that passes through the pullup from $V_{DD}$ will run to ground through the tap.

The only time the term wire can be high is when all poly wires crossing that particular term wire where there are taps are low. Thus, in the AND plane, each term wire implements the logical NOR function; it is high if and only if none of the vertical poly wires is high. If the inputs are inverted before being fed to this plane, the NOR circuit will implement the desired logical AND function.

In the OR plane, the role of rows and columns is reversed, with the term wires becoming poly and the vertical wires running in metal. This is because the OR plane is similar in structure to the AND plane except that it is rotated by 90° clockwise. Each vertical wire in the OR plane begins at the top with a pullup, and at the bottom it is the input to an inverter. Diffusion taps at various places along each vertical wire will connect that wire to ground if the term wire crossing at that point is high. Thus the input to the inverter is high if and only if none of the term wires crossing the taps are high. Therefore, the output of the inverter will be high if one or more of these term wires are high. That is, each vertical wire in the OR plane implements the sum (logical OR) of some subset of the terms represented by the term wires—those that cross the vertical wire where there are taps. This point is illustrated in Figure 8.24. The general procedure for implementing a collection of sums of Boolean functions as a PLA is enumerated below.

1. For each literal used in one or more product terms, we require a vertical wire in the AND plane.
2. For each function to be implemented make one vertical wire in the OR plane.

3. For each unique product term appearing in one or more functions, make one term wire running horizontally across the planes.
4. Each term is the product of certain literals. For each of those literals, tap the term's wire at the points in the AND plane where the wires for each of those literals (i.e., wires carrying signals that are the complements of those literals) cross the term's wire.
5. For each function which is a sum of terms, tap the vertical wire for the sum at the points in the OR plane where the wires for the terms in the sum cross.

### 8.5.1 PLA personality

The essential information about a PLA consists of the following: (a) the names of input variables, that is, the vertical lines in the AND plane and the order of their occurrence, (b) the names for each output wire, that is, the vertical lines in the OR plane and the order of their occurrence, (c) the number of term wires corresponding to the number of unique product terms, and (d) the tap positions. This information is represented in a matrix notation commonly known as *PLA personality*.

The personality consists of two matrices side by side, one for the AND plane and the other for the OR plane. The AND plane matrix consists of 0s, 1s, and 2s. The OR plane matrix consists of only 0s and 1s. The number of rows of both the matrices is equal to the number of unique product terms required to implement all the desired functions. The number of columns of the AND plane is equal to the number of input variables. Similarly the number of columns in the OR plane is equal to the number of output functions. The AND plane matrix is constructed as follows:

1. A '0' means there is a tap under the vertical wire $\overline{x_i}$. This wire is the one that feeds $x_i$ uncomplemented in the AND plane.
2. A '1' means there is a tap under vertical wire for $x_i$, that is, the wire being the one where $x_i$ is complemented before being fed to the AND plane.
3. A '2' means that neither wire for $x_i$ is tapped for the term line under consideration.

In the OR plane a '1' indicates a tap and a '0' indicates no tap in the layout. That is, a '1' indicates that the output function contains the product term represented by that row. The following example will clarify the above ideas.

**Example 8.3**   For the three functions given below, draw the PLA stick diagram†. Also give the personality matrix.

---

†In stick diagrams, wires are represented by lines, and contact-cuts/vias are represented by points.

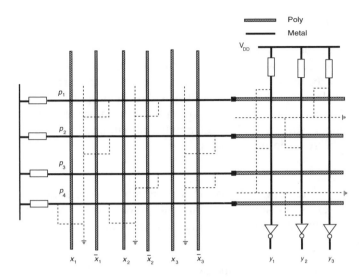

**Figure 8.25** Stick diagram of PLA in Example 8.3.

$$y_1(x_1, x_2, x_3) = x_1 x_2 + x_1 x_3 + x_2 x_3$$
$$y_2(x_1, x_2, x_3) = \overline{x_1}\ \overline{x_2} + x_1 x_3$$
$$y_3(x_1, x_2, x_3) = x_1 x_2$$

SOLUTION  Since the three functions contain four unique product terms, which are $x_1 x_2 = p_1$, $x_1 x_3 = p_2$, $x_2 x_3 = p_3$ and $\overline{x_1}\ \overline{x_2} = p_4$, the PLA layout will have four rows. The number of columns in the AND plane is equal to the number of input variables which is three. Since we are implementing a circuit for three functions, the number of outputs in the OR plane is also three. The PLA layout in the form of a stick diagram is given in Figure 8.25. The personality is given below.

| | | | | | | |
|---|---|---|---|---|---|---|
| $p_1$ | 1 | 1 | 2 | 1 | 0 | 1 |
| $p_2$ | 1 | 2 | 1 | 1 | 1 | 0 |
| $p_3$ | 2 | 1 | 1 | 1 | 0 | 0 |
| $p_4$ | 0 | 0 | 2 | 0 | 1 | 0 |
| | $x_1$ | $x_2$ | $x_3$ | $y_1$ | $y_2$ | $y_3$ |

With this, we conclude the construction procedure of a PLA layout for a given set of functions. In the next section we will see the problem of PLA optimization.

## 8.5.2 Optimization of PLAs

The size of the personality matrix also represents the size of the final PLA layout. In this section we will use the personality representation of the PLA and study algorithms to reduce its size, and hence the area of the final layout. It is clear that the area of a PLA is directly proportional to the size of the PLA personality which depends on (a) the number of product terms, (b) the number of inputs and (c) the number of outputs. The area of a PLA layout can be reduced in two ways:

1. Reduce the number of rows. This is possible by deleting redundant product terms. Deleting a redundant product term corresponds to deleting a row of a personality and hence reducing the size of the PLA layout.

2. Allow two literals to share the same column. This technique is known as '*folding*'. PLA folding consists of allowing AND plane inputs to enter from either the top or the bottom; if the two different inputs arriving from different directions, that is, top or bottom, can share a column then the circuitry of the two columns can be placed in the area of one, thus reducing the width of the PLA layout.

In the following paragraphs we will look at a technique used to determine if a row is redundant. For the sake of notation let us suppose that the inputs to the AND plane are $x_1, \cdots, x_n$ and the outputs from the OR plane are $y_1, \cdots, y_n$. In the PLA personality each term $p$ of $n$-variables represents some points in an $n$-dimensional cube. For example if $x_1$, $x_2$, and $x_3$ are the three inputs of the PLA then the term 120 in the AND plane represents the product term $x_1\overline{x_3}$, which represents the two vertices 100 and 110 of the 3-cube.

### Redundancy check

The first thing to check in reducing the size of the personality is to see if a row is redundant. To find if a row is redundant we must consider every output $y_i$ in which row $r$ has a '1' in the OR plane. Let $S_i$ be the set of other rows that have '1' for output $y_i$. Row $r$ is redundant for output $y_i$ if the set of points covered by one or more points of $S_i$ is a superset of the points covered by $r$. Row $r$ is redundant if and only if it is redundant for all the outputs $y_i$ in which it has a '1'.

**Example 8.4**    For the personality matrix given below determine if row $r_1$ is redundant.

| | | | | | |
|---|---|---|---|---|---|
| $r_1$ | 1 | 1 | 1 | 1 | 1 |
| $r_2$ | 1 | 2 | 1 | 1 | 0 |
| $r_3$ | 0 | 1 | 2 | 1 | 1 |
| $r_4$ | 2 | 1 | 1 | 0 | 1 |
| | $x_1$ | $x_2$ | $x_3$ | $y_1$ | $y_2$ |

SOLUTION    The minterm covered by the term in $r_1$ is 7 (111). For output $y_1$, the other rows that have a '1' in the OR plane are $r_2$ and $r_3$. The minterms covered by $r_2$ (121) are 101 (5) and 111 (7). Therefore the term of $r_1$ (which is 111) is covered by the minterms of $r_2$ and $r_3$ (which are 010, 011, 101, 111). But we cannot say if $r_1$ is redundant until we have repeated the procedure for output $y_2$ since it also has a 1 in the row corresponding to $r_1$.

For the output $y_2$, the rows other than $r_1$ that have a '1' in the OR plane are $r_3$ and $r_4$. The minterms covered by these two product terms are {2,3} and {3,7} respectively. Since the minterm of $r_1$, which is 7, is covered by the minterms of other rows that have a '1' in the same column ($y_i$, $i = 1,2$) of the OR plane for all the outputs, row $r_1$ is redundant.

We can also use the containment condition to determine if the minterms contained in a row being considered for deletion are also contained in the other rows for which the output column contains a 1. This is explained below.

## Containment condition

The containment condition can be mathematically stated as follows. Let $p$ be the product term for the row $r$ whose redundancy is being considered. Focus on those outputs for which this row has a '1'. Let $p_1, p_2, \cdots, p_k$ be the terms of all the *other* rows with '1' in one of those outputs. We can say that row $r$ is redundant if the minterms covered in $p$ are a subset of the minterms covered by $\{p_1, p_2, \cdots p_k\}$. That is, a row $r$ is redundant if

$$p \subseteq p_1 + p_2 + \cdots + p_k \tag{8.5}$$

Another way of testing for the condition of Equation 8.5 is to verify if

$$\overline{p} + p_1 + p_2 + \cdots p_k \tag{8.6}$$

is a tautology.†. The procedure used to test the above condition must

†If for any assignment of input values the expression is always true then the expression is a tautology.

perform two things: (a) determine $\overline{p}$, and (b) test if the sum of products of the complemented terms and the $p_i$'s is a tautology.

If $p$ is expressed as a sequence of 0s, 1s, and 2s then the procedure to find $\overline{p}$ is as follows. For every '0' in position $i$ of $p$, we create a new term with a '1' in that position and '2' elsewhere. The same thing is done for every '1' in position $i$ of $p$, a new term is created with a '0' in that position and 2s elsewhere. No new terms are created for 2s in position $i$. The sum of these newly created terms is the complement of $p$. For example, the complement of the term 120 is 022 + 221. Note that the above procedure is simply the application of the DeMorgan's theorem. The term 120 represents $x_1\overline{x_3}$ and 022+221 represents $\overline{x_1} + x_3$.

Testing for tautology is more involved (Ullman, 1983; Brayton *et al.*, 1984). An heuristic is used which is based on the idea that any function $f(x_1, x_2, \cdots, x_n)$ can be decomposed as follows:

$$f(x_1, x_2, \cdots, x_n) = x_1 f_1(x_2, \cdots, x_n) + \overline{x_1} f_0(x_2, \cdots, x_n) \qquad (8.7)$$

$f_1$ is formed by taking the rows with '1' or '2' in the first column, and $f_0$ is constructed similarly by taking the rows with '0' or '2' in the first column. Thus an $n$-column personality matrix is converted into two $n-1$ column matrices. The original matrix is a tautology if and only if both the new matrices are tautologies. The method can be applied recursively, but fortunately we do not have to extend it until we reach a matrix with one column. There are two useful heuristics that can be used to find the answer to the subproblem (Ullman, 1983). These are:

1. If a matrix has a row with all 2s , then it surely is a tautology, because this one term covers all the points in the Boolean cube.

2. Suppose a matrix has $n$-columns, so its rows are product terms representing points in a Boolean $n$-cube. Each term $t$ with $i_t$ 2s represents $2^{i_t}$ points. The maximum number of points covered by all the terms can be obtained by summing $2^{i_t}$ over all terms $t$. If this number is less than the total number of points in the $n$-cube, that is

$$\Sigma 2^{i_t} < 2^n \qquad (8.8)$$

then we can conclude the matrix is not a tautology. However, if the above inequality fails, then the matrix in question may represent a tautology. Further testing is required. The tautology test can be applied to either the original matrix or the decomposed matrix. If any of the decomposed matrices does not represent a tautology then the original matrix also does not represent a tautology.

The example below will further illustrate the tautology test.

**Example 8.5**  For the personality matrix $M$ given below, determine if it represents a tautology.

$$M = \begin{matrix} 1 & 2 & 2 \\ 2 & 1 & 2 \\ 0 & 1 & 1 \\ 1 & 0 & 2 \end{matrix}$$

SOLUTION    Applying Equation 8.8, there are three literals $x_1$, $x_2$ and $x_3$. Therefore $n = 3$ and $2^n = 8$. However, the sum $\Sigma 2^{i_t} = 4 + 4 + 1 + 2 = 11 > 2^n = 8$. Hence, the matrix may be a tautology.

The above matrix can now be decomposed into two matrices say $M_0$ and $M_1$, where $M_0$ is formed by taking the rows with '0' or '2' in the first column and $M_1$ is formed by taking the rows with '1' or '2' in the first column. The two matrices are given below.

$$M_0 = \begin{matrix} 1 & 2 \\ 1 & 1 \end{matrix} \qquad M_1 = \begin{matrix} 2 & 2 \\ 1 & 2 \\ 0 & 2 \end{matrix}$$

Clearly $M_1$ is a tautology since it contains a row with all 2s (Ullman, 1983; Brayton *et al.*, 1984). $M_0$ cannot be further decomposed, and does not have a row with all 2s. Therefore $M_0$ as well as the original matrix are not tautologies.

**Example 8.6**    For the PLA personality given below determine if row $r_2$ is redundant. Use the containment condition explained above to check for redundancy.

| | $x_1$ | $x_2$ | $x_3$ | $x_4$ | $y_1$ |
|---|---|---|---|---|---|
| $r_1$ | 2 | 1 | 2 | 1 | 1 |
| $r_2$ | 0 | 1 | 1 | 2 | 1 |
| $r_3$ | 0 | 2 | 1 | 0 | 1 |

SOLUTION    To determine if row $r_2$ is redundant, we first find the complement of $r_2$-term (0112). The complement of this term is (1222 + 2022 + 2202). Now we have to check if the matrix formed by the above three terms as three rows and the terms represented by the other rows, that is $r_1$ (2121) and $r_3$ (0210) represents a tautology.

$$\begin{matrix} 1 & 2 & 2 & 2 \\ 2 & 0 & 2 & 2 \\ 2 & 2 & 0 & 2 \\ 2 & 1 & 2 & 1 \\ 0 & 2 & 1 & 0 \end{matrix}$$

Applying the decomposition procedure as in Example 8.5 it is easy to verify that the above matrix represents a tautology. Hence $r_2$ is redundant.

## Raising of terms

Raising of terms consists of taking a term $x_1 f$ (or $\overline{x_1} f$) where $f$ is a product of $n - 1$ other variables besides $x_1$, and see if it can be replaced by $f$. Observe that the points covered by $f$ in the $n$-cube are twice the number of points covered by $x_1 f$. The motivation for doing so is that by increasing the number of points covered by a given product term, there could be a better chance that some other product terms become redundant.

To test if $x_1 f$ can be replaced by $f$ we have to know if the points covered by $\overline{x_1} f$ are also covered by other terms. As before, for each output column $y$, let the set of other available terms with a '1' at this output be $\{p_1, p_2, \cdots, p_k\}$. The containment condition described earlier can be used to check if $\overline{x_1} f \subseteq \{p_1, p_2, \cdots, p_k\}$. Only if the above is true, then the replacement of $x_1 f$ by $f = x_1 f + \overline{x_1} f$ will not change the function.

If there are several output functions then the above procedure must be repeated for each output column in which the row for term $x_1 f$ is a '1'. Raising is represented by replacing a '0' or a '1' in the PLA personality by a '2'. We illustrate this raising technique with the following example.

**Example 8.7**  For the PLA personality shown below verify if the first position of the first row which is a '0' can be raised.

$$
\begin{array}{ccc@{\qquad}c}
0 & 1 & 2 & 1 \\
1 & 1 & 0 & 1 \\
2 & 1 & 1 & 1
\end{array}
$$

SOLUTION  The question being asked is: can the first term 012, be replaced by 212, that is, can the '0' be raised to '2'? This will result in adding the term 112. The term 112 can be added if and only if it is already contained in the function, or if its minterms are contained in the terms covered by the remaining terms. To test this we can apply the containment condition to see if the sum of the complement and the remaining terms is a tautology. That is, if $112 \subseteq 110 + 211$ or if $\overline{112} + 110 + 211$ is a tautology, or if the matrix $M$ below represents a tautology.

$$
M = \begin{array}{ccc}
0 & 2 & 2 \\
2 & 0 & 2 \\
1 & 1 & 0 \\
2 & 1 & 1
\end{array}
$$

|  |  |  | $r_1$ | 0 | 2 | 2 | 1 | 0 |
|---|---|---|---|---|---|---|---|---|
|  |  |  | $r_2$ | 2 | 1 | 2 | 1 | 0 |
| $y_1 =$ | $\overline{x_1} + x_2 + x_3$ |  | $r_3$ | 2 | 2 | 1 | 1 | 1 |
| $y_2 =$ | $x_1\overline{x_2} + x_3$ |  | $r_4$ | 1 | 0 | 2 | 0 | 1 |
|  |  |  |  | $x_1$ | $x_2$ | $x_3$ | $y_1$ | $y_2$ |

**Figure 8.26** Two functions and their PLA personality to illustrate PLA folding.

Once again, applying the technique used in Example 8.5 we find that $M$ represents a tautology. Therefore the term 012 in the first row can be replaced by term 212 by raising the '0'.

## PLA folding

As seen in the previous section, a PLA maps a set of Boolean functions in standard form also known as 2-level sum-of-products form into a regular geometrical structure. The structure follows a fixed architecture in which the connection wires (in metal, poly and diffusion), and power and ground lines are deployed in advance. The possible transistor locations given by the personality matrix depend on the function to be implemented. Because of its regular and fixed architecture, the task of generation of layout cells in the form of a PLA is very simple. Thus, the design time is shortened and the cost is reduced.

In the previous paragraphs, we saw a technique to reduce the area of a PLA layout by deleting redundant rows. In this section we present the next step in reducing the layout area of a PLA. This technique is called *folding*. Folding consists of allowing two signals to share a single column in such a way that some unused space is reclaimed without disturbing the regular structure. The space to be reclaimed in the layout corresponds to 2s in the personality or absence of any transistor in the layout. This is another good reason for using the raising of terms technique explained in the previous subsection.

Consider the two functions and their corresponding PLA personality given in Figure 8.26. For the sake of illustration, let us assume that there are no redundant rows and no terms can be raised. To understand folding, let us concentrate on the AND plane in Figure 8.27. Observe that permuting the rows from $r_1, r_2, r_3, r_4$ to $r_1, r_4, r_3, r_2$, does not disturb the function as long as the taps in the OR plane corresponding to rows of the OR plane matrix are also included in the permutation.

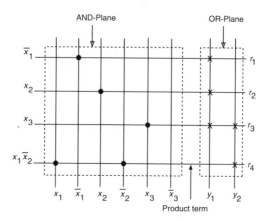

**Figure 8.27** PLA corresponding to personality matrix in Figure 8.26.

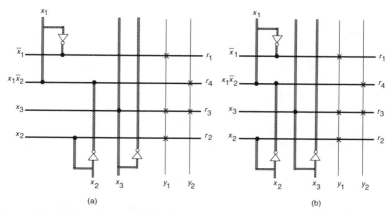

**Figure 8.28** (a) Signal $x_1$ input from top and $x_2$ input from the bottom. (b) Reducing width by shifting circuit of column $x_2$ below $x_1$.

Referring to Figure 8.28(a), we observe that signal $x_1$ is input from the top and signal $x_2$ from bottom. If this is allowed, and if we do not extend the polysilicon wire that carries the signal beyond the last transistor to which it is input, then the circuitry of signal $x_2$ can be shifted below that of signal $x_1$, resulting in the layout compacted by one column. This is shown in Figure 8.28(b). Also observe that if any row has both zeros (or both ones) in the columns being considered for folding, then folding as shown in Figure 8.28(b) is not possible. However, if a row has a '0' in one column and a '1' in another then pairing may be possible.

The folding shown in Figure 8.28 where wire $x_1$ comes over $x_2$ and $\overline{x_1}$ comes over $\overline{x_2}$, that is, where complemented and uncomplemented wires face each other is termed *straight folding*. Sometimes straight folding may not be possible. This situation occurs when any row in the two columns under consideration has both zeros or both ones. In such a situation allowing complemented wires face uncomplemented ones, that is wire $x_1$ facing $\overline{x_2}$ and wire $x_2$ facing $\overline{x_1}$ may allow folding. This type of folding is termed *twisted folding*.

## The algorithm

A graph based algorithm for folding takes a PLA personality as input and produces a pairing of columns indicating which of the two is above the other (Ullman, 1983). It also provides the ordering of the rows that will make this set of column pairs legal.

Let us first consider straight folding. The idea is first to select a pair of columns $c_1$ and $c_2$ in some order. Then assuming that these can be paired (the two columns must not have 1s or 0s in the same row and neither has been paired with another column), we consider pairing $c_1$ and $c_2$ with either one above the other. To determine if $c_1$ and $c_2$ can be paired, a directed graph is constructed. This directed graph represents the constraints on the order of term wires induced by whatever pairing that has been made thus far.

The procedure is to use two nodes $p$ and $q$ for every two columns considered for pairing. Node $p$ is used to represent the constraint that the terms using wire $c_1$ must be above the terms using wire $c_2$, while $q$ represents the constraint that terms using wire $\overline{c_1}$ must be above those using wire $\overline{c_2}$. The procedure applied to construct the directed graph is as follows:

For each row $r$,
1. If it has 1 in column $c_1$, then draw an arc from $r \to p$;
2. If it has 1 in column $c_2$, then draw an arc from $p \to r$;
3. If it has 0 in column $c_1$, then draw an arc from $r \to q$;
4. If it has 0 in column $c_2$, then draw an arc from $q \to r$.

In other words, $p$ is the target of arcs from all rows with '1' in column $c_1$ and source of arcs to all rows with '1' in column $c_2$. Similarly, $q$ is the target of arcs from all rows with '0' in $c_1$ and source of arcs to all rows with '0' in $c_2$.

It is obvious that if both (1) and (2) or both (3) and (4) hold, we cannot make the pairing because a cycle is surely introduced. Therefore for pairing two columns straight, the two bits in any row of the column being considered must not be the same. Having made the above pairing, we check if any cycles have been introduced in the graph. If a cycle is introduced then there is no way we can order the terms and still run the paired columns from top and bottom without occupying the same space.

If cycles are introduced, then the arcs introduced are removed and another pair of columns is considered for pairing. If no cycles are introduced then the pairing is made permanent.

If pairing $c_1$ and $c_2$ straight is not possible, then we may consider pairing them twisted. The procedure used in that case is the same as above except that the roles of $p$ and $q$ are exchanged in (2) and (4). That is,

For each row $r$,
1. If it has 1 in column $c_1$, then draw an arc from $r \to p$;
2. If it has 0 in column $c_2$, then draw an arc from $p \to r$;
3. If it has 0 in column $c_1$, then draw an arc from $r \to q$;
4. If it has 1 in column $c_2$, then draw an arc from $q \to r$.

Again if both (1) and (2) or both (3) and (4) hold then a cycle is introduced and we cannot make the pairing. Therefore, for twisted pairing, the two bits in any row of the two columns being considered must not be different.

Since the possibility of folding two columns together rests entirely on all rows populated on one column being above all rows populated by the other column, $c_i$ and $c_j$ may only be folded together if they are disjoint. If such a fold is actually implemented, the PLA rows must be permuted such that the rows of $R(i)$ are all above those of $R(j)$. This induces a partial ordering on the rows and places constraints on the folding of other columns. A second pair of columns can be simultaneously folded with the first if the second folding pair does not induce constraints that violate the partial ordering induced by the first pair of columns.

A directed edge from vertex $r_i$ to vertex $r_j$ is present in the graph if and only if row $i$ must be above row $j$. The addition of a second folding pair will add directed edges to the graph, and this will violate the partial ordering if and only if the new edges induce a cycle in the graph.

Continuing the above procedure we make a maximal set of pairings that gives us an acyclic graph. The graph represents a partial order on terms wires or rows. We pick a total order consistent with the partial order. This process is known as topological sort. We will now illustrate the entire procedure with the aid of an example.

**Example 8.8** For the PLA personality shown below use the procedure discussed above to fold the AND plane.

| | $x_1$ | $x_2$ | $x_3$ | $x_4$ | | $y_1$ | $y_2$ |
|---|---|---|---|---|---|---|---|
| $r_1$ | 1 | 2 | 1 | 2 | | 1 | 0 |
| $r_2$ | 2 | 0 | 1 | 1 | | 0 | 1 |
| $r_3$ | 0 | 1 | 2 | 1 | | 0 | 1 |
| $r_4$ | 0 | 2 | 0 | 2 | | 1 | 0 |

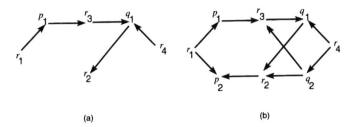

(a)                                    (b)

**Figure 8.29** (a) Directed graph with constraints on order of rows when first two columns considered. (b) Graph with next pair of columns included.

SOLUTION   Let us consider the first two columns $x_1$ (as $c_1$) and $x_2$ (as $c_2$). These two columns are possible candidates for folding. Since no two rows of columns $c_1$ and $c_2$ have both 1s or both 0s, we will attempt to fold them straight.

To do this, we introduce two nodes $p_1$ and $q_1$ in addition to four nodes corresponding to the four rows of the personality. Arcs are then drawn to represent the constraint using the above explained procedure. For example:

In column $c_1$ ($x_1$):
Since we have a '1' in row $r_1$ we draw an arc from $r_1$ to $p_1$. Similarly for zeros in rows $r_3$ and $r_4$ we draw arcs from nodes $r_3$ and $r_4$ to $q_1$.

In column $c_2$ ($x_2$):
Since we have a '1' in row $r_3$ we draw an arc from $p_1$ to $r_3$. And for '0' in $r_2$ we draw an arc from $q_1$ to $r_2$. The constructed graph is shown in Figure 8.29(a). Since the constructed graph has no cycles the pairing is acceptable.

Now we choose the next two columns $x_3$ (as $c_1$) and $x_4$ (as $c_2$). Since these columns have ones in row $r_2$, we cannot fold them straight. We will therefore try to fold them twisted. Two additional nodes $p_2$ and $q_2$ are added and the procedure for twisted folding of the columns as described above is applied. The new graph constructed is shown in Figure 8.29(b). Again we find no cycles and therefore the pairing is acceptable.

Pairing $< x_1, x_2 >$ gives the constraint that $r_1$ must be above $r_3$, and $r_4$ and $r_3$ must be above $r_2$ (see Figure 8.29(a)). Due to the addition of the column pair $< x_3, x_4 >$ new constraints are introduced as shown in Figure 8.29(b). The additional constraint is that row $r_4$ must be above $r_3$. Therefore one possible ordering of rows obtained from the final constructed graph (Figure 8.29(b)) is

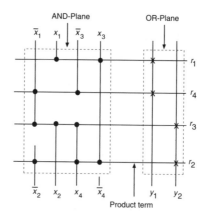

**Figure 8.30** (a) Folded circuit for PLA of Example 8.8.

$$r_1 \text{ above } r_3$$
$$r_4 \text{ above } r_3$$
$$r_3 \text{ above } r_2$$

Therefore the order of rows chosen is $[r_1, r_4, r_3, r_2]$. The folded PLA is shown in Figure 8.30.

With this, we conclude the section on optimization of PLAs. We now discuss some recent work in the areas of module generation and layout optimization.

# 8.6 OTHER APPROACHES AND RECENT WORK

In this chapter, we examined three well known techniques for module generation. In the first approach, we presented the work of Uehara and VanCleemput (1981) in which generation of layouts in standard-cell style from series-parallel CMOS functional cells was presented. Their work relies on an heuristic optimization algorithm that makes use of Euler paths in the graph representation of $n$MOS or $p$MOS circuits. Following their work, an *optimal* non-exhaustive method of minimizing the layout area of CMOS functional cells in the standard-cell style was presented by Maziasz and Hayes (1987). They developed and illustrated a complete graph-theoretic framework for CMOS cell layout. The approach demonstrates a new class of graph-based algebras which characterize this layout problem and is a generalization of the work by Uehara and VanCleemput (1981).

A project developed at AT&T Bell Laboratories called CADRE performs symbolic physical design of full custom CMOS VLSI chips (Ackland *et al.*, 1985). The symbolic design is later converted to geometrical level by a compactor program external to CADRE. The design of a leaf cell generator program called Excellerator (*Expert Cell Generator*) was presented by Poirier (1989). Excellerator fully automates the generation of virtual grid symbolic layouts for CMOS leaf cells. It is intended to fill the role of CADRE's leaf cell agent. The input to Excellerator is a transistor level netlist with optional constraints on layout shape and I/O port positions. The output is a high quality virtual-grid-based layout suitable for use in a two-dimensional tiling methodology. I/O port locations can be optimized. Versatile support for different layout shapes and port locations makes this system ideal for use in a top-down, fully automatic physical design system.

Excellerator's layout structure places the $V_{dd}$ bus along the top and the $V_{ss}$ along the bottom; $p$-type and $n$-type transistors are placed in one or more rows near the $V_{dd}$ and $V_{ss}$ bus respectively, with a gap of at least one grid line separating $p$-transistors from $n$-transistors. Transistors of a symbolic width two or greater are implemented by single transistors connected in parallel. Routing occurs throughout the entire cell area, including the transistor areas. Excellerator has the ability to choose the number of rows for its transistor placement so as to best match a given shape constraint. This, combined with a powerful, yet general router, results in a very flexible layout system. Several layouts of the same circuit may be generated to meet the specified shape (*aspect ratio*) constraints (Poirier, 1989). Although optimized for full-complementary CMOS, Excellerator can handle any type of CMOS circuitry, including domino CMOS. Circuits are not limited to single gates but may constrain multiple compound gates, transmission gates, and feed-throughs.

Yehuda *et al.* (1989) presented a new algorithmic framework for mapping CMOS circuits into area-efficient, high-performance layouts in the style of one-dimensional transistor arrays. They used efficient search techniques and accurate evaluation methods, to quickly traverse the large solution space that is typical to such problems. The quality of designs produced is comparable to handcrafted layouts. Optimized circuits that meet pre-specified layout constraints are generated. Their algorithm has been implemented and used at IBM for cell library generation (Yehuda *et al.*, 1989).

Most of the work presented in this chapter is based on fixed architectures. These impose constraints on the final structure of the layout. This is because of the high complexity of module compilation of flexible architectures. Compilation with flexible architectures provides a good ground for application of artificial intelligence (AI) techniques. Knowledge based approaches have also been attempted (Kim and McDermott, 1986). One system that uses this approach is known as TOPOLOGIZER (Kollaritsch, 1985). This is a rule-based CMOS layout generator. Similar to other approaches, it also uses a style in which all $p$-type transistors are put in rows parallel to the $V_{dd}$ lines and all $n$-type transistors are put in rows parallel to the $V_{ss}$ line. The input consists of a transistor netlist and environment constraints. The environment constraints include layout size (specified by height, width, or aspect ratio). The aspect ratio is defined by the number of rows and columns available for transistor placement. Pin constraints are made up of side, location, layer, and loading. An iterative improvement strategy is used for transistor placement and routing. A gate-matrix-like array of transistor placement having the desired aspect ratio is produced. The routing, however, is not structured as in gate matrix. The output produced by TOPOLOGIZER is a symbolic file of CMOS layout which is design-rule-independent.

In this chapter we also presented the automatic mapping of CMOS functions into a structured array known as gate matrix layout (Wing et al., 1985). Hwang et al. (1987) proposed a new representation of nets for gate matrix layout, called *dynamic-netlists*. Their dynamic-netlist representation is better suited for layout optimization than the traditional *fixed-netlist* since, with it, net-bindings can be delayed until the gate-ordering has been constructed. Based on dynamic-netlists, they developed an efficient modified min-net-cut algorithm to solve the gate ordering problem for gate matrix layout. Through theoretical analysis and experimental results they showed that their approach significantly reduces the number of horizontal tracks and hence the area. The time complexity of their algorithm is $O(N \log N)$, where $N$ is the total number of transistors and gate-net contacts. They also present a modified min-net-cut heuristic for layout minimization.

A new approach to gate matrix layout is presented by Huang and Wing (1989). This approach consists of two stages: the determination of an optimal gate sequence and an assignment of nets to rows such that the nets are realizable. The gate sequence algorithm is based on Asano's approximate search (1981). Modifications are made to it to take into account constraints of transistor sizing, serial subcircuit conflicts, I/O gates, and I/O nets. The net assignment algorithm, called 'zone-net assignment algorithm', assigns nets to a minimum number of rows determined by the gate sequence. It also provides a means to resolve vertical conflicts in the layout. Power connections are made using power nets and possible added power rows. Results of examples show that the new approach can achieve a considerable improvement compared to earlier algorithms while satisfying additional constraints (Huang and Wing, 1989; Wing et al., 1985).

Finally we shed light on some related work in the area of PLA optimization. As discussed, the area of the VLSI layout of a PLA is directly proportional to the size of the PLA personality matrix and can be reduced in several ways.

1. By reducing the number of rows. This is possible by deleting any redundant products terms (Hachtell *et al.*, 1980).
2. By PLA folding (Hachtell *et al.*, 1980).
3. A PLA may be partitioned into two PLAs whose areas sum is less than the area of the original single PLA (Ullman, 1983).
4. Using two to four decoders whose outputs are fed to the AND plane of the PLA. This causes the width of the PLA to remain unchanged, but the number of product terms is considerably reduced (Sasao, 1984; Benten and Sait, 1994).

A restricted definition of a PLA folding, called bipartite folding, was presented by Egan and Liu (1984). The additional constraints of a bipartite folding force the resulting PLA to have a more uniform structure. This structure of a column bipartite folding is then exploited when subsequently folding the rows of the PLA. A column bipartite folding creates fewer constraints upon the ability to fold the rows of the resulting PLA. Thus there is a greater probability of folding the rows. Obviously, the more columns and rows of the PLA that are folded, the lesser the area that is needed to implement the PLA. An efficient branch and bound algorithm which finds an optimal bipartite folding of a PLA was presented (Egan and Liu, 1984). The experimental results show that the size of an optimal bipartite folding compares favourably with the size of folding discovered by known heuristic algorithms. This algorithm can also be used to partition a large PLA into smaller ones (Egan and Liu, 1984).

Lecky *et al.* (1989) presented graph theoretic properties of the PLA folding problem. These properties, not only give insight into the various folding problems, but also provide efficient algorithms for solving them. The work is based on the transformation of the PLA into graphs where cliques (completely connected subgraphs) in the graphs correspond to PLA folding sets (Lecky *et al.*, 1989). Variations of the general folding problem such as bipartite folding and constrained folding are also addressed.

## 8.7 CONCLUSION

In this chapter, we presented an introduction to silicon compilation. Various levels at which hardware can be modelled were examined, namely the behavioural level, the structural level, and the physical level. The main theme of the chapter was the various techniques that can be used to automatically generate VLSI layouts of modules as standard-cells, in gate matrix style, and as PLAs. Graph-based optimization techniques for standard-cell generation, based on the detection of Euler paths, were presented. Optimization of gate matrix was also discussed where the problem reduces to one of column-ordering followed by the use of the left-edge heuristic used in Chapter 7. Finally we presented optimization of PLAs with two possible ways to reduce the size: (1) by deleting redundant product terms thus reducing the height, and (2) by folding columns to reduce the width.

## 8.8 BIBLIOGRAPHIC NOTES

Gate matrix methodology was first proposed by Lopez and Law (1980). The optimization problem and the solution explained in this chapter for gate matrix were formulated by Wing *et al.* (1985). PLA width minimization and folding are due to Hachtell *et al.* (1980), Ullman (1983) and Brayton *et al.* (1984)).

## EXERCISES

**Exercise 8.1**   Explain clearly the differences between:

(1) Behavioural and structural level representation.
(2) Structural and physical level representation.
(3) High level synthesis and silicon compilation.
(4) Logic synthesis and layout synthesis.
(5) Design automation and physical design.

**Exercise 8.2**   Apply the technique discussed in Section 8.3 and implement the layouts of the two functions given below as standard-cells.

$$y_1 = (A \cdot B) + ((C + D) \cdot (E + F)) + (G \cdot H)$$
$$y_2 = A + (B \cdot C) + (D \cdot E)$$

**Exercise 8.3**   For the function given below, draw the CMOS circuit, construct the graph corresponding to it, and find the single Euler path.

$$F = I \cdot (E + (H \cdot (D + (G \cdot (C + (J \cdot (A + B)))))))$$

**Exercise 8.4**   (*) **Programming Exercise:** Write a program to determine Euler paths in graph representations of series/parallel CMOS circuits. Include in your program the heuristic explained in the chapter to reduce the number of edges in graphs.

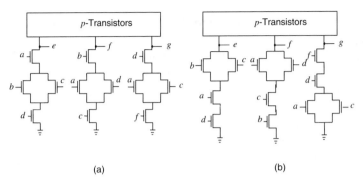

(a)                                              (b)

**Figure 8.31** Circuit of Exercise 8.8.

**Exercise 8.5**   Lay out in gate-matrix style the CMOS circuit of:

(1) the function $X = A \cdot B + C \cdot \overline{D}$.
(2) the function $X = \overline{\overline{E} + A}$, where $A = \overline{B + F}$, and $F = \overline{C + D}$.
(3) the function $X = A \cdot B + B \cdot C + A \cdot C$.
(4) a Half-Adder.
(5) a 4-bit Half-Adder using the layout obtained in Exercise 8.5(4).

**Exercise 8.6   (*) Programming Exercise:** Write a program that will take Boolean function as input and produce an optimal gate-matrix layout representation. Use the functions given in Exercise 8.5 to test your program.

**Exercise 8.7**   For the three input XOR function expressed in standard sum-of-product form

(1) Draw its CMOS circuit.
(2) Draw its gate matrix layout and find the *net-gates* set and the *gate-nets* set.
(3) Apply the technique discussed in Section 8.4 and optimize the gate matrix layout by reducing the number of rows.

**Exercise 8.8**   The Boolean functions implemented by the two circuits given in Figure 8.31 are identical. The circuits differ only in the order of transistors. Using the technique discussed in this chapter implement the circuits as optimal gate matrix layouts. Observe the difference in the number of rows required for each circuit. Comment on your answer.

**Exercise 8.9** Three functions $y_1, y_2, y_3$ are defined on four inputs $x_1, x_2, x_3, x_4$ as follows.
(1) $y_1$ is ON if at least two of the inputs are 0.
(2) $y_2$ is ON if at least two of the inputs are 1.
(3) $y_3$ is ON if exactly two of the inputs are 1.

Show the organization of PLA which implements the three functions $y_1, y_2$, and $y_3$. Use circles to indicate switches. How many product lines did you need?

**Exercise 8.10** (*) **Programming Exercise:** Develop a PLA generator that takes as inputs a PLA personality and produces layouts in $n$MOS/CMOS technology.

**Exercise 8.11** Determine if the matrix $M$ given below is a tautology.

$$M = \begin{matrix} 1 & 2 & 2 & 1 \\ 2 & 1 & 2 & 2 \\ 0 & 1 & 1 & 2 \\ 1 & 2 & 0 & 2 \\ 1 & 0 & 2 & 0 \end{matrix}$$

**Exercise 8.12** (*) **Programming Exercise:** Write a program that will take a PLA personality and

1. raise its terms,
2. determine redundant product terms and delete them, and
3. determine if its AND plane is a tautology.

**Exercise 8.13** Optimize the PLA personality of Exercise 8.9.

**Exercise 8.14** In the implementation of a certain class of problems using PLAs, the inputs are either the literals or their complements, not both. Suggest an efficient algorithm to fold such PLAs. Then apply your algorithm to the PLA personality given below and optimize its area. Show all steps. Do not reduce the number of product terms before folding.

| $p_1$ | 1 | 2 | 2 | 2 | 0 | 1 |
|-------|---|---|---|---|---|---|
| $p_2$ | 2 | 1 | 2 | 1 | 1 | 0 |
| $p_3$ | 1 | 2 | 2 | 1 | 0 | 0 |
| $p_4$ | 2 | 2 | 1 | 2 | 0 | 0 |
| $p_5$ | 2 | 1 | 2 | 2 | 1 | 0 |

$$x_1 \quad \overline{x_2} \quad x_3 \quad \overline{x_4} \quad y_1 \quad y_2$$

**Exercise 8.15**  How will you fold the columns of both the AND plane and the OR plane? Apply your algorithm to the personality given in Exercise 8.14 to include the OR plane personality in folding, and obtain a smaller PLA.

**Exercise 8.16**  Fold the PLA given below.

| $p_1$ | 0 | 2 | 2 | 2 | 0 | 0 | 1 | 0 |
|-------|---|---|---|---|---|---|---|---|
| $p_2$ | 1 | 0 | 0 | 1 | 0 | 1 | 0 | 0 |
| $p_3$ | 1 | 2 | 2 | 0 | 0 | 1 | 1 | 0 |
| $p_4$ | 1 | 0 | 1 | 1 | 1 | 0 | 0 | 1 |
| $p_5$ | 1 | 0 | 2 | 2 | 1 | 0 | 0 | 1 |
| $p_6$ | 0 | 2 | 2 | 0 | 1 | 0 | 1 | 1 |
| $p_7$ | 2 | 1 | 0 | 1 | 0 | 1 | 1 | 1 |

$$x_1 \quad x_2 \quad x_3 \quad x_4 \quad y_1 \quad y_2 \quad y_3 \quad y_4$$

**Exercise 8.17**  Obtain a minimal folded PLA for the following set of equations.

$$y_1 = (A \cdot B \cdot G) + (A \cdot E \cdot F) + (A \cdot B \cdot D \cdot E)$$
$$y_2 = (A \cdot B \cdot G) + (B \cdot C \cdot F \cdot G) + (C \cdot F \cdot H) + (B \cdot G)$$
$$y_3 = (A \cdot E \cdot F) + (C \cdot F \cdot H) + (B \cdot G)$$

# REFERENCES

Ackland, B., A Dickenson, R. Ensor, J. Gabbe, P. Kollaritsch, T. London, C. Poirier, P. Subrahmanyam and H. Watanabe. Cadre – A system of cooperating VLSI design experts. *IEEE Proceedings*, pages 172–174, 1985.

Asano, T. An optimum gate placement algorithm for MOS one-dimensional arrays. *Journal of Digital Systems*, 4:1–27, 1981.

Benten, M. S. T. and Sadiq M. Sait. GAP: A genetic algorithm approach to optimize two-bit decoder PLAs. *International Journal of Electronics*, 76(1):99–106, January 1994.

Brayton, R. K. *et al. Logic Minimization Algorithms for VLSI Synthesis.* Boston, Mass: Kluwer Academic, 1984.

Egan, J. R. and C. L. Liu. Bipartite folding and partitioning of a PLA. *IEEE Transactions on Computer Aided Design*, 3(3):191–199, July 1984.

Fulkerson, D. R. and O. A. Gross. Incidence matrices and internal graphs. *Pacific J. Math*, (3), 1965.

Hachtell, G. D., A. R. Neiston and A. L. Sangiovanni-Vincentelli. Some results in optimum PLA folding. *International Conference on Circuits and Systems*, page 1040, October 1980.

Hashimoto, A. and J. Stevens. Wire routing by optimizing channel assignment within large apertures. *Proceedings of 8th Design Automation Conference*, pages 155–169, 1971.

Huang, S. and O. Wing. Improved gate matrix layout. *IEEE Transactions on Computer Aided Design*, 8(8):875–889, August 1989.

Hwang, D. K., W. K. Fuchs and S. M. Kang. An efficient approach to gate matrix layout. *IEEE Transactions on Computer Aided Design*, 6(5):802–809, September 1987.

Kim, J. and J. McDermott. Computer aids for IC design. *IEEE Software*, pages 38–47, March 1986.

Kollaritsch, P. TOPOLOGIZER: An expert system translator of transistor connectivity to symbolic cell layout. *IEEE Journal of Solid-State Circuits*, SC-3:799–804, June 1985.

Lecky, J. E., O. J. Murphy and R. G. Absher. Graph theoretic algorithms for the PLA folding problem. *IEEE Transactions on Computer Aided Design*, 8(9):1014–1021, September 1989.

Lopez, A. D. and H. S. Law. A dense gate matrix layout method for MOS VLSI. *IEEE Transactions on Electron Devices*, ED-27(8):1671–1675, August 1980.

Maziasz, R. L. and J. P. Hayes. Layout optimization of CMOS functional cells. *Proceedings of 24th Design Automation Conference*, pages 544–551, 1987.

Mead, C. and L. Conway. *Introduction to VLSI Systems.* Reading, MA: Addison-Wesley, 1980.

Nagel, L. W. Spice2: A computer program to simulate semiconductor circuits. *Memo ERL-M520, University of California, Berkeley, CA*, pages 412–417, May 1975.

Ohtsuki, T., H. Mori, E. S. Kuh, T. Kashiwabara, T. Fujisaisa. One dimensional logic gate assignment and interval graphs. *IEEE Transactions on Circuits and Systems*, pages 675–684, September 1979.

Ousterhout, J. K. *et al.* Magic: A VLSI layout system. *Proceedings of 21st Design Automation Conference*, pages 152–159, 1984.

Poirier, C. J. Excellerator: Custom CMOS leaf cell layout generator. *IEEE Transactions on Computer Aided Design*, 8(7):744–755, July 1989.

Sasao, T. Input variable assignment and output phase optimization of PLAs. *IEEE Transactions on Computers*, C-33(10):879–894, October 1984.

Tanaka, T., T. Kobayashi and O. Karatsu. Harp: Fortran to silicon. *IEEE Transactions on Computer Aided Design*, 8(6):649–660, June 1989.

Trickey, H. W. A high level hardware compiler. *IEEE Transactions on Computer Aided Design*, pages 259–269, May 1987.

Uehara, T. and W. M. VanCleemput. Optimal layout of CMOS functional arrays. *IEEE Transactions on Computers*, C-30(5):305–312, May 1981.

Ullman, J. D. *Computational Aspects of VLSI.* Computer Science Press, 1983.

Wing, O., S. Huang and R. Wang. Gate matrix layout. *IEEE Transactions on Computer Aided Design*, pages 220–231, July 1985.

Yehuda, R. B., J. A. Feldman, R. Y. Pinter and S. Wimer. Depth-first-search and dynamic programming algorithm for efficient CMOS cell generation. *IEEE Transactions on Computer Aided Design*, 8(7):737–743, July 1989.

# LAYOUT EDITORS AND COMPACTION

## 9.1 INTRODUCTION

In the previous chapters we discussed various stages of VLSI physical design automation—partitioning, floorplanning, placement, routing, layout generation, etc. Another important constituent in the series is layout editors and layout compaction.

Today, high-level synthesis (HLS) systems are becoming increasingly popular. HLS systems generate a structural level description of the system (for example netlist) from their behavioural models. Physical design is typically the backend of HLS system which takes as input the generated netlist and produces layouts. The physical design stages which generate VLSI layouts use cell libraries and templates. The task of generating libraries and templates at the physical level is fulfilled by layout generators and layout editors. Layout generators make use of a large number of repeated patterns, sometimes also known as leaf cells. Whereas layout generators are used for generating layouts of regular parametrized designs, layout editors are used in the design of layouts of basic cells and layouts of leaf cells.

In this chapter we discuss the capabilities of layout editors, and present some details of a popular public domain layout editor, Magic, as a case study (Ousterhout *et al.*, 1984). The chapter also presents another physical design subproblem, namely, layout compaction. Layout is the final design step, where the attempt is made at improving the area and performance of the layout.

### 9.1.1 Capabilities of layout editors

The capabilities of layout editors have evolved from colour graphic paint tools to those which have knowledge about the electrical implications of the paint. The fabrication process requires the circuit layout to represent appropriate masks. The only feature of interest is the mask geometry. However, creation of mask geometry in exact details as required for the fabrication process is painstaking. Primitive layout systems were text-based and non-graphic. The layouts were drawn on paper and then their coordinates entered via a textual layout description language such as CIF (Mead and Conway, 1980). Disadvantages of text-based layout systems include (a) technology and fabrication process dependence, since the layouts are drawn on a fixed grid and origin, and, (b) the layouts are time consuming to make and difficult to modify.

The second generation layout editors came to the fore with developments in interactive colour graphics. They began with dumb paint programs which could represent geometrical shapes on a fixed grid. The shapes could be rotated, mirrored and scaled. These had associated colours to represent mask layers and allowed hierarchical design (Keller and Newton, 1982; Ousterhout, 1981). Most of the systems restricted the geometry to rectangles (Manhattan geometry), whereas some others allowed arbitrary angles (Boston geometry). Algorithms were developed to manipulate interactively the geometry of the layouts on engineering workstations.

A significant advancement in layout editors came with the concept of scalable layouts based on $\lambda$-grid (See Section 1.3.1). Further, the process specific design rules were also specified in terms of $\lambda$ (Mead and Conway, 1980; Ousterhout, 1981). These however were still unintelligent, in the sense that they had no information about connectivity, electrical properties, or design rules.

Apart from these, the layouts were still drawn as required for mask generation. A more abstract way was to make layout symbolically without using exact layers, in a form somewhat closer to circuit view and then to automatically generate all the layers of the layout for fabrication. A number of systems were developed based on this idea including one in which layouts could be generated from stick diagrams (Williams, 1978).

Present day layout editors are smart and well integrated with the overall design process. In the next section we briefly touch upon some features of a popular public domain layout design system, namely Magic (Ousterhout et al., 1984).

## 9.1.2 Introduction to Magic layout system

Magic came out as a successor to KIC2 and Caesar layout editors (Keller and Newton, 1982; Ousterhout, 1981; Ousterhout *et al.*, 1984). It is an interactive editor for VLSI layouts and runs in a Unix environment under a windowing system such as X11.

Magic was designed with a view that design is iterative in nature and that the designer would need to try out several different alternatives. For this to be feasible, the layout editor must provide (a) ease of design entry (both interactive and non-interactive), (b) on-line and fast DRC, and, (c) a capability for circuit extraction from the layout and its interactive simulation.

A hierarchically designed layout contains cells and subcells. These need to be placed and routed non-interactively to achieve the desired complex layout. In order for the designer to modify the design easily, routing should be fast.

Magic provides all of the above mentioned features. It treats the layouts as circuit objects rather than simple paint layers. It is not bound to any technology or process. Technology dependent knowledge such as design rules, etc., is obtained by Magic from user supplied technology files. To support interactive editing of layouts and their simulation Magic comes with four tools, namely, Box Tool, Wiring Tool, Netlist Tool, and RSIM Tool. The Box tool is the default tool and is used to create and position the boxes of the layout and paint/erase them. The Wiring tool provides an efficient interface to the wiring commands. It is used to connect points using an interactive router. The Netlist tool provides a facility to specify different points on the layout that belong to the same net. It is used to edit netlists interactively. Finally, the RSIM tool provides an interface to the RSIM/IRSIM switch level simulator used when simulating layouts made using Magic (Mayo *et al.*, 1990).

Two important features of Magic that deserve mention are its symbolic layout style and its layout representation. These features are discussed below.

### Logs layout style

Magic represents symbolic layouts using multiple layers. The layouts are in actual size (fully fleshed), however, the system does not represent all the mask layers. For example, layers corresponding to **wells**, **contact-cuts**, etc., are not actually represented. Instead, only those combinations of mask layers which have electrical circuit implications are represented by special

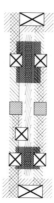

**Figure 9.1** Magic's representation of inverter.

layers. For example, an overlap of poly and diffusion constitutes a region on the *transistor layer*. Therefore, the designer interactively draws his or her layout using the basic interconnect layers such as poly, diffusion, metal, and contacts (not contact-cuts). This design style has been referred to as *logs* (Weste and Eshraghian, 1985). When Magic writes out the CIF file for the foundry, it automatically generates all the mask layers required for fabrication. To do this it requires technology dependent information which it obtains from technology files.

The logs style is much closer to the way a designer looks at the circuit. Moreover, it also makes the task of DRC, routing, compaction, etc., faster. The Magic representation of the inverter circuit is shown in Figure 9.1. Compare this with the mask geometry shown in Figure 1.6 which also contains the actual contact-cuts, wells, etc.

## Layout representation: corner stitching

The heart of the Magic system lies in the innovative data structure used to represent a layout. Magic supports hierarchical representation of designs. A Magic layout cell contains parent geometry and subcells. In each cell the parent geometry is stored in different planes. Those layers which have DRC sensitivity are placed on a single plane. DRC independent layers are on different planes. There is a plane to hold information about subcells. For example, in MOS process, diffusion, polysilicon, and transistor are in the *active plane*. Similarly, metal layers are in the *metal plane*. Each plane is represented using tiles of different materials, that is, the entire plane is made up of tiles. Those regions which do not represent any of the abstract layers are represented by space tiles.

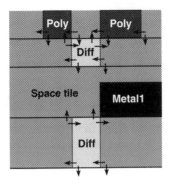

**Figure 9.2**    Corner stitches in the active plane.

The data structure used to represent the layout is termed as corner stitching (Ousterhout, 1984). Each tile on a plane is represented using four corner stitches, two at its lower left corner and two at its upper right corner. The two stitches at lower left corner, namely *bl* and *lb*, denote the neighbouring bottommost tile on the left and leftmost tile at the bottom. Similarly two stitches at upper right corner of tile point to the rightmost tile at top (*rt*) and topmost tile on the right (*tr*). This representation has led to simple search algorithms for the operations required for editing, compaction, layout format translation, etc. The typical search operations involved are point-finding (a point belonging to a certain type of tile), neighbour finding, area search for solid (non-space) tiles, tile insertion/ deletion, compaction, space creation, etc. Figure 9.2 shows the tiles and corner stitches on the active plane of inverter circuit. Although this technique requires three times the storage compared to linked list representation, it allows interactive operation due to fast search algorithms. The representation technique of corner stitching and logs design style enable Magic to provide a number of other facilities discussed below.

**Incremental DRC**: While drawing boxes using the Box tool, Magic runs the DRC program in background (Taylor and Ousterhout, 1984). If the designer activates a change, the DRC stops and lets the command be completed and then restarts.

During the editing of a layout, a designer spends some time thinking, and this time is utilized for DRC. It is the logs design style and corner stitching data structure that make fast DRC possible.

Each cell has *reverify* and *error* planes. When a layout is edited, Magic records the area to be verified in the form of *reverify* tiles. If any DRC violation is found it is immediately shown in the form of error tiles on error planes of each cell. Magic supports hierarchical design rule checking. First, mask information in each cell is checked for any violations, and then each cell and its subcells taken together.

As the layers which have DRC interaction are stored in a single plane, DRC is done on plane by plane basis without the need for considering all planes as in other representations. This feature makes it possible to perform DRC quickly.

Magic's DRC is different from that in other systems. It is edge-based, where edges correspond to those between tiles. The rules are read from a technology file using Magic's DRC language.

**Plowing**: Magic provides the operation of plowing to compact/create space and modify the layout (Scott and Ousterhout, 1984). The user can place a horizontal/vertical line segment (plow) and give the direction and distance for the plow to move. The plow catches edges parallel to it as it moves and carries them along with. It moves edges only, stretching material behind the plow and compressing that in front of it. Plowing maintains connectivity and design rule correctness.

Plowing can be used to compact an entire cell by placing a plow to the left of the cell and moving to the right and then placing a plow at the bottom of the cell and plowing it up. In the next section we shall discuss the topic of compaction, in general.

**Routing**: Interconnecting pins belonging to the same nets can be done either interactively using the wiring tool or by using a non-interactive router. The non-interactive router used by Magic is *Detour* (Hamachi and Ousterhout, 1984). This is a switch-box router based on the greedy channel routing algorithm of Rivest and Fiduccia (1982). Obstacles for routing may be (a) critical nets that have been routed by hand, (b) routed power/ground nets, or (c) hints to the router in the form of partially routed nets.

The router has the ability to avoid obstacles and to consider interactions between nets as channels are routed. It is capable of handling designs not based on fixed routing grid. It modifies the cells by introducing *sidewalks* so that the cells lie on integral grid units.

The router works in three stages. In the first stage called *channel definition*, the empty space is divided into rectangular channels. This is done using the space tiles in the subcell plane. Secondly, *global routing* processes nets sequentially to find the channels to be assigned to each net. The corner stitching representation enables this to be done without creating any new data structure. Finally, in *channel routing* each channel is considered separately and wires are placed to connect the nets.

In channels or portions of channels without obstacles *Detour* produces results comparable to traditional channel routers. In areas containing obstacles, the router either jogs around obstacles, or switches layers and river routes across them, thus combining good features from net-at-a-time router (Lee algorithm) and traditional channel routers (Hamachi and Ousterhout, 1984).

**Circuit extraction:** After the layout is completed, a question that must be answered is, does the layout represent the circuit being designed? To answer this Magic provides a facility for circuit extraction from layout, and simulation of the extracted circuit.

Magic's circuit extractor finds connectivity, transistor dimensions, internodal capacitance and parasitic resistance (Scott and Ousterhout, 1986). The extractor is hierarchical and extracts each cell separately. It has two parts, a basic extractor that considers mask layers and a hierarchical part that considers interaction between a cell and its subcells. The basic extractor extracts nodes, transistors and coupling capacitances. The task of node and transistor extraction are simple due to the corner stitching data structure (a transistor is represented explicitly as a layer type unlike other systems where transistor extraction involves logical operations on layers of diffusion and polysilicon). Then, the hierarchical extractor makes connections between nodes of the cell and its subcells and modifies the estimated capacitances. The task of resistance estimation is done after flattening the hierarchical extracted circuit (Stark and Horowitz, 1987).

Magic also comes with an interface to the switch level simulator, IRSIM. This interface eliminates the need to map node names of circuits to objects in the layout. It allows the user to select nodes in the layout using the mouse and to apply stimuli to them or to display the node values determined by the simulator in the layout itself (Mayo *et al.*, 1990).

A number of features of Magic were discussed in the preceding material. An example showing the use of Magic layout system is presented next.

**Example 9.1** A parity generator for 32-bit words can be realized using a tree architecture as shown in Figure 4.42. This tree can be embedded in a 2-D array shown in Figure 9.3. The boxes labelled '1' are leaf nodes and the one labelled '5' is the root. The leaf cells receive the input and the root node produces the output.

SOLUTION The first step in the design of this circuit is to make the layout of a 2-input XOR gate. A number of XOR gate implementations are possible. The one used here implements the XOR function using two transmission gates and two inverters (multiplexer based

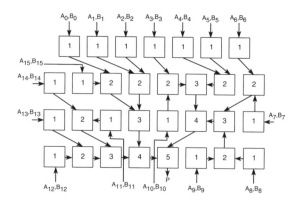

**Figure 9.3** Binary tree embedding on a 2-D array.

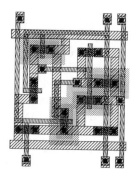

**Figure 9.4** Layout of a 2-bit XOR gate.

implementation). The layout of this can be drawn interactively or by using transmission gates as subcells. Once the layout is complete, with no design rule violations, it can be extracted and simulated to verify its correctness. The layout of the XOR gate is shown in Figure 9.4.

This 2-input XOR gate can now be used to build the layout of the tree. Magic does not provide a placement facility. However, for this example, the *array* construct of Magic can be used to lay out XOR gates as a 2-D array. The cells in the array are systematically numbered by Magic. This can be used to generate the interconnectivity netlist using a simple *C* program. The netlist is an ASCII file with one terminal per line and each net separated from the other using a blank line. The netlist can alternatively be made interactively. Magic's router can now be invoked to automatically route the nets. The layout for the 32-bit parity generator is shown in Figure 9.5.

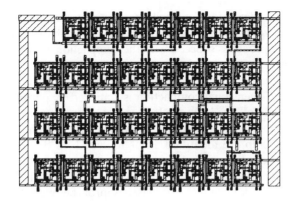

**Figure 9.5**   Magic layout of a 32-bit parity generator.

## 9.2 LAYOUT COMPACTION

When a symbolic or sticks design approach is used, designers create layouts by using abstract objects (such as wires, contacts, transistors) rather than actual mask layout objects. Once the symbolic design is complete, a compaction program is applied to generate the actual masks with the desired spacings consistent with the underlying design rules of the technology. Hence, the designer need not worry about actual sizes or spacings of the layout objects, thus shortening the design turnaround time.

In most general terms compaction can be defined as the process of taking a topological arrangement of geometric objects and attempting to move these objects so as to make the layout as small as possible, while maintaining minimum spacing requirements between the objects. Compaction maintains the underlying circuit topology and enforces design rule correctness. Maintaining the given circuit topology is mandatory in order not to render the previous design steps obsolete.

Compaction is an important step when a symbolic layout approach is adopted. Symbolic design when combined with a good compaction technique can lead to layouts that are as dense as hand-crafted ones.

Besides minimizing the overall chip area (e.g., by removing unnecessary routing spaces), a compactor is a valuable tool when scaling down a design to a new technology. The compactor is usually used to generate different mask layouts corresponding to the same symbolic layout for different technologies. Several surveys of symbolic compaction are available in literature (Boyer, 1988; Ohtsuki, 1986; Preas and Lorenzetti, 1988).

Compaction algorithms are classified according to whether they proceed along one dimension at a time or the two dimensions simultaneously. One-dimensional algorithms compact the layout along the $x$-direction (horizontal), then the $y$-direction (vertical) or vice-versa. Two-dimensional approaches proceed with the compaction along both directions simultaneously.

The one-dimensional approach is the most widely used compaction strategy. Two-dimensional compaction is achieved by performing horizontal compaction, followed by vertical compaction or vice versa. At each compaction phase, wasted layout areas which cut through (horizontally or vertically) the chip are identified and eliminated. In this section we shall describe only one-dimensional compaction.

Compaction algorithms can also be classified based on the model used to abstract the mask layout. Graph-based approaches model the layout as a constraint graph where the vertices represent the layout objects and spacing constraints between pairs of objects are modelled by directed weighted arcs.

Virtual grid-based approaches assume that the objects are laid out on a virtual grid. The layout is compacted along the grid lines while satisfying minimum spacing requirements. Both these approaches are one-dimensional.

The graph-based approach has been the dominating approach since its introduction in the CABBAGE system (Hsueh and Pederson, 1979).

A compaction tool requires three things (Ohtsuki, 1986):
1. the initial layout;
2. technology information (design rules); and
3. a compaction strategy.

## Initial layout representation

The layout may be represented at the mask level (e.g., Magic) or cell level. The layout surface is assumed to be a mask grid, as in Magic, or a virtual grid. For the mask grid representation, the mask objects are basic rectangles aligned on the mask grid lines. The grid unit is the distance between two consecutive mask grid lines, which indicates the smallest distance between any pair of mask objects as allowed by the technology. For the virtual grid representation, the objects are aligned on the virtual grid lines. However, the spacing between the grid lines does not have physical meaning.

At the cell level, the layout is seen as a set of interconnected cells. Each cell is seen as a black box. Only the cell contours and the interconnections between cells are represented.

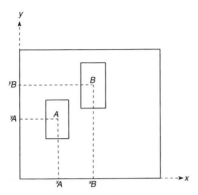

**Figure 9.6**    Spacing constraints between layout objects.

## Technology information

Technology information can be seen as a set of design rules stating spacing constraints between the layout objects. Usually the location of each object is indicated by the $(x, y)$-centre coordinates of the object.

For example, suppose that the centre coordinates of two objects $A$ and $B$ are $(x_A, y_A)$ and $(x_B, y_B)$ (see Figure 9.6). Then a design constraint of the form $x_B - x_A \geq d$ indicates that the compactor must maintain a centre to centre separation between the two blocks equal to at least $d$ units. In general, technology design rules are of three types:

1. minimum spacing rules of the type just discussed;
2. minimum size rules (on the width of objects);
3. minimum overlaps among elements of different mask layers.

These types of rules are defined for each layout object and stored in the technology file or database.

## 9.2.1 Compaction algorithms

The two-dimensional compaction problem has been shown to be NP-complete (Sastry and Parker, 1982). Therefore, although two-dimensional compaction is superior since it better accommodates global tradeoffs in moving an object, most existing approaches are one-dimensional. For example, the plowing technique of the Magic layout system allows one-dimensional compaction only in the direction perpendicular to that of the plow.

One-dimensional compaction algorithms break the two-dimensional problem into two one-dimensional compaction problems. First horizontal (vertical) compaction is performed, followed by vertical (horizontal) compaction. The compaction program identifies movable objects, which are then shifted by the correct amount either horizontally or vertically. As we said, topological relationships between the objects is maintained, i.e., objects are not allowed to jump over each other and connectivity between connected objects must be preserved. All wires are assumed to be stretchable and objects are restricted to have rectangular shapes (and so does the layout surface).

Since horizontal and vertical compactions are similar, we shall describe in this section horizontal compaction only. We will assume that the left boundary of the layout surface is fixed at $x = 0$ and that compaction is performed by shifting horizontally movable elements from right to left. The general compaction procedure has two main steps:

1. The first step consists of searching for the movable objects in the direction of compaction. This step is carried out in consultation with the input description of the layout and the technology files (the design rules).
2. The second step consists of actually performing the compaction. The movable objects are moved while satisfying the design constraints.

These two steps depend on the layout model adopted. Next, we shall briefly describe these for both the virtual grid and constraint graph models.

### 9.2.2 Horizontal virtual grid compaction

A layout of $m \times n$ grid units is represented as a $m \times n$ bitmap $G$. Each bit cell $G(i, j)$ is set either to zero (0) or one (1). A bit cell is set to zero if the corresponding grid cell is occupied by either a block or a vertical wire segment (see Figure 9.7). All other bit cells are set to one. Therefore, vertically connected objects, together with their vertical wire segments will be moved together. On the other hand, horizontal wire segments will be stretched or contracted as the objects they connect are moved. The size of the grid unit is determined by the design rules. Usually, objects are restricted to one unit moves.

One of the search techniques used to identify movable elements is the *cut search* technique (Akers *et al.*, 1970). The cut search technique consists of identifying a set of empty bit cells which cut the layout (the bitmap) in two parts. There can be two kinds of cuts: (a) *simple cut* when the bit cells form a vertical column of the bitmap; and (b) *rift line cut* when the cut is the union of vertical sub-columns joined by horizontal rift lines (see Figure

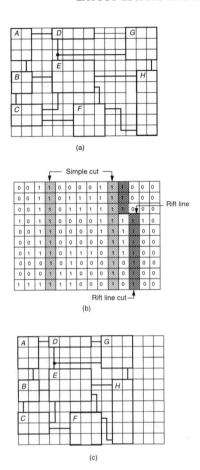

**Figure 9.7** Virtual grid compaction. (a) Initial layout. (b) Corresponding bitmap. (c) Layout after horizontal compaction.

9.7). Therefore, each cut removal results in a 1-grid unit horizontal compaction.

The search for movable objects (identification of the cuts) and the compaction which follows is made easy because of the simple bitmap representation of the layout. However, the grid resolution can be very coarse. You might suggest that you can always reduce the grid unit size. This necessarily improves the grid resolution, but it also leads to unacceptably large bitmaps. Large bitmaps require large memory space and large processing time. The basic cut search technique has been modified to deal with this resolution problem. For example, the interested reader can consult works of Dunlop (1978) and that of Weste (1981).

### 9.2.3 Constraint graph compaction

The constraint graph approach was first described by Hsueh and Pederson (1979). However, constraint graph compaction was first used in the FLOSS system (Cho *et al.*, 1977; Preas and Lorenzetti, 1988). In this approach, the layout topology and spacing constraints are represented by a weighted directed graph called the constraint graph. The construction of this graph is the time consuming step of constraint graph compaction. Among the approaches that have been used to construct this graph are the shadow propagation method and the coarse grid method (Hedges *et al.*, 1985; Hsueh and Pederson, 1979). The shadow propagation method consists of 'shining' a light from behind the group under consideration and identifying all the groups that will be covered by the shadow of that group. For the coarse grid method, the layout is represented as a bitmap. In this case, identification of group adjacencies is straightforward.

Each node in the graph represents a group of blocks that are connected by vertical wire segments. The edges of the graph connect groups that have spacing constraints, i.e., there is an arc $(g_i, g_j)$ with weight $d_{i,j}$ if and only if groups $g_i$ and $g_j$ are adjacent and are constrained to remain some distance $d_{i,j}$ apart. Lower bound distance constraints are represented by positive weights on the corresponding edges, while upper bound constraints are represented by negative weights. For example, if two groups $g_i$ and $g_j$ must be kept at least $d_{i,j}$ units apart, then the vertices corresponding to these two groups will be connected with the weighted directed edge $(g_i, g_j)$, with weight $+d_{i,j}$ (refer to Figure 9.8(a)). On the other hand, constraints used to maintain connectivity between two elements are stated as upper bounds and therefore are modelled by two directed arcs with negative weights. This is illustrated in Figure 9.8(b), where we have a contact cut $w_1$ and a wire $w_2$ that must remain connected. If the centre to centre distance between these two elements is $d$, then the elements will remain connected if the following inequality remains satisfied $|x_1 - x_2| \leq d$. This constraint is represented by two directed edges with negative weights set to $-d$ (see Figure 9.8(b)). The negative weights indicate that the centre of either element can be to the left of the centre of the other element.

The constraint that two elements $a$ and $b$ must be kept a fixed distance $d$ units apart is represented by two directed edges with equal and opposite weights, i.e., an edge $(a, b)$ with weight $d$ and an edge $(b, a)$ with weight $-d$.

To complete the construction of the constraint graph, two extra vertices are added: a source vertex $L$ representing the leftmost position of the layout, and a sink vertex $R$ representing the layout rightmost position. For each element $e$ adjacent to the leftmost side, a directed arc $(L, e)$ is added with positive weight set to the required distance from the layout leftmost side to the centre of the element. Similarly, for each element $e$

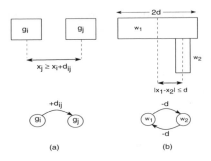

**Figure 9.8**  Graph modelling of spacing and connectivity constraints. (a) Spacing constraint. (b) Connectivity constraint.

adjacent to the rightmost side, a directed arc $(e, R)$ is added with positive weight set to the required distance from the centre of the element to the layout rightmost side.

Once the construction step is complete, the constraint graph is used to assign $x$-locations to the vertices so as to satisfy the constraints and minimize the layout width. This is achieved by assigning $x$-values to the vertices so as to minimize the length of the longest $L$ to $R$ path(s) (the critical path(s)). The length of a path is equal to the sum of the weights of those arcs traversed by the path.

In the remainder of this section, we shall assume that the constraint graph does not have any edges with negative weights (the graph is therefore acyclic).

Central to the constraint graph approach is the critical path algorithm. This algorithm is used to find for each vertex $v$, the range of $x$-locations $[l(v), r(v)]$ that can be tolerated by that vertex.

Let $LengthFromL(v)$ be the length of the longest path from the source $L$ to vertex $v$. Similarly, define $LengthToR(v)$ to be the length of the longest path from vertex $v$ to the sink $R$. Therefore, the range of movement tolerance for each vertex $v$ can be defined as follows:

$$l(v) = LengthFromL(v)$$
$$r(v) = LengthFromL(R) - LengthToR(v)$$

It should be obvious to the reader (see Exercises 9.15–9.16) that $l(L)$ $= r(L) = 0$, and $l(R) = r(R) = LengthFromL(R)$, which is the length of the longest path from the source $L$ to the sink $R$. Also, based on the principle of optimality, for each vertex $v$ on one of the critical path(s), $l(v) = r(v)$.

For acyclic graphs, the critical path problem is straightforward. A semi-formal description of such an algorithm is given in Figure 9.9.

**Algorithm** Critical-Path($L, R, G$);
  (* $G$: Constraint graph *)
  (* $L$: source vertex of graph $G$ *)
  (* $R$: sink vertex of graph $G$ *)
  (* $LengthFromL(i)$: length of longest path from $L$ to vertex $i$ *)
  (* $LengthToR(i)$: length of longest path from $i$ to $R$ *)
  (* $d_{i,j}$: weight of edge $(i, j)$ (spacing constraint between nodes $i$ and $j$) *)
  (* $\Gamma^{+}(i)$: successor vertices of $i$ *)
  (* $\Gamma^{-}(i)$: predecessor vertices of $i$ *)
**Begin**
1.  (* Initialization: levelize the graph *)
    Let $K$ be the number of levels, and $V_i$ the vertices in level $i$;
    therefore, $V_1 = \{L\}$ and $V_K = \{R\}$;
2.  (* Forward trace *)
    $LengthFromL(L) = 0$;
    **For** $k = 2$ **To** $K$ **Do**
      **ForEach** $i \in V_k$ **Do**
        $LengthFromL(i) \leftarrow \max_{j \in \Gamma^{-}(i)}\{LengthFromL(j) + d_{j,i}\}$
      **EndForEach**
    **EndFor**;
3.  (* Backward trace and compute tolerance ranges *)
    $LengthToR(R) = 0$;
    **For** $k = K - 1$ **DownTo** $1$ **Do**
      **ForEach** $i \in V_k$ **Do**
        $LengthToR(i) \leftarrow \max_{j \in \Gamma^{+}(i)}\{LengthToR(j) + d_{i,j}\}$;
        $l(i) = LengthFromL(i)$;
        $r(i) = LengthFromL(R) - LengthToR(i)$
      **EndForEach**
    **EndFor**;
**End**.

**Figure 9.9** Algorithm to determine range of tolerance for all groups.

Once the moving tolerances for all vertices have been computed, the next task is to assign locations to the corresponding elements consistently with their tolerance ranges. To do that, we need to have a moving strategy.

There are three general moving strategies: the minimum, the maximum, and the optimum strategies.

The minimum moving strategy assigns each element closest to the right boundary of the chip, i.e., for each vertex $v$, $x(v) = r(v)$. This is the strategy used in the plowing technique of Magic.

The maximum moving strategy assigns elements closest to the left boundary of the chip, i.e., for each vertex $v$, $x(v) = l(v)$. Both these strategies are simple and lead to layouts that are densely packed on either side of the chip. This is undesirable since it leads to an increase in the overall wire length.

An optimum moving strategy attempts to assign $x$-locations to elements according to some optimization criterion. A possible criterion would be to assign each element to the middle of its range of tolerance, i.e., for each vertex $v$, $x(v) = \frac{l(v)+r(v)}{2}$. Another possible criterion is to use a force-directed approach. In this case, the criterion would be to minimize the sum of the forces acting on each element. The objective minimized in this case is the sum of the squares of the $x$-distances between the element and its neighbours.

The general steps of the constraint graph compaction approach are summarized in Figure 9.10. Next, we shall illustrate this approach with an example.

**Algorithm** Constraint_Graph_Compaction;
   1. Construct the constraint graph $G(V, E)$;
   2. Apply the critical path algorithm and find for each vertex $v, [l(v), r(v)]$;
   3. Move each element to within its range of tolerance;

**End.**

**Figure 9.10**  Constraint graph compaction.

**Example 9.2**  Given the layout of Figure 9.11, perform horizontal compaction using the constraint graph approach. Assume that the minimum horizontal centre to centre spacings between the individual elements are as follows, $d_{A,D} = 5$, $d_{A,E} = 5$, $d_{C,E} = 5$, $d_{C,F} = 5$, $d_{D,G} = 4$, $d_{D,H} = 4$, $d_{E,H} = 4$, $d_{E,I} = 5$, and $d_{F,I} = 5$. Assume further that the elements have the following widths, $w_A = 4$, $w_B = 6$, $w_C = 4$, $w_D = 4$, $w_E = 4$, $w_F = 4$, $w_G = 2$, $w_H = 2$, and $w_I = 4$.

SOLUTION  The constraint graph is given in Figure 9.11(b). Since blocks $B$ and $C$ are connected by vertical wires, they are grouped together. The same applies to blocks $D$ and $E$, and blocks $G$ and $H$. In this example, we assume that the minimum distance between two adjacent blocks is equal to 1.

The minimum distance $d_{i,j}$ between two groups $g_i$ and $g_j$ is defined as follows,

$$d_{i,j} = \max_{a \in g_i \& b \in g_j} \{d_{a,b}\} \qquad (9.1)$$

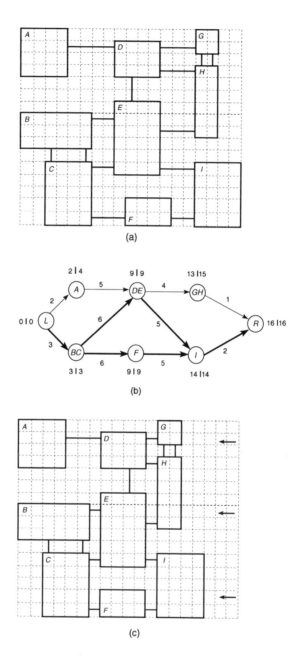

**Figure 9.11** Illustration of constraint graph compaction. (a) Initial layout. (b) Constraint graph where each vertex is labelled by $l(v)|r(v)$. (c) Compacted layout.

Therefore, $d_{A,DE} = \max\{5,5\} = 5$. Similarly, $d_{L,A} = 2$, $d_{L,BC} = 3$, $d_{BC,DE} = 6$, $d_{BC,F} = 6$, $d_{DE,GH} = 4$, $d_{DE,I} = 5$, $d_{F,I} = 5$, $d_{GH,R} = 1$, and $d_{I,R} = 2$.

The graph has five levels. Let $V_i$ be the vertices of level $i$. Then, $V_1 = \{L\}$, $V_2 = \{A, BC\}$, $V_3 = \{DE, F\}$, $V_4 = \{GH, I\}$, and $V_5 = \{R\}$. Then, a forward trace of the graph is performed where, for each vertex $i$, the values $LengthFromL(i)$ are determined as described in the algorithm of Figure 9.9. For this example, we obtain the following:

$level1$ :
$LengthFromL(L) = 0.$

$level2$ :
$LengthFromL(A) = LengthFromL(L) + d_{L,A} = 0 + 2 = 2;$
$LengthFromL(BC) = LengthFromL(L) + d_{L,BC} = 0 + 3 = 3.$

$level3$ :
$LengthFromL(DE) = \max\{LengthFromL(A) + d_{A,DE};$
$\quad LengthFromL(BC) + d_{BC,DE}\} = \max\{7,9\} = 9;$
$LengthFromL(F) = \max\{LengthFromL(BC) + d_{BC,F}\} = 9.$

$level4$ :
$LengthFromL(GH) = LengthFromL(DE) + d_{DE,GH} = 13;$
$LengthFromL(I) = \max\{LengthFromL(DE) + d_{DE,I};$
$\quad LengthFromL(F) + d_{F,I}\} = \max\{14, 14\} = 14.$

$level5$ :
$LengthFromL(R) = \max\{LengthFromL(GH) + d_{GH,R};$
$\quad LengthFromL(I) + d_{I,R}\} = \max\{14, 16\} = 16.$

The next step is to perform a backward trace of the graph. During this step, for each vertex $v$, the values $LengthToR(v)$, $l(v)$, and $r(v)$ are computed as described in the algorithm of Figure 9.9. Hence, the following is obtained for all vertices of the graph:

*vertex R:* $\quad LengthToR(R) = 0;$ $\quad l(R) = 16,$ $\quad r(R) = 16 - 0 = 16;$
*vertex GH:* $LengthToR(GH) = 1;$ $\quad l(GH) = 13,$ $\quad r(GH) = 16 - 1 = 15;$
*vertex I:* $\quad LengthToR(I) = 2;$ $\quad l(I) = 14,$ $\quad r(I) = 16 - 2 = 14;$
*vertex DE:* $LengthToR(DE) = 7;$ $\quad l(DE) = 9,$ $\quad r(DE) = 16 - 7 = 9;$

> vertex F:     LengthToR(F) = 7;     l(F) = 9,     r(F) = 16−7 = 9;
> vertex A:     LengthToR(A) = 12;     l(A) = 2,     r(A) = 16−12 = 4;
> vertex BC:     LengthToR(BC) = 13;     l(BC) = 3,     r(BC) = 16−13 = 3;
> vertex L:     LengthToR(L) = 16;     l(L) = 0,     r(L) = 16−16 = 0.

The final step of the algorithm is to find the $x$-location of the centre of each group. Let $x_{min}(i)$, $x_{max}(i)$, and $x_{opt}(i)$ be the $x$-location of the centre of group $i$ according to the minimum, maximum, and optimum strategies respectively. For this example, assuming that $x_{opt}(i) = \frac{l(i)+r(i)}{2}$, these variables will be assigned the following values:

> vertex $R$:     $x_{min}(R) = 16$;     $x_{max}(R) = 16$,     $x_{opt}(R) = 16$;
> vertex $GH$:     $x_{min}(GH) = 15$;     $x_{max}(GH) = 13$,     $x_{opt}(GH) = 14$;
> vertex $I$:     $x_{min}(I) = 14$;     $x_{max}(I) = 14$,     $x_{opt}(I) = 14$;
> vertex $DE$:     $x_{min}(DE) = 9$;     $x_{max}(DE) = 9$,     $x_{opt}(DE) = 9$;
> vertex $F$:     $x_{min}(F) = 9$;     $x_{max}(F) = 9$,     $x_{opt}(F) = 9$;
> vertex $A$:     $x_{min}(A) = 4$;     $x_{max}(A) = 2$,     $x_{opt}(A) = 3$;
> vertex $BC$:     $x_{min}(BC) = 3$;     $x_{max}(BC) = 3$,     $x_{opt}(BC) = 3$;
> vertex $L$:     $x_{min}(L) = 0$;     $x_{max}(L) = 0$,     $x_{opt}(L) = 0$.

Figure 9.11(c) gives the resulting compacted layout when a horizontal maximum moving strategy is adopted.


## 9.3 OTHER APPROACHES AND RECENT WORK

In this chapter, we described traditional and most widely used approaches to leaf cell compaction. Our objective in this chapter was to explain the basic goal and techniques used to compact layouts and reduce the required chip area. Compaction dates back to the early seventies. After nearly twenty five years of efforts, compaction is now a mature field of design automation. A nice survey paper on compaction was published by Boyer (1988). Wolf and Dunlop wrote a good description of compaction approaches and algorithms in Chapter 6 of the book edited by Preas and Lorenzetti (1988). Lengauer (1990) dedicated a chapter to compaction.

There are several other compaction approaches. For example, two-dimensional compaction was formulated as a mathematical optimization problem (Watanabe, 1984). The general combinatorial optimization technique, simulated annealing, has also been applied to perform compaction (Mosteller *et al.*, 1987).

Area minimization is not the only important optimization criterion during compaction. With the increasing performance requirement (circuit speed), it is becoming imperative to minimize the interconnect delays. It is important that the compactor be aware of the critical wires so that time critical wires are not stretched. Numerous approaches to optimize other cost measures have been reported (Shiele, 1983; Maley, 1985; Lakhani and Varadarajan, 1987; Crocker *et al.*, 1987; Wolf *et al.*, 1988). These approaches, in addition to optimizing the area, perform also wirelength minimization. This is of crucial importance when the constraint graph approach is used. The reason is that for the constraint graph approach, the range of movement tolerance of each group gives freedom of movement to the group, which could lead to harmful consequences to the performance speed of the circuit.

Compaction for improved performance is one of the many active research areas in layout compaction.

## 9.4 CONCLUSION

Layout editors are essential to VLSI design. Usually they work with symbolic objects which are rectangles close to those in actual masks, or symbols such as wires, vias, contacts, transistors, etc. In this chapter we presented the Magic layout system which includes among other features an efficient layout editor, an interactive design rule checker, and a compactor. The Magic layout editor works with basic tiles/rectangles. In this chapter we presented briefly the capabilities of this popular and widely available layout system. Magic was implemented by Ousterhout *et al.*, (1984) at the University of Berkeley, and is available in public domain.†

Compaction is the last step in the design process. The handcrafting of designs using layout editors is a very tedious process. Layouts produced by such editors generally obey design rules, however, they are not area efficient. Even when an automatic layout approach is used, the resulting layout usually contains unnecessary empty spaces. Compaction is needed to remove any unnecessary wasted area without violating design rules.

Even for designs of reasonably small size, manual compaction is very difficult. Two major techniques widely used for automatic compaction are the grid-based approach and the graph-based approach. However, the most attractive approach is the graph-based. Both techniques were presented in this chapter and illustrated with examples.

---

† To obtain Magic via FTP contact 'magic@decwrl.dec.com'. Magic may also be obtained on magnetic tapes from 'EECS/ERL Industrial Liaison Program, 479 Cory Hall, University of California at Berkeley, Berkeley, CA, 94720, USA'.

## 9.5 BIBLIOGRAPHIC NOTES

There are several papers describing various aspects of the Magic layout system (Ousterhout et al., 1984). These aspects are the internal data structure (corner stitching) used by Magic (Ousterhout, 1984), the interactive design rule checking (Taylor and Ousterhout, 1984), the plowing technique used to stretch or compact layouts (Scott and Ousterhout, 1984), and the routing algorithm (Hamachi and Ousterhout, 1984).

Two-dimensional compaction is an NP-hard problem (Sastry and Parker, 1982). For this reason, heuristic techniques are employed which consist of performing one-dimensional horizontal (vertical) compaction followed by one-dimensional vertical (horizontal) compaction. The most widely used approaches for one-dimensional compaction are the constraint-graph approach and the virtual grid approach (Boyer, 1988). The most attractive and flexible of the two is the graph-based approach. Several experimental studies were conducted to compare both approaches in terms of running time and quality of solution (layout area after compaction) (Wolf, 1985; Eichenberger, 1986; Lo et al., 1987). The experiments indicated that both approaches have comparable run times. However, the constraint-graph approach consistently produced smaller layouts (15 per cent to 30 per cent smaller) than those obtained by the virtual grid approach. An experiment to study the sensitivity of the compaction algorithm to small changes in the layout was also conducted (Wolf, 1985). Results of the experiment showed that the virtual grid approach produced a wide range of layout sizes. On the other hand, the graph-based approach was much less sensitive to layout changes and consistently produced layouts of comparable sizes.

Although all experiments are in favour of the constraint graph approach, the virtual grid approach is still widely used. The reason is that the virtual grid approach is much easier to implement. We believe that both approaches will remain dominant in the coming years over other compaction approaches, with a slight shift towards graph-based techniques. A very good discussion on performance aspects of these two approaches is provided in Chapter 6 of the book edited by Preas and Lorenzetti (1988).

In this chapter, we considered only the problem of leaf-cell (flat cells with no hierarchy) compaction. Leaf-cell compaction is the backbone of hierarchical compaction. Hierarchical compactors allow the objects to be wires, transistors, or cells of any complexity (Ohtsuki, 1986; Preas and Lorenzetti, 1988; Boyer, 1988). A cell is modelled as a black box, i.e., only the cell contour is of interest to the hierarchical compactor.

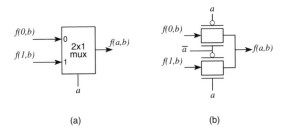

**Figure 9.12**    A $2 \times 1$ multiplexer used as a universal module, $f(a, b) = \bar{a} \cdot f(0, b) + a \cdot f(1, b)$
(a) Block diagram. (b) Circuit diagram.

# EXERCISES

Note:
For Exercises 9.1 to 9.9 design the layout using Magic's layout editor (or any other layout editor of similar capability available in your laboratory). Then, extract the circuit from the layout, and simulate the extracted circuit to verify the design.

**Exercise 9.1**    A $2 \times 1$ multiplexer realized using transmission gates can be used as a universal logic module. The block diagram and the circuit are illustrated in Figure 9.12.

1. Design the layout of a $2 \times 1$ multiplexer (Figure 9.12) using CMOS transmission gates.
2. Test the $2 \times 1$ multiplexer to generate 2-variable Boolean functions.

**Exercise 9.2    Programming Exercise:**    Write a program to generate an $n \times 1$ multiplexer using $2 \times 1$ multiplexers. You may use any technique for placement. The program must generate the netlist which will be used by the routing tool to automatically interconnect multiplexers.

**Exercise 9.3**    Design a full-adder using only $2 \times 1$ multiplexers. Using this module, hierarchically design a 4-bit ripple carry adder. The full-adder cell may be designed so that by abutting four such cells the 4-bit adder is synthesized.

**Exercise 9.4**    Design a 4-bit adder using only $2 \times 1$ multiplexers. You may use as many $2 \times 1$ multiplexers as required. Place them in rows as in standard-cell design methodology. The netlist must be generated automatically. Use Magic's router to interconnect the required nets.

| $C_1$ | $C_2$ | Function |
|-------|-------|----------|
| 0 | 0 | AND |
| 0 | 1 | OR |
| 1 | 0 | XOR |
| 1 | 1 | XNOR |

**Table 9.1 Logic unit function table.**

Compare the area and performance of your design with the one in Exercise 9.3.

**Exercise 9.5** Modify the design of full-adder in Exercise 9.3 to work as an adder/subtractor. An additional control line is required to choose between the operation add or subtract. Use this design to generate the layout of a 4-bit arithmetic unit.

**Exercise 9.6** Design the layout of a logical unit with two control lines $C_1, C_2$, that will perform the logical operations given in Table 9.1.

**Exercise 9.7** Integrate the designs of the arithmetic unit designed in Exercise 9.5 and the logic unit designed in Exercise in 9.6 to build an ALU.

**Exercise 9.8** Design a controller for a digital system that is used to multiply two 4-bit numbers using the ALU designed in Exercise 9.7.

**Exercise 9.9** (*) **Programming Exercise:** Write a program that will read a PLA personality and produce the layout of the PLA.

**Exercise 9.10**
(a) For the layout given in Figure 9.7(c), perform a vertical compaction using virtual grid approach.
(b) For the layout given in Figure 9.7(a), perform a vertical compaction followed by a horizontal compaction using the virtual grid approach.
(c) Compare the compacted layouts resulting from (b) with that resulting from (a).
(d) Discuss your results.

**Exercise 9.11**

(a) For the layout given in Figure 9.7(a), perform a horizontal compaction using constraint graph approach.

(b) Compare and discuss the resulting compacted layout with that obtained from the virtual grid approach (see Figure 9.7).

**Exercise 9.12**

(a) For the horizontal compacted layout given in Figure 9.11(c), perform a vertical compaction using constraint graph approach.

(b) Perform a horizontal compaction on the layout resulting from part (a).

(c) Using the constraint graph approach, perform a vertical compaction followed by horizontal compaction on the layout of Figure 9.11(a).

(d) Compare and discuss the resulting layouts from part (a) and part (c).

**Exercise 9.13**   Give a layout example for which horizontal compaction followed by vertical compaction leads to a different compacted layout than if vertical compaction is performed first followed by horizontal compaction.

**Exercise 9.14**   Compare and discuss the time complexity of compaction algorithms using the virtual grid and constraint-graph approaches.

**Exercise 9.15**   Show that, for each vertex $v$ in the constraint graph, $r(v) = LengthFromL(R) - LengthToR(v)$, i.e., moving the element by $d \leq r(v)$ will not create a longer path than the current longest path.

**Exercise 9.16**   Show that, for any vertex $v$ on the critical path(s), $l(v) = r(v)$.

**Exercise 9.17**   Discuss why virtual grid compaction is more sensitive to layout changes than graph-based compaction.

**Exercise 9.18**   (*) **Programming Exercise:** Write a program that will read a layout described in the CIF language and performs one-dimensional compaction using the virtual grid approach. The direction of compaction should be a user specified parameter. The output should be given in the CIF language.

**Exercise 9.19**   **(\*) Programming Exercise:** Write a program that will accept as input a layout described in the CIF language and performs one-dimensional compaction using the constraint-graph approach. The direction of compaction should be a user specified parameter. The output should be given in the CIF language.

# REFERENCES

Akers, S. B., J. M. Geyer and D. L. Roberts. IC mask layout with a single conduct layer. *Proceedings of 7th Annual Design Automation Workshop*, pages 7–16, 1970.

Boyer, D. G. Symbolic layout compaction review. *Proceedings of 25th Design Automation Conference*, pages 383–389, 1988.

Cho, Y. E., A. J. Korenjak and D. E. Stockton. Floss: An approach to automated layout for high-volume designs. *Proceedings of 14th Design Automation Conference*, pages 138–141, 1977.

Crocker, W. H., C. Y. Lo and R. Vadarahan. Macs: A module assembly and compaction system. *Proceedings of IEEE International Conference on Computer Design*, pages 205–208, 1987.

Dunlop, A. E. Slip: Symbolic layout of integrated circuits with compaction. *Computer Aided Design*, 10(6):387–391, June 1978.

Eichenberger, P. *Fast Symbolic Layout Translation for Custom VLSI Integrated Circuits*. Ph.D. Thesis, Stanford University, 1986.

Hamachi, G. T. and J. K. Ousterhout. A switchbox router with obstacle avoidance. *Proceedings of 21st Design Automation Conference*, pages 173–179, 1984.

Hedges, T. W., W. Dawson and Y. E. Cho. Bitmap graph build algorithm for compaction. *Digest of International Conference on Computer Aided Design*, pages 340–342, 1985.

Hsueh, M. Y. and D. D. Pederson. Computer-aided layout of LSI circuit building blocks. *Proceedings of IEEE International Symposium on Circuits and Systems*, pages 474–477, 1979.

Keller, K. H. and A. R. Newton. KIC2: A low cost editor for integrated circuit design. *Proceedings of Spring COMPCON*, pages 305–306, 1982.

Lakhani, G. and R. Varadarajan. A wire-length minimization algorithm for circuit layout. *Proceedings of International Conference on Circuits and Systems*, pages 276–279, 1987.

Lengauer, T. *Combinatorial Algorithms for Integrated Circuit Layout*. John Wiley & Sons, 1990.

Lo, C. Y., R. Varadarajan and W. H. Crocker. Compaction with performance optimization. *Proceedings of International Conference on Circuits and Systems*, pages 514–517, 1987.

Maley, F. M. Compaction with automatic jog introduction. *Chapel Hill VLSI Conference*, pages 261–283, 1985.

Mayo, R. N., M. H. Arnold, W. S. Scott, D. Stark, G. T. Hamachi. *1990 DECWRL/Livermore Magic Release*. Digital Western Research Laboratory, 1990.

Mead, C. and L. Conway. *Introduction to VLSI Systems*. Reading, MA: Addison-Wesley, 1980.

Mosteller, R. C., A. H. Frey and R. Suaya. 2-D compaction: A Monte Carlo method. *Proceedings of Conference on Advanced Research in VLSI*, MIT Press, pages 173–197, 1987.

Ohtsuki, T. *Advances in CAD for VLSI: Layout Design and Verification, Chapter 6*. North Holland, 1986.

Ousterhout, J. K. Caesar: An interactive editor for VLSI layouts. *VLSI Design*, 2(4):34–38, 1981.

Ousterhout, J. K. Corner stitching: A data structuring technique for VLSI layout tools. *IEEE Transactions on Computer Aided Design*, 3(1):87–100, 1984.

Ousterhout, J. K. *et al.* Magic: A VLSI layout system. *Proceedings of 21st Design Automation Conference*, pages 152–159, 1984.

Preas, B. and M. Lorenzetti. *Physical Design Automation of VLSI Systems, Chapter 6.* Benjamin Cummings, 1988.

Rivest, R. L. and C. M. Fiduccia. A greedy channel router. *Proceedings of 19th Design Automation Conference*, pages 418-424, 1982.

Sastry, S. and A. Parker. The complexity of two-dimensional compaction of VLSI layouts. *Proceedings of IEEE International Conference on Circuits and Computers*, pages 402–406, 1982.

Scott, W. S. and J. K. Ousterhout. Plowing: Interactive stretching and compaction in Magic. *Proceedings of 21st Design Automation Conference*, pages 166–172, 1984.

Scott, W. S. and J. K. Ousterhout. Magic's circuit extractor. *IEEE Design and Test*, pages 24–34, 1986.

Shiele, W. Improved compaction with minimized wire length. *Proceedings of 20th Design Automation Conference*, pages 121–127, 1983.

Stark, D. and M. Horowitz. REDS: Resistance extraction for digital simulation. *Proceedings of 24th Design Automation Conference*, pages 570–573, 1987.

Taylor, G. S. and J. K. Ousterhout. Magic's incremental design-rule checker. *Proceedings of 21st Design Automation Conference*, pages 160–165, 1984.

Watanabe, H. *IC Layout Generation and Compaction using Mathematical Optimization.* Ph.D. Thesis, University of Rochester, 1984.

Weste, N. Virtual grid symbolic layout. *Proceedings of 18th Design Automation Conference*, pages 225–233, 1981.

Weste, N. and K. Eshraghian. *Principles of CMOS VLSI Design, A Systems Perspective.* Addison-Wesley, 1985.

Williams, J. D. STICKS, a graphical compiler for high level LSI design. *Proc. 1978 National Computer Conference, AFIP Press*, pages 289–295, May 1978.

Wolf, W. An experimental comparison of 1-D compaction algorithms. *Chapel Hill VLSI Conference*, pages 165–180, 1985.

Wolf, W. H., R. G. Mathews, J. A. Newkirk and R. W. Dutton. Algorithms for optimizing two-dimensional symbolic layout compaction. *IEEE Transactions on Computer Aided Design*, CAD-7(4):451–466, April 1988.

# APPENDIX A

This appendix introduces the notations and formalisms used throughout this book. The first section introduces basic concepts from graph theory. The second section describes the notation used to specify the computational complexity of algorithms. The final section introduces terms such as 'hard', 'easy', and 'intractable', which are used to specify the complexity of problems. The idea of NP-completeness is discussed with examples.

## A.1 GRAPH THEORY

A graph $G = (V, E)$ consists of a set of nodes $V$ and a set of edges $E$. An edge is a pair of nodes from $V$; we use an unordered pair $(i, j)$, $i \in V$ and $j \in V$, if the edge is undirected. The corresponding graph is an undirected graph. An edge is an ordered pair of nodes $< i, j >$ when a direction is associated with the edge. Graphs with directed edges are called digraphs. When we refer to a 'graph' without a qualifier, we mean undirected graphs. A graph can have at most $\binom{|V|}{2}$ edges, i.e., an edge between every pair of nodes. A graph is said to be *complete* if it has all the possible edges. A graph is called *dense* if the number of edges $|E|$ is $O(|V|^2)$. *Sparse* graphs have $O(|V|)$ edges.

A *path* in a graph is a set of edges of the form $(i, j), (j, k), (k, l) \cdots$. A graph is *connected* if there is a path from any vertex to any other vertex. Given a graph $G = (V, E)$, a *subgraph* $H = (U, F)$ can be constructed by setting $U \subset V$ and selecting $F$ to contain all those edges in $E$ which are incident on the nodes of $U$. Given a graph $G$, a *clique* is a complete subgraph of $G$. The size of a clique is the number of nodes in the clique. For example, an edge is a clique of size 2. A clique is a *maximal size clique* (or simply *maximal clique*) if it is not a subgraph of another clique. A maximum clique is a clique whose size is the largest possible. A maximum clique is necessarily a maximal clique, but not vice versa. A *clique partition*, or a *clique cover*, is a set of cliques $C_1, C_2, \cdots, C_r$ satisfying two conditions: (1) for $i \neq j$, the node set of $C_i$ does not overlap with the node set of $j$ and (2) the union of the node sets of $C_i$ is the set $V$. The number $r$ is known as the size of the clique cover and ranges from 1 (for a complete graph) to $|V|$ (for a graph that has no edges). The *clique number* of a graph $G$ is the size of the smallest possible clique cover of $G$.

Given a graph $G$, a set of nodes in $G$ is said to be *independent* if no pair of nodes in the set share an edge. A *maximal independent set* is an independent set of nodes, which is not a subset of another independent set. A *proper node colouring* of $G$ is obtained by associating a colour with each node in the graph, such that no two nodes which are directly connected by an edge are assigned the same colour. The minimum number of colours required to properly colour a graph is known as the *chromatic number* of the graph. For further definitions in graph theory, the reader may refer to Reingold *et al.* (1977).

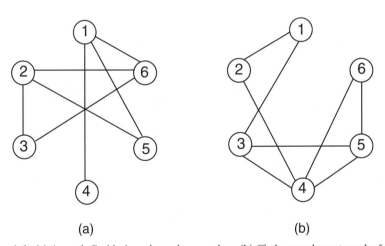

(a)                                    (b)

**Figure A.1**  (a) A graph $G$ with six nodes and seven edges. (b) $G'$, the complement graph of $G$.

**Example A.1**  Consider the graph in Figure A.1(a). The nodes 2,3,6 form a clique; this is also a maximal clique, since there is no larger clique that contains these nodes. There is no clique of size 4 in this graph. Therefore, {2,3,6} is also a maximum clique. {1,2,5} is another maximum clique in the graph. A clique cover of size 3 is obtained by considering the cliques $C_1 = \{2, 3, 6\}$, $C_2 = \{1, 5\}$, and $C_3 = \{4\}$. The reader can verify that this is also a minimum sized clique cover. Therefore, the clique number of the graph is 3.

A proper colouring of the graph is obtained by using three colours: colour nodes 3 and 5 using red paint, nodes 2 and 4 using blue paint, and nodes 1 and 6 using green paint. It is easy to verify that there cannot exist a 2-colouring of this graph; the nodes $\{2,3,6\}$ form a clique of size 3, and must all be painted using different colours. Therefore, the chromatic number of the graph is 3.

In a directed graph, a path is said to exist from node $i$ to node $j$ if it is possible to reach node $j$ starting from node $i$ by following a set of unidirectional edges. A *cycle* or a *circuit* is a closed path; in other words, a set of nodes $i_1, i_2, \cdots, i_k$ form a cycle if it is possible to start from $i_1$ and visit $i_2, i_3, \cdots, i_k$ and return to $i_1$ by following a set of unidirectional edges. A directed graph that has no cycles is said to be *acyclic*. A *Eulerean path* in a graph is a path that includes every edge in the graph exactly once. Similarly, a Eulerean circuit is a cycle which includes each edge exactly once. Note that a Eulerean circuit may visit the same node several times. A *Hamiltonian circuit* is a cycle which includes each node in the graph exactly once.

## A.2 COMPLEXITY OF ALGORITHMS

Two important ways to characterize the performance of an algorithm are its *space complexity* and *time complexity*. Space complexity refers to the amount of core memory (main memory) taken up by the algorithm. Similarly, time complexity refers to the amount of CPU-time required to execute the algorithm. Sometimes, the time complexity of an algorithm is simply referred to as its complexity. The space and time complexities depend on the problem size as well as the target machine. One way to describe the performance of an algorithm is to quote its running time (in seconds) and its memory usage (in megabytes) on the machine that was used; but this method has some limitations. It does not bring out the growth rate of the time and space complexities with the problem size. Secondly, the numbers quoted refer to specific machines and hence are not universal. In order to overcome these problems, the analysis of the time complexity of an algorithm is restricted to determining an expression of the number of steps needed as a function of the problem size. Since the step count measure is somewhat coarse, one does not aim at obtaining an exact step count. Instead, one attempts only to get asymptotic bounds on the step count (Sahni and Bhatt, 1980). Asymptotic analysis makes use of the Big-Oh, Big-Omega and Big-Theta notations. These notations help describe the complexity of an algorithm as a function of the problem size.

## A.2.1 Big-Omega notation

We say that $f(n) = \Omega(g(n))$ if there exist constants $n_0$ and $c$ such that we have $f(n) \geq c \cdot g(n)$ for all $n > n_0$. Another way to say the same thing is to specify that $f(n)$ is *lower bounded* by $g(n)$. This notation is useful when we speak of *problem* complexity. For example, it is well known that at least $n \log_2 n$ comparisons are necessary to sort $n$ keys. The same statement can be rephrased as 'a comparison-based algorithm for sorting $n$ keys requires $\Omega(n \log n)$ time'.

## A.2.2 Big-Oh notation

We say that $f(n) = O(g(n))$ if there exist constants $n_0$ and $c$ such that for all $n > n_0$, we have $f(n) \leq c \cdot g(n)$. Alternatively, we say that $f(n)$ is *upper bounded* by $g(n)$. The Big-Oh notation is used to describe the space and time complexity of *algorithms*. For example, consider the 'bubble sort' algorithm to sort $n$ real numbers.

> **Example A.2**  The procedure 'BubbleSort' shown below requires $O(n)$ storage (for the array A) and $O(n^2)$ running time (two nested **for** loops). The above statement should be taken to mean that the BubbleSort procedure requires no more than linear amount of storage and no more than a quadratic number of steps to solve the sorting problem. In this sense, the following statement is also equally true: the procedure BubbleSort takes $O(n^2)$ storage and $O(n^3)$ running time! This is because the Big-Oh notation only captures the concept of 'upper bound'. However, in order to be informative, it is customary to choose $g(n)$ to be as small a function of $n$ as one can come up with, such that $f(n) = O(g(n))$. Hence, if $f(n) = a \cdot n + b$, we will state that $f(n) = O(n)$ and not $O(n^k), k > 1$.

**Algorithm** BubbleSort $(A[1:n])$;
**Begin** /* Sort Array $A[1:n]$ in ascending order */
    **var** $i, j$;
      **For** $i = 1$ **To** $n$ **Do**
        **For** $j = i + 1$ **To** $n$ **Do**
          **If** $A[i] > A[j]$ **Then**
              swap $(A[i], A[j])$;
**End Algorithm**;

The Big-Oh notation can also be used to describe the complexity of a particular step in an algorithm. For example, the 'swap' operation in the bubble sort procedure requires a constant number of steps; an alternate way to describe this is to say that swapping requires $O(1)$ time.

### A.2.3 Big-Theta notation

We say that $f(n) = \Theta(g(n))$ if $f(n) = \Omega(g(n))$ **and** $f(n) = O(g(n))$. $\Theta(g(n))$ is known as the *tight bound* for $f(n)$. In Example A.2 above, we say that the BubbleSort procedure requires $\Theta(n)$ storage and $\Theta(n^2)$ running time.

## A.3 HARD PROBLEMS VERSUS EASY PROBLEMS

An algorithm is said to be a *polynomial-time* algorithm if its time complexity is $\Theta(p(n))$, where $n$ is the problem size and $p(n)$ is a polynomial function of $n$. The BubbleSort algorithm of Example A.2 is a polynomial-time algorithm. In contrast to a polynomial-time algorithm, an *exponential-time* algorithm is one whose time complexity is $\Theta(a^n)$, where $a$ is a real constant larger than 1. A problem is said to be *tractable* (or *easy*) if there exists a polynomial-time algorithm to solve the problem. From Example A.2 above, we may conclude that sorting of real numbers is tractable, since the BubbleSort algorithm solves it in $O(n^2)$ time.

Unfortunately, there are problems of great practical importance that are not computationally easy. In other words, polynomial-time algorithms have not been discovered to solve these problems. The bad news is, it is unlikely that a polynomial-time algorithm will ever be discovered to solve any of these problems. Such problems are also as known as 'hard problems' or 'intractable problems'. Examples of hard problems are the travelling salesperson problem, satisfiability, graph partitioning, Steiner tree, quadratic assignment, etc. (Garey and Johnson, 1979). Most design automation problems are hard problems, such as global routing, two-layer channel routing, floorplanning, placement, etc. (Sahni and Bhatt, 1980).

**Example A.3** Consider the satisfiability problem (SAT). Given a circuit with $n$ inputs and a single output; each input can be set to 0 or 1. The question is, *is it possible to set the inputs such that the output is driven to logic 1?* This problem commonly arises in the testing of combinational circuits, where it is desired to find an input test pattern that will drive the output to a specified state. Since each of the $n$ inputs can take two values, there are $2^n$ input patterns. An algorithm which applies all input patterns to solve the satisfiability problem will require

$O(2^n)$ time. This is an exponential-time algorithm. It turns out that all known algorithms for the problem require exponential time in the worst case. In electronic testing, the D-algorithm is commonly used to generate test patterns for combinational circuits (Abramovici *et al.*, 1990; Fujiwara, 1985). The D-algorithm is based on the concept of *backtracking*. While it is quick for most circuits, the D-algorithm is known to take exponential time for certain classes of circuits.

**Example A.4**   The travelling salesperson problem (TSP) is one of the oldest problems known to be computationally hard. There are many variations of the problem; in this example, a version known as the 'Euclidean TSP' is described. The objective of the problem is to find the shortest Hamiltonian tour of $n$ cities. The input to the problem is given as an $n \times n$ matrix $C$. Entry $C_{ij}$ of the matrix represents the Euclidean distance from city $i$ to city $j$. It is assumed that the salesperson can travel from any city $i$ to any other city $j$. A Hamiltonian tour begins at some city $a$, includes each of the remaining cities exactly once, and returns to the city $a$. The tour can be represented by a permutation $T$ of $n$ cities; $T_i$ is the $i$th city visited. The cost of the tour $T$ is given by

$$c(T) = (\sum_{i=1}^{n-1} C_{T_i T_{i+1}}) + C_{T_n T_1} \qquad (A.1)$$

Given $n$ cities, there exist $\frac{(n-1)!}{2}$ Hamiltonian tours. To see why, observe that a permutation $T$ of $n$ cities is completely specified by the first $n-1$ cities. Further, a permutation $T$ and the reversal of the permutation $T$ represent the same tour; this accounts for the factor 2 in the denominator. When $n$ is large, the number of tours is exponential. Literally hundreds of heuristic algorithms have been developed to solve the TSP (Lawler, 1992; Reingold *et al.*, 1977; Manber, 1989).

### A.3.1 NP-complete problems and reduction

It is known that many hard problems are 'equivalent' in the following sense: if someone were to discover a polynomial-time algorithm to solve one of these problems, then the same algorithm can be modified to solve all these problems in polynomial-time. These problems are called NP-complete problems. The term 'NP' needs some explanation. It stands for 'nondeterministic polynomial.' A problem is said to be NP if a nondeterministic (or *randomized*) algorithm can solve the problem in polynomial-time. The word 'complete' refers to the above notion of equivalence—if we were to construct an imaginary graph with NP-complete problems for its nodes and join two nodes if the corresponding problems are equivalent, then we will end up with a complete graph.

How is one NP-complete problem equivalent to another? To understand this, we need the notion of reducibility. Let $P_1$ be an NP-complete problem. $P_1$ is said to be reducible to $P_2$ if, using polynomial amount of work, $P_1$ can be made to look like $P_2$. If $P_1$ can be reduced to $P_2$ then essentially $P_1$ is 'equivalent' to $P_2$. In other words, if someone were to discover an algorithm for solving $P_2$ in polynomial-time, then we can use the same algorithm to solve $P_1$ in polynomial-time—simply reduce $P_1$ to $P_2$, and solve $P_2$.

**Example A.5**  Consider the 'maximum independent set' problem (MIS) for graphs. Given a graph $G$ on $n$ nodes, it is required to identify the largest subset of nodes which are independent of one another, i.e., do not share edges. To illustrate, a graph with six nodes and seven edges is shown in Figure A.1(a). By inspection, a maximum independent set in this graph is $\{4,5,6\}$. Another maximum independent set is $\{3,4,5\}$. It will be shown that MIS is reducible to the 'maximum clique' problem (MC), which identifies the largest clique in the graph. Incidentally, the largest clique in the graph is the largest subset of nodes such that every pair of nodes in the subset is connected by an edge.

The MIS problem can be reduced to the MC problem by constructing the complement graph of $G$. Let $K_n = (V, E_n)$ be the complete graph on $n$ nodes. The complement graph of $G$ is defined as $G' = (V, E_n - E)$. Creating $G'$ can be completed in polynomial time, since a graph on $n$ nodes can have at most $\binom{n(n-1)}{2}$ edges. Figure A.1(b) shows the complement graph of the graph in A.1(a). Note that it has eight edges. Together, the two graphs of Figure A.1 have 15 edges. The reader may verify that $K_6$ also has 15 edges. It is easy to see that the maximum clique in $G'$ corresponds to the maximum independent set in $G$. For example, $\{4,5,6\}$ in Figure A.1(b) is a maximum clique. Thus, the problems MIS and MC are equivalent in complexity.

One other comment is in order. This regards the proof of NP-completeness of a problem $Q$. To show that $Q$ is NP-complete, two things must be accomplished:

1. Prove that $Q$ is NP.

2. Prove that a known NP-complete problem $Q'$ is polynomially reducible to $Q$.

# REFERENCES

Abramovici, M., M. A. Breuer and A. D. Friedman. *Digital System Testing and Testable Design*. Potomac, Md: Computer Science Press. 1990.

Fujiwara, H. *Logic Testing and Design for Testability*. The MIT Press. 1985.

Garey, M. R. and D. S. Johnson. Computers and Intractability: A Guide to the theory of NP-Completeness. W. H. Freeman and Company, New York, 1979.

Lawler, E. L. *A Traveling Salesman Problem: A guided tour of Combinatorial Optimization*. A Wiley-Interscience Publication, NY, Wiley, 1992.

Manber, U. *Introduction to Algorithms. A Creative Approach*. Addison-Wesley, 1989.

Reingold, E. M., J. Nievergelt and N. Deo. *Combinatorial Algorithms: Theory and Practice*. Prentice-Hall, 1977.

Sahni, S. and A. Bhatt. The complexity of design automation problems. In *Proc. of the Design Automation Conference*, pages 402–410, 1980.

# INDEX